T0140941

Gas Network Optimization by MINLP

vorgelegt von
Dipl.-Math. Jesco Humpola
aus Melle

von der Fakultät II – Mathematik und Naturwissenschaften
der Technischen Universität Berlin
zur Erlangung des akademischen Grades

Doktor der Naturwissenschaften
– Dr. rer. nat. –

genehmigte Dissertation

Promotionsausschuss:

Vorsitzender:	Prof. Dr. Peter Bank
Gutachter:	Prof. Dr. Dr. h.c. mult. Martin Grötschel
Gutachter:	Prof. Dr. Thorsten Koch
Gutachter:	Prof. Andrea Lodi, Ph.D.

Tag der wissenschaftlichen Aussprache: 06. November 2014

Berlin 2014
D 83

Bildnachweis Einband Vorderseite:
Mit freundlicher Genehmigung von Open Grid Europe GmbH, Essen.

Bibliografische Information der Deutschen Nationalbibliothek:

Die Deutsche Nationalbibliothek verzeichnet diese Publikation in der Deutschen Nationalbibliografie; detaillierte bibliografische Daten sind im Internet über http://dnb.d-nb.de abrufbar.

ISBN: 978-3-8325-4505-5

Logos Verlag Berlin GmbH
Comeniushof, Gubener Str. 47,
10243 Berlin
Tel.: +49 (0)30 42 85 10 90
Fax: +49 (0)30 42 85 10 92
INTERNET: http://www.logos-verlag.de

Abstract

One quarter of Europe's energy demand is provided by natural gas distributed through a vast pipeline network covering the whole of Europe. At a cost of 1 million Euros per kilometer the extension of the European pipeline network is already a multi billion Euro business. The challenging question is how to expand and operate the network in order to facilitate the transportation of specified gas quantities at minimum cost. This task can be formulated as a mathematical optimization problem that reflects to real-world instances of enormous size and complexity. The aim of this thesis is the development of novel theory and optimization algorithms which make it possible to solve these problems.

Gas network topology optimization problems can be modeled as nonlinear mixed-integer programs (MINLPs). Such an MINLP gives rise to a so-called *active transmission problem* (ATP), a continuous nonlinear non-convex feasibility problem which emerges from the MINLP model by fixing all integral variables. The key to solving the ATP as well as the overall gas network topology optimization problem and the main contribution of this thesis is a novel *domain relaxation* of the variable bounds and constraints in combination with a penalization in the objective function. In case the domain relaxation does not yield a primal feasible solution for the ATP we offer novel sufficient conditions for proving the infeasibility of the ATP. These conditions can be expressed in the form of an MILP, i.e., the infeasibility of a non-convex NLP can be certified by solving an MILP. These results provide an efficient bounding procedure in a branch-and-bound algorithm.

If the gas network consists only of pipes and valves, the ATP turns into a *passive transmission problem* (PTP). Although its constraints are non-convex, its domain relaxation can be proven to be convex. Consequently, the feasibility of the PTP can be checked directly in an efficient way. Another advantage of the passive case is that the solution of the domain relaxation gives rise to a cutting plane for the overall topology optimization problem that expresses the infeasibility of the PTP. This cut is obtained by a Benders argument from the Lagrange function of the domain relaxation augmented by a specially tailored *pc-regularization*. These cuts provide tight lower bounds for the passive gas network topology optimization problem.

The domain relaxation does not only provide certificates of infeasibility and cutting planes, it can also be used to construct feasible primal solutions. We make use of parametric sensitivity analysis in order to identify binary variables to be switched based on dual information. This approach allows for the first time to compute directly MINLP solutions for large-scale gas network topology optimization problems.

All the research in this thesis has been realized within the collaborative research project "Forschungskooperation Netzoptimierung (ForNe)". The developed software is in use by the cooperation partner Open Grid Europe GmbH.

Parts of this thesis have been published in book chapters, journal articles and technical reports. An overview of the topics and solution approaches within the research project is given by Martin et al. (2011) and Fügenschuh et al. (2013). Gas network operation approaches and solution methods are described in detail by Pfetsch et al. (2014) and with a special focus on topology optimization in Fügenschuh et al. (2011). The primal heuristic presented in this thesis is published by Humpola et al. (2015a). The method for pruning nodes of the branch-and-bound tree for an approximation of the original problem is described in Humpola and Fügenschuh (2015) and Humpola et al. (2015b) and Humpola and Serrano (2017). The Benders like inequality is introduced by Humpola et al. (2016).

Zusammenfassung

Ein Viertel des europäischen Energiebedarfs wird durch Gas gedeckt, das durch ein europaweites Pipelinesystem verteilt wird. Aufgrund von Ausbaukosten von 1 Mio. Euro pro Kilometer ist der Netzausbau ein Milliardenunterfangen. Die größte Herausforderung besteht darin zu entscheiden, wie das Netzwerk kostengünstig ausgebaut und genutzt werden kann, um notwendige Gasmengen zu transportieren. Diese Aufgabe kann mit Hilfe eines mathematischen Optimierungsproblems formuliert werden, wobei anwendungsnahe Instanzen eine enorme Größe und Komplexität aufweisen. Ziel der vorliegenden Arbeit ist die Entwicklung neuer mathematischer Theorien und damit einhergehender Optimierungsalgorithmen, die es ermöglichen, derartige Probleme zu lösen.

Die Optimierung der Topologie eines Gasnetzwerks kann mit Hilfe eines nichtlinearen gemischt-ganzzahligen Programms (MINLP) modelliert werden. Durch Fixierung aller ganzzahligen Variablen ergibt sich ein kontinuierliches Zulässigkeitsproblem, das als *aktives Transmissionsproblem* (ATP) bezeichnet wird. Die zentrale Methode um dieses ATP zu lösen, ist eine neuartige *Relaxierung*, welche Variablenschranken und einige Nebenbedingungen relaxiert und in der Zielfunktion bestraft. Diese Relaxierung bildet den Kern der in dieser Arbeit vorgestellten Theorie und ermöglicht so die effiziente Lösung der Topologieoptimierung eines Gasnetzwerkes. Für den Fall, dass die Relaxierung keine Primallösung für das ATP liefert, ist es uns gelungen, hinreichende Bedingungen für die Unzulässigkeit des ATP zu formulieren, die durch ein MILP dargestellt werden. Kurz gefasst kann die Unzulässigkeit eines nicht-konvexen NLP durch Lösung eines MILP bewiesen werden. Beide Methoden liefern effiziente Schranken in einem branch-and-bound Lösungsverfahren.

Besteht ein Gasnetzwerk nur aus Rohren und Schiebern, dann wird das ATP als *passives Transmissionsproblem* (PTP) bezeichnet. Obwohl die Nebenbedingungen des PTP nicht konvex sind, konnten wir zeigen, dass seine Relaxierung konvex ist. Daher kann die Unzulässigkeit des PTP direkt auf effiziente Weise geprüft werden. Außerdem können mit Hilfe der Relaxierung in diesem speziellen Fall Schnittebenen für das Topologieoptimierungsproblem aufgestellt werden. Diese repräsentieren die

Unzulässigkeit des PTP und folgen aus der Lagrange Funktion der Relaxierung zusammen mit einer speziellen Erweiterung, der sogenannten *pc-Regularisierung*.

Abgesehen von den genannten Klassifizierungen kann die Relaxierung auch genutzt werden, um primale Lösungen zu konstruieren. Hier nutzen wir die parametrische Sensitivitätsanalyse, um mit Hilfe dualer Informationen Binärvariablen des ATP zu identifizieren, deren Werte angepasst werden müssen. Dieser Ansatz erlaubt es zum ersten Mal, direkt MINLP Lösungen für das Topologieoptimierungsproblem realer Gasnetzwerke zu berechnen.

Die Resultate dieser Arbeit wurden im Rahmen des Forschungsprojekts "Forschungskooperation Netzoptimierung (ForNe)" erarbeitet. Die entwickelte Software wird vom Kooperationspartner Open Grid Europe GmbH aktiv genutzt.

Teile dieser Arbeit sind in Buchkapiteln, Journalen und technischen Berichten publiziert. Eine Übersicht über die Themen und Lösungsansätze im ForNe-Projekt veröffentlichten Martin u. a. (2011) und Fügenschuh u. a. (2013). Für Lösungsmethoden für die operative Nutzung von Gasnetzwerken verweisen wir auf Pfetsch u. a. (2014). Ansätze für eine Topologieoptimierung wurden von Fügenschuh u. a. (2011) beschrieben. Die in dieser Arbeit präsentierte primale Heuristik ist publiziert von Humpola u. a. (2015a). Die genannte Methode, um Knoten innerhalb des branch-and-bound Baums abzuschneiden, wurde für eine Approximation des Topologieproblems von Humpola und Fügenschuh (2015) und Humpola u. a. (2015b) und Humpola und Serrano (2017) beschrieben. Ein Bericht über die Ungleichungen nach Benders ist in Humpola u. a. (2016) nachzulesen.

Acknowledgements

First of all, I would like to thank my supervisor Prof. Dr. Dr. h.c. mult. Martin Grötschel for enabling me to work in such a fascinating field of work, making it possible for me to apply my theoretical advances to real life scenario.

In addition I wish to thank my second supervisor Prof. Dr. Thorsten Koch. He encouraged me to attend diverse mathematical conferences, resulting in profitable discussions, novel ideas and a meeting with Prof. Andrea Lodi, who later agreed to be one of my supervisors. Moreover, from very early on, he put his trust in me and encouraged me to follow my own ideas.

Furthermore I would also like to thank Prof. Andrea Lodi for inviting me to Bologna, having long and fruitful discussions on further applications of the research of this thesis and for including my defense into his busy schedule.

I have to thank Prof. Dr. Ralf Borndörfer for the constructive discussions we had over the years and all valuable insights and ideas he shared with me. Thank you for supporting and motivating me and for providing me with the last finishing touches and suggestions for this thesis.

I am especially obliged to Prof. Dr. Armin Fügenschuh, Dr. Thomas Lehmann and Dr. Nam Dung Hoang for careful reading this thesis and for their diverse criticism and suggestions. I have to emphasize Prof. Dr. Armin Fügenschuh for his willingness to support my publication ideas, enhancing them and to make our final results so engaging to read.

I am very grateful to my colleague Felipe Serrano for his essential comments on Chapter 6 which resulted in the rewriting and rethinking of the whole chapter.

Thanks must go to my colleagues of the working group “Energy” at ZIB, Dr. Benjamin Hiller, Ralf Lenz, Jonas Schweiger, Robert Schwarz. I always appreciated our vital discussions and the pleasant working atmosphere. I also like to thank them for commenting on parts of this thesis.

I would also like to thank our industry partners, in particular Dr. Klaus Spreckelsen, Dr. Lars Huke and his team from Open Grid Europe GmbH. They made it possible to test the algorithms developed in this thesis within real-life applications by providing the data.

I wish to thank all my former and fellow colleagues of the ForNe project team and the SCIP group, especially Prof. Dr. Marc E. Pfetsch for all the hours of fertile discussions we had.

Thank you to Dr. Radoslava Mirkov for all the advice, friendship and inspring conservations we had over the past years. I am happy I came to Berlin for my PhD.

Also I have to give many thanks to Zara, Daniel, Nadine and my mother for all the proof reading and rephrasing of my thesis.

I have to express my biggest gratitude to my family, especially to my parents for supporting me in my studies. Because of you I became the person I am today!

Last but not least, I want to thank Nadine for all her patience and her constant trust in me.

Contents

Chapter 1

Introduction: Gas Network Optimization

Natural gas is a nontoxic, odorless, transparent, and flammable gas that originates from underground deposits. Today natural gas is mainly used for heating private houses and office buildings, for the generation of electrical power, as fuel for vehicles, and for several reactions in chemical process engineering. Natural gas usage represents one quarter of the world's energy demand (BGR 2013). It must be transported from the deposits to the customers, sometimes over distances of thousands of kilometers. For very long distances it is more economic to cool the gas down to $-160°$ so that it becomes liquid and can be transported by ships (see Cerbe 2008). For shorter distances or for the delivery to the end customers large pipeline systems are used. Existing gas networks have usually grown over time. In Germany, the high-pressure pipeline system was built by gas supply companies. It has a size of approximately 35 000 km (FNBGas 2013).

Historically these companies were both gas traders and gas network operators. They purchased gas from other suppliers and operated the necessary infrastructure to transport the gas from those suppliers to their own customers. During the liberalization of the German gas market these business functions were separated by regulatory authorities (GasNZV 2005). Nowadays, there are companies that trade gas and others whose sole task is the operation of gas networks for the transportation of gas. One of the requirements set by the regulatory authorities is that every trader can use the network infrastructure to transport gas. Open access to these gas networks has to be granted to all the trading companies free of any discrimination. This means that the gas supplies and demands cannot be fully controlled by the network operator. Therefore the network operator is required to have a high degree of operational flexibility. The majority of network management is carried out manually with the aid of simulation software. There is a need to develop a more automated process in order to cope with anticipated challenges associated with the addition of more traders accessing the network. Here we have developed mathematical optimization methods to improve the network operation and to enhance the cost effectiveness of investments in the infrastructure.

1.1 Optimization Tasks

The physics of a *gas transport network*, which is used for the transportation of gas, can roughly be described as follows: Most of the network elements are *pipelines*. A gas flow through a pipeline (pipe) is induced if the gas has different *pressures* at the end nodes of the pipe. Usually gas pipelines can withstand nominal pressures of 16 bar up to 100 bar. Typically at long transport distances of about 100 km to 150 km the gas pressure gets too low which is technically not feasible. In situations where lack of pressure is an issue *compressors* are used to increase the pressure again. High gas pressures can also be problematical in parts of the network. Therefore in order to protect the network it can be necessary to reduce the pressure by using *control valves*. This is particularly important when the network includes older pipelines which have a lower pressure limit. If parts of the network need to be deactivated, then *valves* allow to split the network into physically independent subnetworks.

A gas network is mathematically modeled by a *directed graph*, refer to Korte and Vygen (2007) for the notations in graph theory. This graph consists of *nodes* together with connecting *arcs*. The end points of an arc are nodes. Each node corresponds to a geographical position. Each *arc* models a network element. We use a directed graph in order to distinguish between gas flow in the direction of an arc and in the opposite direction. A network element can either be a pipeline, a valve, a compressor or a control valve. All network elements determine a specific relation between the flow through the element and the pressure at the end nodes. Pipelines are called *passive network elements*, while valves, compressors and control valves are called *active network elements*. Pipelines have a unique relation between the flow through a pipeline and the pressures at its end nodes. This is different for active elements. Their physical behavior can be influenced by the network operator, for instance, a valve can be *open* or *closed*. An open valve means that the pressures at the end nodes are equal, while closed means that there is no gas flow through the valve. Compressors and control valves can be open or closed. This means that each active network element has different *operation modes* or rather *configurations* available. An accurate description of every network element is given in Section 3.1.

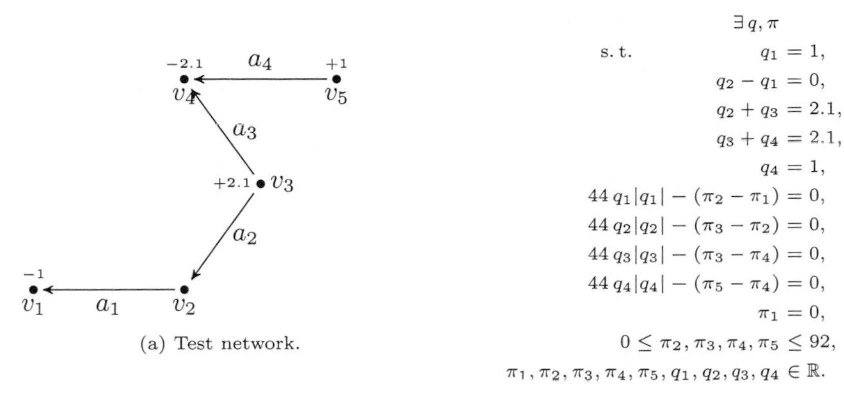

(a) Test network.

(b) Nomination validation problem for the network shown in Figure 1.1a (left picture).

Figure 1.1: An example of a test network and a model of the corresponding nomination validation problem as used in this thesis. The instance has two entries at node v_3 with flow amount +2.1 and v_5 with flow amount +1. Two exits are at node v_1 with flow amount −1 and v_4 with flow amount −2.1. Node v_2 is a transmission node. These node flows imply the arc flows 1 for a_1 and a_2 and a_4 and 1.1 for a_3. Every arc is a pipeline which means a unique relation between the arc flow and the pressures at the end nodes. The pressure at node v_1 is fixed to zero which implies that any other node pressure is fixed. Hence the pressure at node v_5 is also fixed. The pressure bounds of the pipes imply that the pressure at node v_5 violates its upper bound. So the nomination validation problem is infeasible for this instance.

We distinguish between two different optimization tasks. Given is a *nomination* that specifies for each single node the amount of gas that enters or leaves the network there. The network in combination with the nomination defines an *instance*. In order to operate a gas transport network the task is to compute a configuration of valves, control valves and compressors, a pressure for each node, and a flow for each element. This computation requires that the flow specified by the nomination is transported through the network and all technical and physical as well as legal constraints are fulfilled. The arising problem is a feasibility problem which is called *nomination validation problem* of the specified instance. Figure 1.1 shows an example. A simple tree network together with a nomination is shown in 1.1a and the associated nomination validation problem is infeasible. This means that it is not possible to transport gas through the network according to the specified nomination. A model of this problem as used in this thesis is shown in 1.1b and will be explained in detail in Chapter 3. Due to the infeasibility of the nomination the network needs to be extended. In principal all network elements, i.e., pipelines, valves, control valves and compressors, can be added to the gas network. The additional elements are called *extensions*. The task is to compute cost-optimal extensions in order to transport the gas defined by the nomination through the network. The arising problem is called *topology expansion problem*. Its objective is to minimize the building costs of the

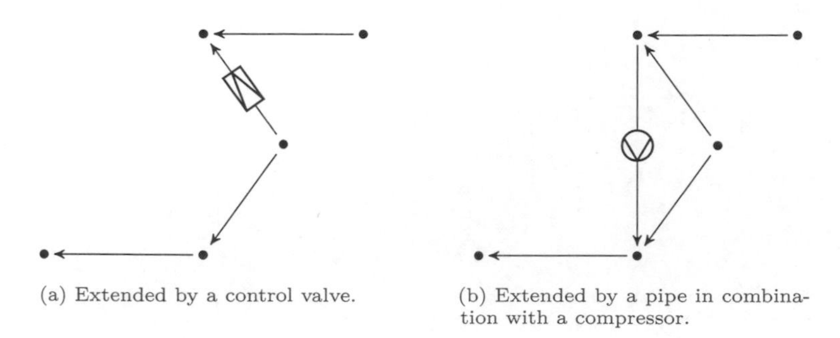

(a) Extended by a control valve.

(b) Extended by a pipe in combination with a compressor.

Figure 1.2: Extended versions of the network shown in Figure 1.1a. The discussion in Figure 1.1 explains that the nomination validation problem 1.1b is infeasible because the pressure at node v_5 violates its upper pressure bound. Decreasing the pressure in node v_4 is a possible adaptation in order to reduce the pressure in node v_5. This adjustment is achieved by the proposed extensions in 1.2a and 1.2b.

additional network elements. Note that operation costs of the network elements are assumed to be a constant term in this objective which is not taken into account for the optimization. They are caused by compressors which consume energy for the compression of the gas or pipelines which have to be maintained regularly, for example. Figure 1.2 shows two different suitable extensions for the network in Figure 1.1a. We refer to Figure 1.3 for a visualization of the effect of extending a gas network. More precisely the node pressures are shown before and after extending the network. The networks shown so far are only small examples. A real-world gas network with approximately 4000 arcs for which we want to solve the nomination validation problem is shown Figure 1.4.

The building costs of a new network element depend on its type. For building a pipeline the costs are mainly made up of the material costs, the construction costs and costs of getting permission rights for the use of land. Construction costs for compressors are different as a compressor is typically built at a single place. We assume that the construction costs also contain a term representing the operation costs of the specific network element.

1.2 Previous Work

A general survey over the application of optimization methods in the natural gas industry is given by Zheng et al. (2010). The surveys by Shaw (1994) and Ríos-Mercado and Borraz-Sánchez (2012) are closer to the problems we are studying. A monograph outlining the earlier state-of-the-art is described in the book of Osiadacz (1987).

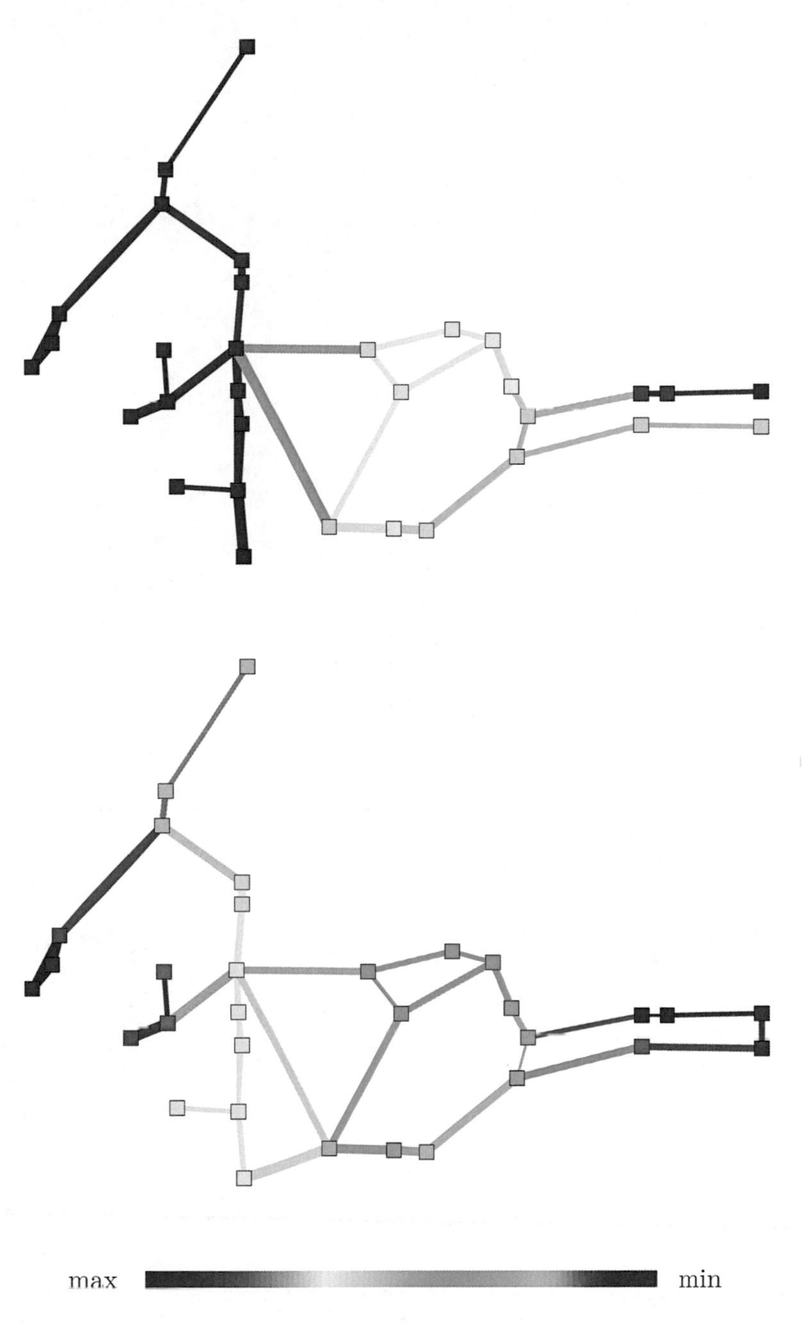

Figure 1.3: Element flow and node pressure corresponding to a realization of a nomination in a test network consisting of pipelines only. The line width represents the flow value (the thicker the more flow), while its color depicts the mean value of the node pressures at both end nodes. The nodes are depicted by squares and the node colors represent the node pressures. In our test case we added two pipelines to the network which results in a different flow and pressure distribution (see the lower picture). In both pictures the maximum of the node pressures corresponds to the color red and the minimum to the color blue.

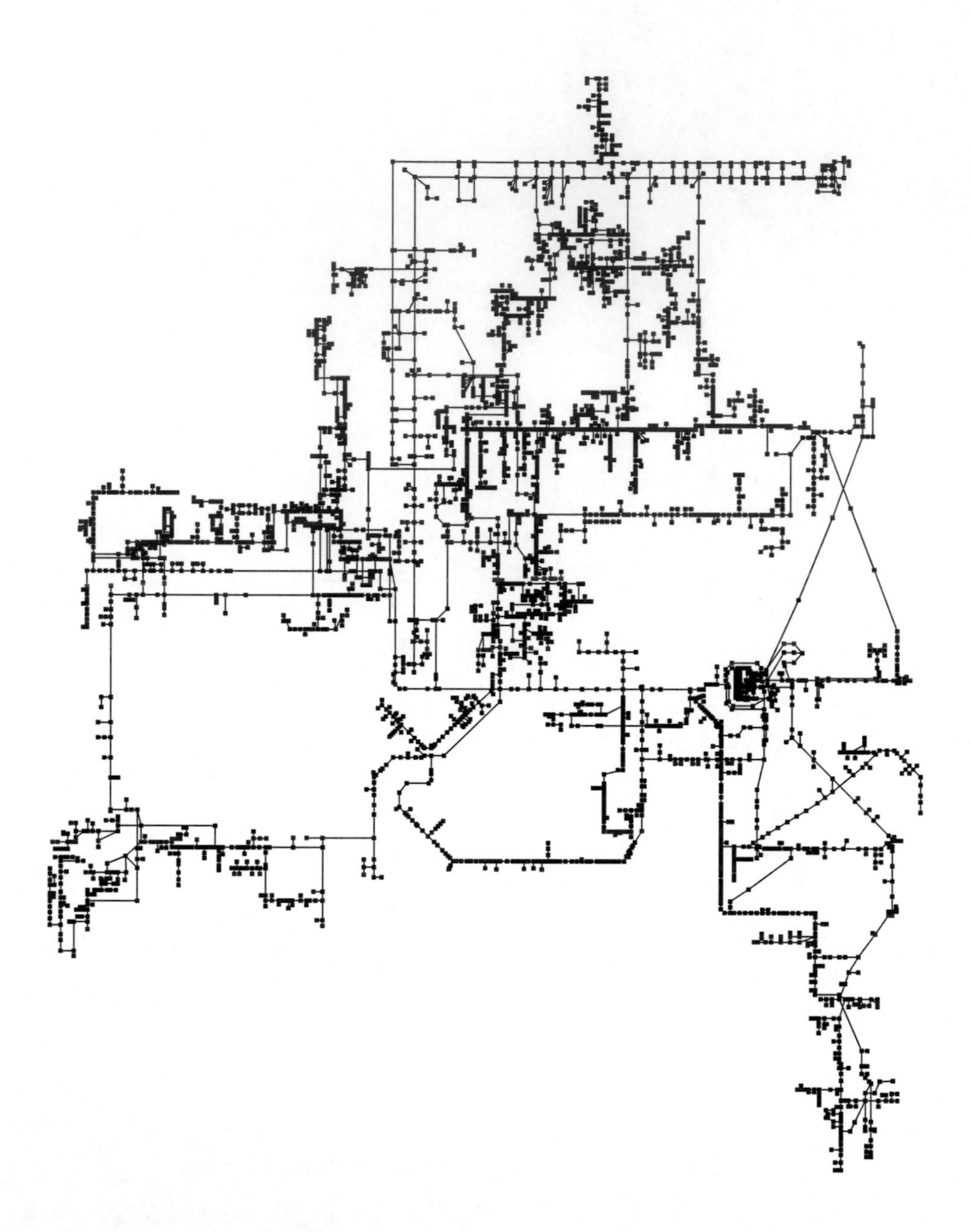

Figure 1.4: The real-world network **net7** provided by the cooperation partner Open Grid Europe GmbH. The task is to solve the nomination validation problem for different instances of this network. Here state-of-the-art solvers show a poor performance as discussed in Section 1.4.

Several models and solution approaches exist for the nomination validation problem. One class of approaches is based on solving a *mixed-integer linear program* (MILP). The main difficulty then is to get an adequate model for the nonlinearities. For the stationary case, Möller (2004) used piecewise-linear approximations; see also Martin et al. (2006). A similar approach using different linearizations was used by Tomasgard et al. (2007) and Nørstebø et al. (2010).

Among the many approaches based on solving a *nonlinear program* (NLP), we mention Percell and Ryan (1987), who use a gradient-descent based method and De Wolf and Smeers (2000), who consider sequential linear programming approaches. Based on a fine simulation model, Jeníček (1993) and Vostrý (1993) use subgradient-based methods. Also sequential quadratic programming (Furey 1993; Ehrhardt and Steinbach 2005; Ehrhardt and Steinbach 2004) and interior point methods (Steinbach 2007) have been considered. Bonnans et al. (2011) use interval analysis techniques to analyze an approximated model of the Belgian network with 20 pipes. Approaches that rely on network reduction techniques are discussed in Hamam and Brameller (1971), Mallinson et al. (1993), Wu et al. (2000), and Ríos-Mercado et al. (2002).

Another widely used approach is dynamic programming. A survey for the work up until 1998 is written by Carter (1998). The first application were so-called gun-barrel networks, i.e., straight line networks with compressors (Wong and Larson 1968b; Wong and Larson 1968a). The method was later extended to more complex network topologies; see e.g., Lall and Percell (1990), Gilmour et al. (1989), and Zimmer (1975). Further extensions of this approach were proposed by Borráz-Sánchez and Ríos-Mercado (2004).

Not surprisingly, many purely heuristic approaches have been developed. We mention applications of simulated annealing by Wright et al. (1998) (specifically for the optimization of compressor operations) and Mahlke et al. (2007), tabu search by Borraz-Sánchez and Ríos-Mercado (2004), general expert systems by Sun et al. (2000), genetic algorithms by Li et al. (2011), and ant colony optimization by Chebouba et al. (2009). Kim et al. (2000) and Ríos-Mercado et al. (2006) present a two stage iterative heuristic to minimize fuel cost of the compressors.

For the topology expansion problem there exist several approaches to improve the topology of a gas network which mainly consist of various heuristic and local optimization methods. Boyd et al. (1994) apply a genetic algorithm to solve a pipe-sizing problem for a network with 25 nodes and 25 pipes, each of which could have one of six possible diameters. André et al. (2006) consider a similar problem and present a heuristic method based on relaxations and local optima. Castillo and Gonzáleza (1998) also apply a genetic algorithm for finding a tree topology solution for a network problem with up to 21 nodes and 20 arcs. In addition to pipes compressors can also be placed in the network. Mariani et al. (1997) describe the design problem of a natural gas pipeline. They present a set of parameters to evaluate the quality of the transportation system. Based on these ones they evaluate a number of potential configurations to identify the best among them. Osiadacz and Górecki (1995) formulate a network design problem for a given topology as a nonlinear optimization problem, for which they iteratively compute a local optimum. For a given topology the diameter of the pipes is a free design variable. Their method is applied to a network with up to 108 pipes and 83 nodes. De Wolf and Smeers (1996) also use a nonlinear formulation, which is then solved heuristically. For a given topology with up to 30 arcs and nodes they can determine optimized diameters of pipes. Further approaches are available from Hansen et al. (1991), Bonnans et al. (2011), Babonneau et al. (2012), and Zhang and Zhu (1996). A method for a complete redesign of a gas transport network is presented by Hübner (2009).

1.3 Solution Approach

Within the joint research project "Forschungskooperation Netzoptimierung (ForNe)" in cooperation with our industry partner Open Grid Europe GmbH (OGE) we have developed a method for the nomination validation as well as the topology expansion problem. OGE is a company which actually operates and maintains the largest gas transport network in Germany. Such a company operating a gas network is also called *transmission system operator* (TSO).

The previous approach of OGE for solving the nomination validation problem was as follows: Given a nomination experts used simulation software like Simone (LIWACOM 2005) and PSIGanesi (Scheibe and Weimann 1999) to compute flows for each element and pressures for each node such that the flow specified by the nomination is transported through the network. This computation needs to be as such that physical, technical and legal constraints are fulfilled. The software simulates the gas physics and characteristics of the active and passive elements of the network. Therefor a set of input parameters must be available which require expert knowledge. For solving the topology expansion problem the gas network was manually extended first, while the nomination validation problem on the extended network was then solved in a second step. These two steps were iterated while the number and size of extensions in the first step were consecutively reduced.

An aim of the research project ForNe was to develop mathematical optimization techniques which aid these processes. The physics of gas networks are described by nonlinear equations. Additionally discrete decisions are necessary because active elements have different configurations as they can, for instance, be open or closed. A possible approach is to state the nomination validation as well as the topology expansion problem as a *mixed-integer nonlinear program* (MINLP). A formal definition of an MINLP is given in Section 2.1.

The general approach in the research project for solving the nomination validation problem was to split the solution process into two steps. In a first step different approaches and MINLP models were developed for computing reasonable discrete decisions, i.e., configurations of compressors, valves and control valves, for a given nomination. Here physical constraints are simplified. Once these decisions are computed, an NLP is solved in a second step. This NLP models the physical and technical constraints of all network elements in detail. A solution for this NLP yields a result which is in precision comparable to the manual approach using the simulation software. For more details and a precise description of this two stage solution approach we refer to Koch et al. (2015). All these models focus on the stationary case meaning transient gas flows are not considered. In addition it should be noted that the gas flow through a pipeline is modeled independently of time. This is due to uncertain gas consumption in the long run. Here especially the nominations for the topology optimization problem can only be specified roughly and independent of time.

Let us briefly summarize the first stage MINLP models and solution approaches described in Koch et al. (2015). Geißler (2011) presents a solution method which bases on MIP relaxations. All nonlinearities are modeled by discretization techniques while the user can give a predefined maximal approximation error. Based on this model, Morsi (2013) presents a network decomposition approach. Furthermore there are presented two heuristic strategies. The first one is based on network reductions (see Stangl 2014) and the second one on an MPEC solution approach. For the second stage NLP model we refer to Schmidt et al. (2014).

The strategy that we follow in this thesis is to provide an MINLP model for the first stage. We do not differ between the nomination validation and the topology expansion problem. Instead we consider both problems as a *topology optimization problem* and model this by an MINLP. Therefor we assume that the network operator OGE provides a predefined set of additional network elements together with installation positions. Then the topology optimization problem is to compute a cost-optimal selection in order to transport the flow defined by the nomination through the network. This means that the nomination validation problem is equal to the topology optimization problem without objective function for a fixed selection.

BARON			ANTIGONE			SCIP			SCIP + heuristic		
feas	infeas	no sol	feas	infeas	no sol	feas	infeas	no sol	feas	infeas	no sol
-	3	27	-	28	2	1	-	29	18	-	12

Table 1.1: Computational study of the nomination validation problem for 30 nominations on the network net7 shown in Figure 1.4. For all instances we set a time limit of 14 400 s. The computational results of the last column "SCIP + heuristic" are obtained by a specially tailored heuristic presented in Chapter 7. The last column is used only to demonstrate that at least 18 instances are feasible.

Note that a cost-optimal solution for the topology optimization problem is a *global optimal solution* which means that any other feasible solution does not improve the objective function value.

1.4 Computational Study

We focused on the nomination validation problem for the network shown in Figure 1.4. It was provided by our cooperation partner OGE. We applied state-of-the-art MINLP solvers to the mixed-integer nonlinear program that we introduce in Chapter 3 as a model of the nomination validation problem. Our aim was to get an impression of the solving performance of the solvers. For a computational study we consider 30 different feasible nominations. The arising MINLPs consist of 1478 binary and 21 957 continuous variables. There are 3757 nonlinear and 31 931 linear constraints. All nonlinear constraints are of the same type. They consist of two different continuous variables z_1 and z_2 and write as $z_1|z_1| = z_2$. Among the linear constraints there are 3406 "bigM"-constraints meaning that they are either active or inactive dependent on the value of a binary variable.

For solving these instances we applied the state-of-the-art MINLP solvers SCIP (Achterberg 2009; Vigerske 2012), BARON (Tawarmalani and Sahinidis 2005) and ANTIGONE (Misener and Floudas 2014). SCIP was used in combination with CPLEX 12.1 (CPLEX) as linear programming solver and IPOPT 3.10 (Wächter and Biegler 2006) as nonlinear solver. The version of BARON that we used is 12.7.3, the version of ANTIGONE is 1.1.

A summary of our computational results is shown in Table 1.1. Detailed results are shown in the right column of Table A.1 – A.3. It turns out that neither BARON nor ANTIGONE were able to compute any feasible solution within a time limit of 4 hours. ANTIGONE detects 28 instances to be infeasible while at least 18 out of 30 instances are feasible. SCIP cannot compute any feasible solution for 29 instances. The results of the last column of Table 1.1 are obtained by a specially tailored heuristic which is described in Chapter 7.

Overall the performance of state-of-the-art MINLP solvers is not appropriate for solving the topology optimization problem (3.2.1). Therefore the aim of this thesis is to present different methods for improving the performance of SCIP for solving (3.2.1) on real-world gas networks.

1.5 Outline of the Thesis

The outline of the following chapters is as follows: In Chapter 2 we give a brief summary of solving MINLP by branch-and-bound, separation, and spatial branching in general. We refer to the MINLP solver SCIP which enables this approach. In Chapter 3 we provide a detailed description of the topology optimization problem (3.2.1). We also identify the *passive transmission problem* (3.4.1) and the *active transmission problem* (3.4.2) as subproblems of the topology optimization problem (3.2.1). The networks and corresponding nominations which we consider for our computational studies are presented. We consider different networks which are either obtained from literature, or were generated manually representing realistic networks, or are based on real-world data. These networks differ according to size and types of network elements that are included. We split the networks into different groups where we distinguish between the elements that are contained. For extensions that are pipes we differentiate whether the pipe would follow an existing pipe or not. In the first case the extension is called a *loop*. Now we distinguish between three types of network:

1. networks that only consist of pipelines and valves,

2. networks that only consist of pipelines, loops and valves,

3. networks that contain active elements.

Note that this is a hierarchical classification. The latter group always contains the former groups. The subsequent Chapters 4 – 6 focus on a specialized solution approach for each of these network types. When using a gas network for our computational studies we apply the solution methods presented for the smallest set in which it is contained following the aforementioned classification.

In Chapter 4 we focus solely on the first type of network, namely those that only contain pipes and valves. We describe and compare novel solution methods for solving the passive transmission problem as part of the topology optimization problem (3.2.1) to global optimality. The solution methods consist of solving convex relaxations of the passive transmission problem. These methods are further integrated into the solver SCIP that we use for solving (3.2.1). This integration allows the number of globally solved instances within our given time frame to increase by 29 %. On average the run time is reduced by 72 % for those instances already solvable by SCIP.

In Chapter 5 we focus on the second and more complicated type of network namely those that additionally contain loops. We derive an improved Benders cut (Geoffrion 1972) for the topology optimization problem from the dual solution of the relaxations of the passive transmission problem. This cut represents the infeasibility of the current passive transmission problem and allows us to speed up the overall solution process when added to the problem formulation (3.2.1). Using this strategy we are able to increase the number of globally solved instances by approximately 13 %. For those instances which are already solvable by SCIP the run time is reduced by approximately 33 %.

In Chapter 6 we focus on the third type of network which consists of all types of network elements. We introduce a special tailored algorithm for solving the active transmission problem (3.4.2) which is part of the topology optimization problem (3.2.1). This algorithm consists of a primal heuristic and sufficient conditions of infeasibility for (3.4.2). We integrate this algorithm into the solver SCIP for solving problem (3.2.1). Thereby the number of globally solved topology expansion instances is increased by 20 %.

With the above described methods we are still not able to solve the real-world instances used for the computational study in the previous Section 1.4. These instances are feasibility instances. Therefor in Chapter 7 we introduce a primal heuristic for the topology optimization problem which is integrated into the solver SCIP. Here we make use of the dual solution of a relaxation of the active transmission problem (3.4.2). It allows to solve approximately 60 % (18 out of 30) of the initially discussed nomination validation instances on `net7` in Section 1.1. Recall that only approximately 3 % of these instances (1 out of 30) were solvable using state-of-the-art MINLP solvers.

Chapter 8 contains concluding remarks and ideas for future research. Moreover we present additional benefits resulting from our methods which are especially useful for a TSO.

Chapter 2

Solving Mixed-Integer Nonlinear Optimization Problems

In this chapter we roughly describe a solution method for an MINLP such as the topology optimization problem (3.2.1). We start with a mathematical definition of MINLP in Section 2.1. Then, in Section 2.2, we briefly explain the solution methods of SCIP for solving an MINLP, see Achterberg (2009) and Vigerske (2012). For a survey of different solution methods we refer to Belotti et al. (2012) and for different MINLP solvers to Bussieck and Vigerske (2010), D'Ambrosio and Lodi (2011), and D'Ambrosio and Lodi (2013). In Section 2.3 we explain the difference between *convex* and *non-convex* MINLP and the impact on the aforementioned solution methods. Afterwards we present the necessary conditions for optimality of a primal solution in Section 2.4.

From the computational study in Section 1.4 we concluded that we have to improve the solution approach for solving the topology optimization (3.2.1) problem in several directions. As discussed in the previous chapter we will present different results that allow us to prune nodes of the branch-and-bound tree and further a primal heuristic. For an implementation of these methods SCIP provides a flexible framework. In Section 2.5 we explain our adaptations of SCIP for solving the topology optimization problem.

2.1 Definition of MINLP

Nonlinear optimization problems containing discrete and continuous variables are called mixed-integer nonlinear programs (MINLPs). A general MINLP can be formulated as

$$\min\{f(x) \mid x \in X\} \tag{2.1.1a}$$

with

$$X := \{x \in [\underline{x}, \overline{x}] \mid g(x) \leq 0, h(x) = 0, x_i \in \mathbb{Z}, i \in I\}. \tag{2.1.1b}$$

Here $\underline{x}, \overline{x} \in \mathbb{R}^n$ determine the *lower and upper bounds* on the variables, $I \subseteq \{1, \ldots, n\}$ denotes the set of variables with *integrality requirement*, $f : [\underline{x}, \overline{x}] \to \mathbb{R}$ is the *objective function*, and $g : [\underline{x}, \overline{x}] \to \mathbb{R}^m$ and $h : [\underline{x}, \overline{x}] \to \mathbb{R}^\ell$ are the *constraint functions*. The set X is called *feasible set* of (2.1.1). We assume $f(x)$, $g(x)$ and $h(x)$ to be at least continuous. Further we assume that $f(x)$ is linear. This is obtained by shifting a nonlinear objective function to the constraints while setting $f(x) = z$ for a new continuous variable z. This goes along with replacing the objective "min $f(x)$" by "min z".

A point $x \in X$ is called *local optimum*, if there exists an $\epsilon > 0$ such that for all $y \in X$ with $\|x - y\| < \epsilon$ we have $f(x) \leq f(y)$. A local optimum whose objective function value equals the optimal value is called a *global optimum*. Note that, due to continuity of $f(x)$ and $g(x)$, there always exists a global optimum of (2.1.1), if its optimal value is finite.

MINLP problems arise in many fields such as energy production and distribution, logistics, engineering design, manufacturing, and chemical and biological sciences, see Floudas (1995), Grossmann and Kravanja (1997), Tawarmalani and Sahinidis (2002), Pintér (2006), and Ahadi-Oskui et al. (2010).

2.2 Details on SCIP for Solving MINLP

Below we describe the solution technique for solving MINLP (2.1.1) as implemented in SCIP, see Achterberg (2004) and Vigerske (2012). By "solving" we mean to compute a feasible solution for an approximation of a given instance of the problem together with a computational proof of its global optimality. The solution is globally optimal for an approximation of the problem and not for the original version due to numerical reasons.

The solution process in SCIP is as follows: The MINLP (2.1.1) is first relaxed to a *mixed-integer linear program* (MILP) and further to a *linear program* (LP). For a detailed introduction into linear and integer programming and combinatorial optimization see for example Nemhauser and Wolsey (1989). Recall that $X \subseteq \mathbb{R}^n$ is the feasible set of (2.1.1). A linear *outer approximation* of the feasible set X is computed such that

$$X \subseteq \{x \in [\underline{x}, \overline{x}] \,|\, Dx \leq d\}$$

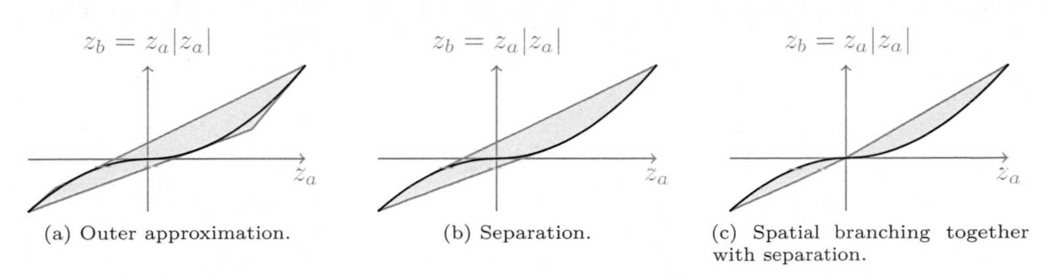

(a) Outer approximation.

(b) Separation.

(c) Spatial branching together with separation.

Figure 2.1: Example of outer approximation, separation, and spatial branching when handling the nonlinear relation $z_b = z_a|z_a|$ between two continuous variables z_a and z_b.

for a suitable matrix $D \in \mathbb{R}^{m \times n}$ and vector $d \in \mathbb{R}^m$. As an example for a linear outer approximation consider the nonlinear constraint

$$z_a|z_a| = z_b, \tag{2.2.1}$$

for two continuous variables z_a, z_b, see Fügenschuh et al. (2010). An outer approximation of (2.2.1) with four different linear inequalities is shown in Figure 2.1a, for instance. The LP relaxation of MINLP (2.1.1) then writes as

$$\min\{f(x) \mid Dx \leq d, x \in [\underline{x}, \overline{x}]\}. \tag{2.2.2}$$

Recall that f is a linear function. LP (2.2.2) is solved in practice by the dual simplex algorithm (see Dantzig 1951). The obtained solution value defines a lower bound on the optimal value of the original MINLP (2.1.1). In case this solution fulfills all constraints of (2.1.1), it is a proven global optimal MINLP solution. However, this rarely happens in practice. Hence either *cutting planes* are added to strengthen the relaxation or a *branching* on a variable is performed as described below.

By adding cutting planes the linear relaxation is improved. For a suitable matrix $D' \in \mathbb{R}^{m' \times n}$ having more rows than D, i.e., $m' > m$, and $d' \in \mathbb{R}^{m'}$, the linear relaxation of the feasible set X is improved by

$$X \subseteq \{x \in [\underline{x}, \overline{x}] \mid D'x \leq d'\} \subset \{x \in [\underline{x}, \overline{x}] \mid Dx \leq d\}.$$

The feasible space which remains after adding all available cutting planes is visualized in Figure 2.1b for the nonlinear function (2.2.1). Adding cutting planes always yields a tighter LP relaxation of the original problem while the relaxation is still a linear program. For more details on separation we refer to Nemhauser and Wolsey (1989).

A branching is either performed on an integer or on a continuous variable. In both cases the domain of a variable is restricted and a new subproblem is created.

Branching on an integral variable x_i means to subdivide the previous linear relaxation into two parts

$$X \subseteq \left\{x \in [\underline{x}, \overline{x}] \,\middle|\, Dx \le d,\ x_i \le y\right\} \cup \left\{x \in [\underline{x}, \overline{x}] \,\middle|\, Dx \le d,\ x_i \ge y+1\right\}.$$

for an integral value y. When branching on a continuous variable we speak of *spatial branching*. Spatial branching on the continuous variable x_i of the solution x^* to the linear relaxation refers to subdividing the previous linear relaxation into two parts

$$X \subseteq \left\{x \in [\underline{x}, \overline{x}] \,\middle|\, Dx \le d,\ x_i \le x_i^*\right\} \cup \left\{x \in [\underline{x}, \overline{x}] \,\middle|\, Dx \le d,\ x_i \ge x_i^*\right\}.$$

For each part of the relaxation a subproblem is created and a tighter outer approximation can be computed due to tighter variable bounds, see Figure 2.1c for example. Spatial branching thus improves the relaxation, in particular, in places where the functions cannot be properly approximated by cutting planes. A *branching tree* is used for managing the different subproblems. The initial LP relaxation (2.2.2) is associated with the root of the branching tree. Whenever a problem corresponding to a node of this tree is split in subproblems, each of these is associated with a child node.

Branching is pursued until all integral variables take integral values and the outer approximation is "close enough" to the feasible region. If a subproblem is infeasible or its optimal value is larger than the best available solution so far, then the subproblem needs not to be investigated furthermore.

The methods above are summarized as branch-and-bound, separation and spatial branching. This way, global bounds on the objective function can be computed and the problem can be solved to global optimality up to a certain accuracy. For more details on branch-and-bound for MILP refer to Nemhauser and Wolsey (1989).

SCIP provides different *constraint handlers*. They allow to deal with different types of constraints of an MINLP. For instance, the nonlinear constraint (2.2.1) is handled by the so called "cons_abspower" constraint handler as described above. A constraint of the form $x = 1 \Rightarrow g(z) \le 0$, where $g(z)$ is a linear function and x a binary variable, is called *indicator constraint* and handled by the "cons_indicator" constraint handler. Roughly spoken, the constraint is modeled by the constraint handler as

$$g(z) \le M\,(1 - x),$$

where $M \in \mathbb{R}_{\ge 0}$ is a suitable constant which is called "bigM". Additionally there are some auxiliaries implemented to handle numerical intractability. We refer to these indicator constraints because they are used for modeling the different modes of active elements in our model (3.2.1) for the topology optimization problem. The "cons_nonlinear" constraint handler is the most general MINLP constraint handler of SCIP for nonlinear constraints, see Vigerske (2012).

Apart from constraint handlers SCIP has several other functionalities for solving MINLPs. There are for example *primal heuristics*, for computing primal feasible solutions. Five heuristics are specifically tailored towards mixed-integer nonlinear programming problems: Undercover, RENS, nonlinear versions of RINS and Crossover, see Danna et al. (2004) and Berthold (2014). Further an NLP local search heuristic (Vigerske 2012), and a nonlinear diving heuristic (Bonami and Gonçalves 2012). Other heuristics are MILP heuristics that are applied to the linear outer approximation plus the integrality constraints. *Propagators* are used to strengthen variable bounds, different *branching strategies* allow one to decide which variable branching is performed, and *node selection strategies* are available, which select nodes of the branching tree to be considered next. Again we refer to Vigerske (2012) and Achterberg (2009) for details.

2.3 Convex and Non-Convex MINLP

In general, when solving an MINLP of the form (2.1.1) we differ between *convex* and *non-convex* MINLP. The optimization problem (2.1.1) is convex if the constraint functions as well as the objective function are convex. A constraint function g_j : $[\underline{x}, \overline{x}] \rightarrow \mathbb{R}$ is convex if

$$g_j(\lambda x_1) + g_j((1-\lambda)x_2) \leq \lambda g_j(x_1) + (1-\lambda)g_j(x_2) \quad \forall \lambda \in [0,1] \tag{2.3.1}$$

holds for any two points $x_1, x_2 \in [\underline{x}, \overline{x}]$. Consequently (2.1.1) is non-convex if any constraint function g_j or h_j does not fulfill the condition (2.3.1). A convex MINLP has the important property that for fixed integral variables $x_i, i \in I$, every local minimum of (2.1.1) is global (see Boyd and Vandenberghe 2004, Section 4.2.2). For the special case of convex MINLP there exist several solvers, see Bonami et al. (2008), Abhishek et al. (2010), IBM (CPLEX), and FICO (2009).

An example for a non-convex MINLP is given in Figure 2.2. The optimization problem as shown in 2.2b is a continuous non-convex optimization problem because the feasible set as shown in Figure 2.2a is not convex. We conclude from Figure 2.2a that the non-convex feasible set can be divided into two parts, one convex part with $x \leq 0$ and another part with $x \geq 0$. This additional information allows splitting problem 2.2b into two problems, while the convex one yields a global optimal solution. Note that the information about a smart split of the feasible set is typically not

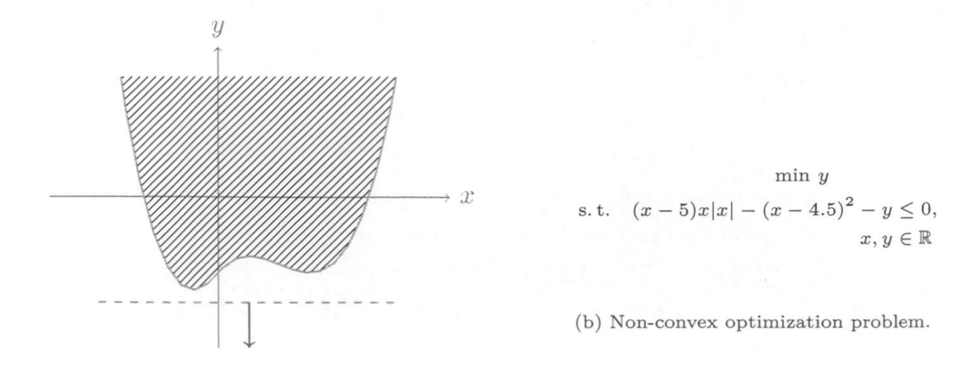

(a) Non-convex feasible set.

(b) Non-convex optimization problem.

Figure 2.2: Example of a non-convex continuous optimization problem. The feasible set as shown in the left figure is non-convex. The objective function of the optimization problem shown in the right part is linear.

available. Hence the solution methods for convex MINLP cannot be applied directly for solving non-convex MINLP. For more details on solution methods for MINLP we refer to Vigerske (2012). We briefly sketch two methods.

A method for solving a non-convex MINLP is to apply the branch-and-bound, separation, and spatial branching technique described in Section 2.2 as implemented in SCIP. Here branching on integer as well as spatial branching on continuous variables needs to be performed, see Figure 2.1 for instance. This is different for a convex MINLP. Because of the convexity of the constraints, there is no need to apply spatial branching. Any tangential hyperplane of a convex function yields a globally valid inequality for MINLP (2.1.1). Hence all nodes of the branching tree only arise from branching on integral variables. Therefore non-convex MINLP are difficult to solve in comparison to convex MINLP. This in turn means, the more information is known about the solution space of an MINLP, the more the solving performance can be improved. Concerning the example shown in Figure 2.2 this means that by splitting the problem 2.2b into two MINLP, a convex and a non-convex one, possibly time consuming spatial branching can be avoided.

A different solution approach for convex MINLP is to apply an NLP based branch-and-bound method. Here a convex NLP is solved instead of an LP at each node of the branching tree. This way even separation is not necessary. This solution method can be extended to non-convex MINLP by using *convex underestimators* for non-convex functions. A convex underestimator for $g(x)$ is a convex function $\tilde{g}(x)$ with $\tilde{g}(x) \leq g(x)$ on the domain of g. Note that $\tilde{g}$ is not required to be linear. Using convex underestimators for non-convex MINLP, i.e., for g and h, goes along with spatial branching, i.e., branching on continuous variables, see Vigerske (2012).

2.4 Necessary Conditions for Local Optimality

After fixing all integral variables of MINLP (2.1.1), a continuous nonlinear optimization problem (NLP) remains. We write this abbreviated as

$$\min\{f(z) : g_i(z) \leq 0, h_j(z) = 0, z \in \mathbb{R}^n\} \tag{2.4.1}$$

where f, g_i ($i = 1, \dots, m$), and h_j ($j = 1, \dots, \ell$) are continuous functions and z represents the remaining unfixed variables. If these functions are continuously differentiable, then the *Karush-Kuhn-Tucker conditions* of this optimization problem yield a necessary condition for a feasible solution of (2.4.1) to be locally optimal, see Conn et al. (2000). These conditions are defined as follows:

Definition 2.4.1:
Let the functions f, g, h of the optimization problem (2.4.1) be continuously differentiable. The conditions

$$\nabla_z L(z, \mu, \lambda) = 0, \tag{2.4.2a}$$
$$g_i(z) \leq 0, \quad \forall i = 1, \dots, m, \tag{2.4.2b}$$
$$h_j(z) = 0, \quad \forall j = 1, \dots, \ell, \tag{2.4.2c}$$
$$\lambda_i \geq 0, \quad \forall i = 1, \dots, m, \tag{2.4.2d}$$
$$\lambda_i \, g_i(z) = 0, \quad \forall i = 1, \dots, m, \tag{2.4.2e}$$

with the Lagrange function $L : \mathbb{R}^n \times \mathbb{R}^\ell \times \mathbb{R}^m \to \mathbb{R}$ defined as

$$L(z, \mu, \lambda) = f(z) + \sum_{i=1}^{m} \lambda_i \, g_i(z) + \sum_{j=1}^{\ell} \mu_j \, h_j(z) = 0$$

*are called **Karush-Kuhn-Tucker conditions** (KKT conditions as abbreviation) of the continuous nonlinear optimization problem (2.4.1). Every vector (z^*, μ^*, λ^*) which fulfills the KKT conditions is called **KKT point**. The components of μ^* and λ^* are called Lagrange multipliers.*

Theorem 2.4.2 (Karush (2014) and Kuhn and Tucker (1951)):
Let the functions f, g, h of the nonlinear optimization problem (2.4.1) *be continuously differentiable functions. Let z^* be a local optimum of* (2.4.1). *If*

- *either the functions g and h are linear*
- *or the gradients of those constraints, which are fulfilled with equality, are linearly independent*

then there exist Lagrange multipliers μ^*, λ^* such that the vector (z^*, μ^*, λ^*) is a KKT point of (2.4.1).

For a feasible solution z^* of (2.4.1) and Lagrange multipliers μ^*, λ^* such that (z^*, μ^*, λ^*) is a KKT point we call the multipliers the *dual part* of the KKT point, and also *dual solution.* When solving (2.4.1) by IPOPT (Wächter and Biegler 2006) for instance, the solver does not only return a primal feasible solution but also a dual solution, and both together form a KKT point.

2.5 A Specially Tailored Adaptation of SCIP

We are going to solve the topology optimization problem (3.2.1) as stated in Chapter 3 by SCIP. As described in Section 2.2, SCIP solves this problem by branch-and-bound, separation and spatial branching. However, its solving performance for real-world instances is insufficient as demonstrated in Section 1.4.

The implementation framework of SCIP allows to influence the solution process in different directions. In the following we describe important functionalities of this MINLP solver that we are going to exploit in this thesis. We also shortly refer to each chapter where this functionality is used in order to improve the solving performance of (3.2.1).

SCIP allows tracking the branching tree. For every node of this tree the user has the possibility to influence the solving behavior. So it is for instance possible to prune a node manually. We only prune a node if we can guarantee that the corresponding problem of this node is infeasible. Therefor we call different routines (as described in Chapter 4 and Chapter 6) in order to detect infeasibility of the current node. One of these routines for example consists of solving a convex relaxation of the current node. As described in Section 2.3 every local optimum of a convex optimization problem is a global optimum. Using the solver IPOPT which only computes a local optimum we are able to solve globally the problem associated with the current node of the branching tree. If we detect infeasibility, then we prune the node. If these routines yield a primal feasible solution, then we add it to the solution pool of SCIP. If possible, then SCIP itself prunes the current node.

As mentioned and described in Section 2.4 the nonlinear solver IPOPT computes a local optimal primal solution together with a corresponding dual solution. Primal and dual solution together form a KKT point, i.e., they fulfill the KKT conditions (2.4.2). These conditions have a practical interpretation for our application, as discussed in Chapter 4. This interpretation allows to derive inequalities as described in Chapter 5, which are valid for the topology optimization problem (3.2.1). We add these inequalities to the *cut pool* of SCIP, while SCIP itself manages the handling of these inequalities.

SCIP also allows to include primal heuristics. These are handled similarly to the already contained heuristics in terms of frequency and order. We complement the heuristics of SCIP by another one which is specially tailored to the topology optimization problem (3.2.1), see Chapter 7.

Chapter 3

An MINLP Model for Gas Network Topology Optimization

In this section we present an MINLP model for the topology optimization problem. Therefor we first describe the technical background of gas networks in Section 3.1. Then the MINLP model is explained in Section 3.2 followed by a complexity analysis in Section 3.3. Two different subproblems of the MINLP are defined in Section 3.4. They form the origin of the discussions in the ongoing parts in this thesis. Finally the computational setup is described in Section 3.5 and different (real-world) instances are presented which we consider for our computations in the next chapters.

3.1 Technical Background

Let us give a mathematical description of the physical and technical properties of a gas transport network. In terms of the notation in the context of graph theory we refer to Korte and Vygen (2007).

We use a directed graph $G = (V, A)$ to model the network, where V denotes the set of nodes and $A \subseteq V \times V$ the set of arcs. Apart from the arc models which are described later we introduce the following generic variables and constraints. We have variables for the flow through each arc a that are bounded by $\underline{q}_a$ and $\overline{q}_a$:

$$q_a \in [\underline{q}_a, \overline{q}_a] \qquad \forall\, a \in A.$$

A positive value corresponds to a flow in the direction of the arc, and a negative value is a flow in the opposite direction. A gas flow through a pipe is induced if the gas has different pressures at the end nodes of the pipe. Hence each node of the gas transport network is associated with a pressure value. We model these pressures, but, instead of using variables which model the pressures directly, we model the squared pressure at each node v by

$$\pi_v \in [\underline{\pi}_v, \overline{\pi}_v] \qquad \forall\, v \in V.$$

Here $\underline{\pi}_v$ and $\overline{\pi}_v$ are the specified bounds for the variable. We also say that this variable π_v models the *node potential* at node v, and $\underline{\pi}_v, \overline{\pi}_v$ are the *node potential bounds*. Furthermore we add the pressure variable p_v when needed.

The arcs of the gas transport network (V, A) represent the various elements. Recall that we distinguish between passive network elements, namely pipes, whose behavior cannot be influenced by the network operator, and active network elements, namely valves, control valves, and compressors. In the following we describe each element individually. For a more detailed description we also refer to Fügenschuh et al. (2015).

Pipes

The majority of the arcs in a gas transport network are pipes. A precise description of the model of a pipe is given in Pfetsch et al. (2014), Koch et al. (2015), and Geißler (2011). We give a brief summary. A pipe is specified by its *length* L, *diameter* D and *roughness* k of the pipe wall. We assume all pipes to be straight and of cylindrical shape and restrict to the modeling of *one-dimensional* flow in the pipe direction x. Under these assumptions the mass flow q is related to *gas density* ρ and *velocity* v via

$$q = A\,\rho\,v, \tag{3.1.1}$$

where $A = D^2\pi/4$ denotes the constant *cross-sectional area* of the pipe. As pipes in Germany are usually at least one meter below the ground it is reasonable to assume the *temperature* T to be constant. In such a situation isothermal flow is an appropriate model. Now the gas flow in such a pipe is described by the following set of nonlinear, hyperbolic partial differential equations (see Feistauer 1993; Lurie 2008), often referred to as Euler Equations:

$$\frac{\partial \rho}{\partial t} + \frac{1}{A}\frac{\partial q}{\partial x} = 0, \tag{3.1.2a}$$

$$\frac{1}{A}\frac{\partial q}{\partial t} + \frac{\partial p}{\partial x} + \frac{1}{A}\frac{\partial (q\,v)}{\partial x} + g\,\rho\,s + \lambda\frac{|v|\;v}{2D}\rho = 0. \tag{3.1.2b}$$

The *continuity equation* (3.1.2a) and the *momentum equation* (3.1.2b) describe the *conservation of mass* and the *conservation of momentum*, respectively. Here,

$p = p(x,t)$, $q = q(x,t)$, $v = v(x,t)$ are pressure, mass flow and velocity in the direction of the pipe depending on *time* t. The constant g denotes the *gravitational acceleration* (with standard value $9.806\,65\,\mathrm{m\,s^{-2}}$), and $s \in [-1,1]$ denotes the constant *slope* of the pipe. Furthermore ρ denotes the *gas density* and λ the *friction factor*.

Since our model is intended to solve the topology expansion problem we have to construct it from a planner's perspective. Expansion problems are typically considered for the long-term network planning. In contrast to, e.g., real-time optimal control problems for gas networks, transient effects are neglected. Here a stationary model for the gas flow is reasonable where all time derivatives are zero. In this case the continuity equation (3.1.2a) simplifies to

$$\frac{\partial q}{\partial x} = 0, \tag{3.1.3}$$

which means that the gas flow is constant within a pipe.

The momentum equation (3.1.2b) relates all forces acting on gas particles to each other. Investigating the addends on the left-hand side of the momentum equation from left to right, the first term represents the flow rate change over time. The second term is the pressure gradient, followed by the so-called impact pressure. The fourth term represents the impact of gravitational forces which are influenced by the slope of the pipe. Finally, the last term is the most important one. It represents the friction forces acting on the gas particles due to pipe walls. These forces are responsible for the major part of pressure drop within pipes. As the time derivatives are zero, the first addend can be neglected. The third term $\partial_x(qv)/A$ contributes less than $1\,\%$ to the sum of all terms under normal operating conditions, see Wilkinson et al. (1964). So we assume that the term can be neglected, too. In conclusion, (3.1.2b) can be written as

$$\frac{\partial p}{\partial x} + g\,\rho\,s + \lambda \frac{|v|\;v}{2D}\rho = 0. \tag{3.1.4}$$

To calculate the friction factor λ we use the formula of Nikuradse (see Nikuradse 1933; Nikuradse 1950; Mischner 2012)

$$\lambda = \left(2\log_{10}\left(\frac{D}{k}\right) + 1.138\right)^{-2},$$

which is suitable on large transport networks where we typically have to deal with highly turbulent flows. Moreover, for the derivation of an algebraic pressure loss equation, we have to introduce the equation of state linking gas pressure and density

$$\rho = \frac{\rho_0 z_0 T_0}{p_0} \frac{p}{z(p,T)T}. \tag{3.1.5}$$

Here T_0 and p_0 are the *norm temperature* and *norm pressure*. The *compressibility factor* $z(p,T)$ characterizes the deviation of a real gas from ideal gas. We assume that this factor can be approximated by a suitable constant along the entire pipe which we denote z_m. The *compressibility factor under norm conditions* is denoted by z_0 and ρ_0 is the gas density under norm conditions.

Lemma 3.1.1:
The solution $p(x)$ to (3.1.4) with initial value $p(0) = p_{\text{in}}$ is given by

$$p(x)^2 = \left(p_{\text{in}}^2 - \tilde{\Lambda}\,|q|\,q \frac{e^{\tilde{S}x} - 1}{\tilde{S}}\right) e^{-\tilde{S}x} \tag{3.1.6}$$

with

$$\tilde{S} := 2\,g\,s \frac{\rho_0 z_0 T_0}{z_m\,T\,p_0}, \quad \tilde{\Lambda} := \lambda \frac{p_0\,z_m\,T}{\rho_0 z_0 T_0\,A^2\,D}.$$

Proof. In (3.1.4), we replace the gas velocity v by the mass flow q using (3.1.1) and the gas density ρ by the pressure p using the equation of state (3.1.5). This yields

$$\frac{\partial p}{\partial x} + g \frac{\rho_0 z_0 T_0\,p}{p_0\,z_m\,T} s + \lambda \frac{|q|\,q}{2\,A^2\,D} \frac{p_0\,z_m\,T}{\rho_0 z_0 T_0\,p} = 0,$$

where we use the assumption that the gas temperature and compressibility factor are constants T and z_m, respectively. Multiplication by $2\,p$ leads to

$$\frac{\partial}{\partial x} p^2 + \tilde{S}\,p^2 = -\tilde{\Lambda}\,|q|\,q.$$

By substituting $y = p^2$ we obtain the first-order linear ordinary differential equation (ODE)

$$\frac{\partial}{\partial x} y + \tilde{S}\,y = -\tilde{\Lambda}\,|q|\,q, \quad y(0) = p_{\text{in}}^2.$$

This ODE can be solved analytically by "variation of constants" and we arrive at

$$y(x) = p(x)^2 = \left(-\tilde{\Lambda}\,|q|\,q\,\frac{1}{\tilde{S}} e^{\tilde{S}x} + p_{\text{in}}^2 + \tilde{\Lambda}\,|q|\,q\,\frac{1}{\tilde{S}}\right) e^{-\tilde{S}x},$$

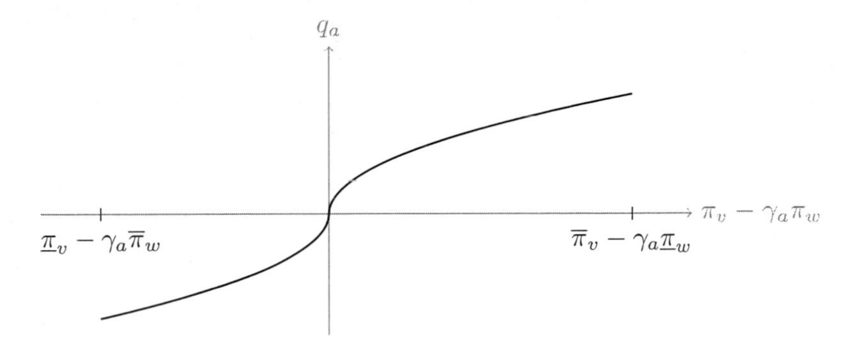

Figure 3.1: Nonlinear relation $\alpha_a \, q_a |q_a|^{k_a} = \pi_v - \gamma_a \pi_w$ of arc flow and node potential differences of a pipe $a = (v, w)$ for $\alpha_a = 1$, $k_a = 1$, and $\gamma_a = 1$. Increasing the arc flow goes along with increasing the difference of the node potentials π_v and π_w at the end nodes v and w.

where the last two terms in parentheses represent the integration constant obtained from the initial value $y(0) = p_{\text{in}}^2$. This concludes the proof for $s \neq 0$. For the other case $s = 0$ we proceed by taking the limit for $s \to 0$ (equivalently $\tilde{S} \to 0$) in (3.1.6) using l'Hôpital's rule. □

By evaluating the solution of (3.1.6) at $x = L$ (with $p(L) = p_{\text{out}}$) and fixing the notation $\Lambda := \tilde{\Lambda} \, L$ and $S := \tilde{S} \, L$, we finally obtain a well-known relationship of inlet and outlet pressures and the mass flow through the pipe, see, e.g., Lurie (2008):

$$p_{\text{out}}^2 = \left(p_{\text{in}}^2 - \Lambda \, |q| \, q \, \frac{e^S - 1}{S} \right) e^{-S} \tag{3.1.7}$$

with

$$\Lambda = \lambda \frac{p_0 \, z_m \, T \, L}{\rho_0 z_0 T_0 \, A^2 D}, \qquad S = 2 \, g \, s \frac{\rho_0 z_0 T_0 \, L}{p_0 \, z_m \, T}. \tag{3.1.8}$$

Recall that the variable $q_a \in \mathbb{R}$ represents the arc flow for a pipe $a = (v, w) \in A$, where a positive value is a flow from v to w, and a negative value is a flow in the opposite direction from w to v. Furthermore the variables π_v, π_w model the node potential values. The fundamental equation we use for a pipe $a = (v, w)$ is

$$\alpha_a \, q_a |q_a|^{k_a} = \pi_v - \gamma_a \pi_w. \tag{3.1.9}$$

Here $\alpha_a \in \mathbb{R}_{\geq 0}$, $k_a \in \mathbb{R}_{\geq 0}$ and $\gamma_a \in \mathbb{R}_{\geq 0} \setminus \{0\}$ are constants that subsume all physical properties of the pipe, the flow, and the interactions of the flow with the pipe. According to (3.1.7) and (3.1.8) we set

$$k_a := 1, \qquad \alpha_a := \Lambda \frac{e^S - 1}{S}, \qquad \gamma_a := e^S.$$

The constant γ_a in particular represents the height difference between nodes v and w. If some pipelines $a_1, \ldots, a_n$ form a circle, it is assumed that $\gamma_{a_1} \cdot \ldots \cdot \gamma_{a_n} = 1$. This reflects that there is no height difference when traversing a circuit completely. If $a = (v, w)$ is an arc and $a' = (w, v)$ is its anti-parallel counterpart, then we assume that the constants γ_a are such that $\gamma_a = \gamma_{a'}^{-1}$. Although each arc in principle might have a different value for k_a it is natural to assume that all pipes have the same constant in a gas network. However, this equality is no assumption for our proceeding work. A visualization of equation (3.1.9) is given in Figure 3.1.

Valves

A valve is installed in the network to separate or to join two independent nodes. The spatial dimension of a valve is small in comparison to the pipes. For a discrete decision valves allow either being open or closed. In our model the node potential values are identified when the valve is open ($\pi_v = \pi_w$). If the valve is closed then they are decoupled and the flow is forced to zero ($q_a = 0$). A binary variable $x_a \in \{0, 1\}$ distinguishes between these states. With this variable we use the following constraints for a valve $a = (v, w)$:

$$
\begin{aligned}
x_a = 0 \Rightarrow q_a &= 0, \\
x_a = 1 \Rightarrow \pi_v &= \pi_w, \\
q_a, \pi_v, \pi_w &\in \mathbb{R}, \\
x_a &\in \{0, 1\}.
\end{aligned}
$$

Compressors

At certain locations in gas transport networks it is necessary to increase the gas pressure. For example, if the pressure is too low after a transport distance of 100 km to 150 km, then compressors might be used to increase it again. For the mathematical description of such an active network element, various models exist in the literature, see Carter et al. (1993), Carter (1996), and Wu et al. (2000) for instance. We follow the approach of De Wolf and Smeers (2000), and make use of the following formulation for a compressor $a = (v, w)$:

$$\alpha_a \, q_a |q_a|^{k_a} \geq \pi_v - \pi_w. \tag{3.1.10}$$

Here $\alpha_a \in \mathbb{R}_{\geq 0}$ and $k_a \in \mathbb{R}_{\geq 0}$ are constants. We introduce a slack variable $y_a \in [\underline{y}_a, \overline{y}_a]$ where $\underline{y}_a$ and $\overline{y}_a$ are lower and upper bounds. Then we rewrite inequality (3.1.10) as equality

$$\alpha_a \, q_a |q_a|^{k_a} - \beta_a y_a = \pi_v - \pi_w \tag{3.1.11}$$

where $\beta_a \in \mathbb{R}$ is the weight of the slack variable y_a.

In practice we need further restrictions such as a minimal and a maximal pressure difference or a restriction of the pressure difference that depends on the flow. Therefor we allow a linear inequality system coupling the flow q_a and the pressures p_v, p_w such that

$$A_a \left(q_a, p_v, p_w\right)^T \leq b_a. \tag{3.1.12}$$

Here it holds $A_a \in \mathbb{R}^{\nu_a \times 3}$ and $b_a \in \mathbb{R}^{\nu_a}$ for some value $\nu_a \in \mathbb{N}$. When the slack variable y_a is unbounded, then the only restrictions for the compressor are given by constraint (3.1.12). Hence (3.1.12) specifies the *operating range* of the compressor. The flow can only go in positive direction through a compressor, hence a corresponding lower bound needs to be set by this linear inequality system, i.e., $q_a \geq 0$.

The previous description belongs to the *active mode* of a compressor. Additionally, a compressor has other *operation modes* and can be in *bypass* and *closed* mode. These modes are identical with the modes of a valve. We introduce three binary variables $x_{a,0}, x_{a,1}, x_{a,2} \in \{0, 1\}$ which model the closed, bypass and active mode respectively. With these decision variables we use the following constraints and variable bounds for a compressor $a = (v, w)$:

$$\begin{aligned}
x_{a,0} + x_{a,1} + x_{a,2} &= 1, \\
p_v^2 &= \pi_v, \\
p_w^2 &= \pi_w, \\
\underline{y}_a \leq y_a &\leq \overline{y}_a, \\
q_a, y_a, \pi_v, \pi_w, p_v, p_w &\in \mathbb{R}, \\
x_a &\in \{0, 1\},
\end{aligned}$$

$$\begin{aligned}
&x_{a,0} = 1 \Rightarrow q_a = 0, \\
&x_{a,1} = 1 \Rightarrow \pi_v = \pi_w \\
&x_{a,2} = 1 \Rightarrow \alpha_a \, q_a |q_a|^{k_a} - \beta_a y_a = \pi_v - \pi_w, \quad A_a \left(q_a, p_v, p_w\right)^T \leq b_a.
\end{aligned} \tag{3.1.13}$$

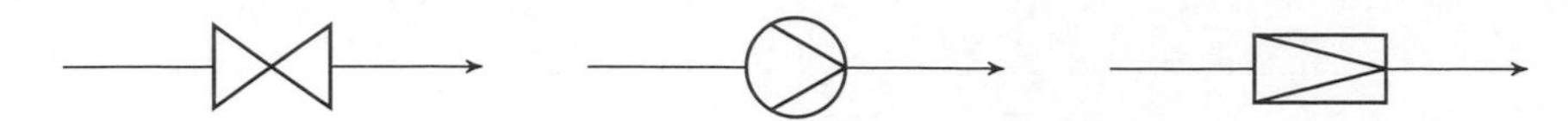

Figure 3.2: Technical arc symbols of a valve, a compressor and a control valve.

Control Valves

It can be necessary to reduce the pressure along an arc $a = (v, w)$ in the network in order to protect parts of the network from high pressures due to pressure limits of older pipelines. Technically this reduction is realized by a control valve that reduces the gas pressure. A control valve $a = (v, w)$ is inverse to a compressor. Hence we model it similarly to a compressor by

$$\alpha_a \, q_a |q_a|^{k_a} - \beta_a y_a = \pi_v - \pi_w. \tag{3.1.14}$$

Here $\alpha_a \in \mathbb{R}_{\geq 0}$, $k_a \in \mathbb{R}_{\geq 0}$, and $\beta_a \in \mathbb{R}$ are constants. The difference between a compressor and a control valve is either the sign of β_a or the bounds on y_a. Note that the flow direction through a control valve is also fixed by setting the lower bound to zero, i.e., $q_a \geq 0$.

Similar to a compressor, the feasible region of a control valve is restricted by additional constraints like (3.1.12). For instance we set a minimal and maximal pressure difference between the end nodes v and w. Furthermore bypass and closed mode are available. Concluding, we model a control valve $a = (v, w)$ by (3.1.13) which is exactly the same as the model of a compressor. For a more detailed description of a control valve we refer to Cerbe (2008).

3.2 An MINLP Model

In the previous section we presented models for the different network elements pipelines, valves, control valves and compressors. Let us integrate these results into a mixed-integer nonlinear programming model for the topology optimization problem. Recall that the problem is to compute a cost-efficient selection of network elements that have to be added to the network in order to make the nomination feasible which is specified by the vector d.

Multigraph

We model the topology optimization problem on a multigraph. This multigraph is obtained as follows: Given the directed graph $G = (V, A)$ which represents the original gas transport network, we define an extended set of arcs $A_X \subseteq V \times V \times \mathbb{N}_{\geq 0}$

where each arc $(v, w, i) \in A_X$ represents the arc $a = (v, w) \in V \times V$ together with index i. This set A_X contains all "original" arcs from A with the additional index 1, that is, $(a, 1) \in A_X$ for all $a \in A$. Each valve $a \in A$ is *additionally* represented by the arc $(a, 0) \in A_X$ to indicate the status that the valve is closed. Hence the arc $(a, 1)$ is associated with an open valve. Each arc $a \in A$ that represents a control valve or a compressor is additionally represented by the arcs $(a, 0), (a, 2) \in A_X$ to indicate the state that the active element is closed or active, respectively. The arc $(a, 1)$ is associated with the state bypass.

Furthermore the extended arc set A_X contains possible new network elements (pipes, valves, compressors, or control valves), where in principle a new element can be built between any pair of existing nodes $v, w \in V, v \neq w$. A possible new extension between nodes v and w is represented by two arcs: $(v, w, 0)$ to indicate in the model below that the element is not built and $(v, w, 2)$ to model that the new element is built. Here we assume that a new active element is only in active mode. This is a reasonable assumption because the bypass mode of a compressor or a control valve equals the open mode of a valve which we refer to as its active mode. Furthermore the closed mode of these elements means that they are not added to the network.

Several loops for an existing arc $a \in A$ are represented by (a, i) for index $i \in \{2, 3, \ldots\}$ to model a given set of design parameters. We always assume that arc (a, i) contains the capacity of $(a, 1)$ for $i \geq 2$. In the model below we select exactly one of the arcs $(a, 1), (a, 2), \ldots$ to be allowed to transport flow.

By (V, A_X) we denote the gas transport network together with its possible extensions. Note that (V, A_X) is a graph with multiple parallel arcs and hence a multigraph. The ongoing model is defined on this multigraph.

Variables

Let us introduce the following variables. The flow on arc $(a, i) \in A_X, i \neq 0$ is denoted by $q_{a,i} \in \mathbb{R}$, where a positive value means that the flow is heading in the same direction as the arc, and a negative value indicates the opposite direction. Note that we do not add a flow variable for an arc $(a, 0) \in A_X$ because this arc corresponds to the closed state of an active element or indicates that an extension is not added to the network. The potential value of a node $v \in V$ is modeled by $\pi_v \in \mathbb{R}$. The pressure itself is modeled by $p_v \in \mathbb{R}$. The variable $y_{a,i} \in \mathbb{R}$ specifies the slack component in (3.1.11) and (3.1.14). For pipelines and valves this variable is fixed to zero. We introduce a binary decision variable $x_{a,i} \in \{0, 1\}$ for each arc $(a, i) \in A_X$, where $x_{a,i} = 1$ represents the decision that arc (a, i) is used (i.e., a necessary condition for a nonzero flow).

The bounds of these variables are given as parameters. For each node $v \in V$ we have lower and upper bounds on the node potential, $\underline{\pi}_v, \overline{\pi}_v \in \mathbb{R}$ with $\underline{\pi}_v \leq \overline{\pi}_v$. For each arc $(a,i) \in A_X$ we have lower and upper bounds on the flow, $\underline{q}_{a,i}, \overline{q}_{a,i} \in \mathbb{R}$ with $\underline{q}_{a,i} \leq \overline{q}_{a,i}$. Furthermore bounds on the weighted slack variable $\underline{y}_{a,i}, \overline{y}_{a,i} \in \mathbb{R}$ with $\underline{y}_{a,i} \leq \overline{y}_{a,i}$ are given.

Constraints

We use the following constraints for modeling the topology optimization problem. The node flow is specified by the nomination d. To ensure a consistent flow model, flow conservation constraints

$$\sum_{\substack{(a,i)\in\delta^+_{A_X}(v),\\ i\neq 0}} q_{a,i} - \sum_{\substack{(a,i)\in\delta^-_{A_X}(v),\\ i\neq 0}} q_{a,i} = d_v \qquad \forall\, v \in V$$

state that the balance of entering and leaving flows has to match exactly the nominated amount d_v at each node $v \in V$. Furthermore each arc flow variable is coupled with the node potential values at its end nodes. More precisely, each arc $(a,i) = (v,w,i) \in A_X, i \neq 0$, which represents either a pipeline, a valve, a control valve or a compressor, is modeled by this basic equation:

$$x_{a,i} = 1 \Rightarrow \alpha_{a,i}\, q_{a,i}|q_{a,i}|^{k_a} - \beta_{a,i} y_{a,i} - (\pi_v - \gamma_a \pi_w) = 0.$$

It is then specified by appropriate parameters $\alpha_{a,i}, \beta_{a,i}, \gamma_a$ and k_a as described in the previous Section 3.1. We have an arc coefficient $\alpha_{a,i} \in \mathbb{R}_{\geq 0}$, a weighting factor $\beta_{a,i} \in \mathbb{R}_{\geq 0}$ for the slack variable, a coefficient $\gamma_a \in \mathbb{R}_{\geq 0}\setminus\{0\}$, and the power $k_a \in \mathbb{R}_{\geq 0}$. If $\alpha_{a,i} = 0$ for an arc $(a,i) = (v,w,i)$ then we assume $\beta_{a,i} = 0$ and $\gamma_a = 1$ such that the arc is modeled by $\pi_v = \pi_w$. For γ_a we have the following conditions: Let (V,E) be the undirected version of (V,A) obtained by removing the orientation of the arcs $a \in A'$. This way each arc $a \in A$ uniquely corresponds to an edge in $e \in E$ and vice versa. Consider a circuit in (V,E). Let $v_1, \ldots, v_n$ be the nodes of this circuit with $v_n = v_1$, $e_1, \ldots, e_{n-1}$ the edges and $a_1, \ldots, a_{n-1}$ the corresponding arcs. Then we assume

$$\left(\prod_{\substack{i=1\\ a_i=(v_i,v_{i+1})}}^{n-1} \gamma_a\right)\left(\prod_{\substack{i=1\\ a_i=(v_{i+1},v_i)}}^{n-1} \gamma_a^{-1}\right) = 0.$$

This condition ensures a zero height difference in total when traversing the edges of the circuit in clockwise direction. Note that those corresponding arcs a_i which are not oriented in clockwise direction, i.e., $a_i = (v_{i+1}, v_i)$, are traversed in backward direction. Here the value $\gamma_{a_i}^{-1}$ is regarded.

For those arcs $(a,i) \in A_X$ which are representing active compressors or control valves we assume additional data being given as a $\nu_a \times 3$-dimensional matrix A_a and ν_a-dimensional vector b_a (here ν_a is the number of linear constraints necessary to describe the operating range). We add the constraint

$$x_{a,i} = 1 \Rightarrow A_a \left(q_{a,i}, p_v, p_w\right)^T \le b_a.$$

For all the other arcs we set $A_a = 0$ and $b_a = 0$.

To complete the model for arcs, we ensure that the flow is fixed to zero in the case that arc (a,i) is not used, i.e.,

$$x_{a,i} = 0 \Rightarrow q_{a,i} = 0.$$

Furthermore we add the constraint

$$\sum_{i:(a,i)\in A_X} x_{a,i} = 1 \qquad \forall a \in A$$

to ensure that only one state of the arc is selected.

For technical reasons only a subset of all possible discrete settings is allowed. This subset is given by $\mathcal{X} \subseteq \{0,1\}^{A_X}$. The set $\mathcal{X}$ is described by linear inequalities in general, i.e.,

$$\mathcal{X} = \left\{x \in \{0,1\}^{A_X} \mid Lx \le t\right\},$$

where L is a matrix of integers with $|A_X|$ columns and t is vector of integers such that the length of t is equal to the number of rows of L.

Objective

A cost coefficient $c_{a,i} \in \mathbb{R}_{\ge 0}$ for each arc $(a,i) \in A_X$ reflects the costs of using arc (a,i). We set $c_{a,1} := 0$ for all existing arcs $a \in A$ and $c_{a,0} := 0$ for every active element $a \in A$. Furthermore we set $c_{a,2} := 0$ for all compressors and control valves $a \in A$. The objective function then is

$$\min \sum_{(a,i)\in A_X} c_{a,i}\, x_{a,i}.$$

MINLP Model

We summarize the previous variables and constraints. The following nonlinear non-convex mixed-integer program with indicator constraints is used for modeling the topology optimization problem:

$$\min \sum_{(a,i)\in A_X} c_{a,i}\, x_{a,i} \tag{3.2.1a}$$

s. t.

$$x_{a,i} = 1 \Rightarrow \alpha_{a,i}\, q_{a,i}|q_{a,i}|^{k_a} - \beta_{a,i} y_{a,i} - (\pi_v - \gamma_a \pi_w) = 0 \qquad \forall\, (a,i) \in A_X,\ a = (v,w), i \neq 0, \tag{3.2.1b}$$

$$x_{a,i} = 1 \Rightarrow A_a\, (q_{a,i}, p_v, p_w)^T \leq b_a \qquad \forall\, (a,i) \in A_X,\ a = (v,w), i \geq 2, \tag{3.2.1c}$$

$$x_{a,i} = 0 \Rightarrow q_{a,i} = 0 \qquad \forall\, (a,i) \in A_X, i \neq 0, \tag{3.2.1d}$$

$$\sum_{i:(a,i)\in A_X} x_{a,i} = 1 \qquad \forall\, a \in A, \tag{3.2.1e}$$

$$\sum_{\substack{(a,i)\in\delta^+_{A_X}(v),\\ i\neq 0}} q_{a,i} - \sum_{\substack{(a,i)\in\delta^-_{A_X}(v),\\ i\neq 0}} q_{a,i} = d_v \qquad \forall\, v \in V, \tag{3.2.1f}$$

$$p_v|p_v| - \pi_v = 0 \qquad \forall\, v \in V, \tag{3.2.1g}$$

$$L\,x \leq t, \tag{3.2.1h}$$

$$\pi_v \leq \overline{\pi}_v \qquad \forall\, v \in V, \tag{3.2.1i}$$

$$\pi_v \geq \underline{\pi}_v \qquad \forall\, v \in V, \tag{3.2.1j}$$

$$q_{a,i} \leq \overline{q}_{a,i} \qquad \forall\, (a,i) \in A_X, i \neq 0, \tag{3.2.1k}$$

$$q_{a,i} \geq \underline{q}_{a,i} \qquad \forall\, (a,i) \in A_X, i \neq 0, \tag{3.2.1l}$$

$$x_{a,i} = 1 \;\Rightarrow\; y_{a,i} \leq \overline{y}_{a,i} \qquad \forall\, (a,i) \in A_X, i \neq 0, \tag{3.2.1m}$$

$$x_{a,i} = 1 \;\Rightarrow\; y_{a,i} \geq \underline{y}_{a,i} \qquad \forall\, (a,i) \in A_X, i \neq 0, \tag{3.2.1n}$$

$$p_v, \pi_v \in \mathbb{R} \qquad \forall\, v \in V, \tag{3.2.1o}$$

$$q_{a,i}, y_{a,i} \in \mathbb{R} \qquad \forall\, (a,i) \in A_X, i \neq 0, \tag{3.2.1p}$$

$$x_{a,i} \in \{0,1\}\ \forall\, (a,i) \in A_X. \tag{3.2.1q}$$

Hereinafter we refer to this model as topology optimization problem (3.2.1). The aforementioned model can also be used in the context of water network optimization, see Bragalli et al. (2012), Burgschweiger et al. (2009), and Gleixner et al. (2012). There the task is to operate the network and to compute a selection of pipe diameters such that the specified nomination is feasible in the network.

Example

In Figure 3.3 we show an example of a small network to demonstrate our notation. In the first part 3.3a, the original network (V, A) is shown. It contains five pipes and one control valve. In 3.3b the arc flow and node potential variables are shown. The last part 3.3c of the figure shows the decision variables x. Simple arcs, such as $(2,1),(5,1),(6,1)$ which correspond to the original arcs $2,5$ and 6 represent pipelines of the network. In particular, these pipelines are not extendible via loops, i.e., by adding parallel pipes. A valve is shown in arc pair $(4,0),(4,1)$. Note that there is **no** flow variable $q_{4,0}$. Multiple arcs such as $(1,1),(1,2)$ and $(3,1),(3,2),(3,3)$ represent each a pipeline (which is $(1,1)$ and $(3,1)$), together with one or two possible loop extensions, respectively.

3.3 Complexity Analysis

Let us characterize the complexity of the topology optimization problem (3.2.1). We reduce the NP-hard problem 3SAT, see Korte and Vygen (2007) for details, to the topology optimization problem (3.2.1).

3SAT is defined as follows: A collection of clauses $\mathcal{Z} = \{Z_1, Z_2, \ldots, Z_r\}$ in boolean variables $x_1, x_2, \ldots, x_s$ is given. Each clause Z_i is a disjunction of three literals $l_{i,1}, l_{i,2}, l_{i,3}$ where a literal $l_{i,j}$ is either a variable x_k or its negation $\overline{x}_k$ for some $k \in \{1, \ldots, s\}$. The clause Z_i is satisfied if at least one of its literals is "true". We denote by $x_{i,j}$ the boolean variable of literal $l_{i,j}$, i.e., if $l_{i,j} = x_k$ or $l_{i,j} = \overline{x}_k$, then $x_{i,j} = x_k$. The problem is to determine whether $\mathcal{Z}$ is satisfiable, that is, whether there is a truth assignment to the variables $x_1, x_2, \ldots, x_s$, which simultaneously satisfies all the clauses in $\mathcal{Z}$.

Consider an instance of 3SAT. We define a corresponding instance of the topology optimization problem. It does not contain any extension arcs, i.e., the objective function equals zero and every feasible solution is an optimal one. The transmission network (V, A) of the instance is shown in Figure 3.4. It has $14r$ vertices, where each clause $Z_1, \ldots, Z_r$ is associated with 14 vertices. Each arc $a \in A$ of the network is either a valve, or a pipe with constants $\alpha_a = \beta_a = 0$ and $\gamma_a = 1$, i.e., the pipe $a = (v, w)$ is modeled by $\pi_v = \pi_w$. The lower flow bound on each arc is zero, i.e., we

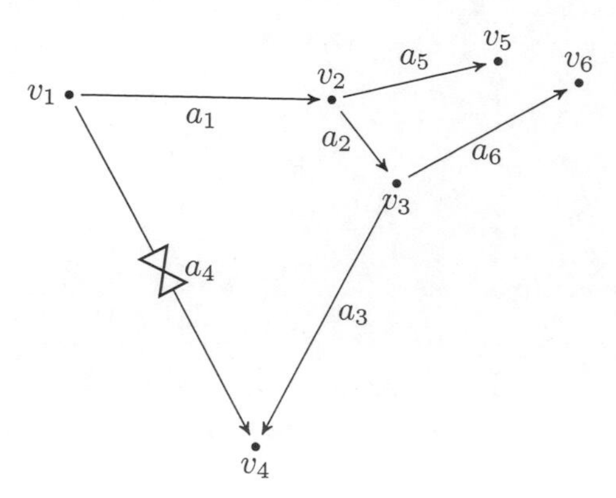

(a) Original network with 1 valve and 5 pipes.

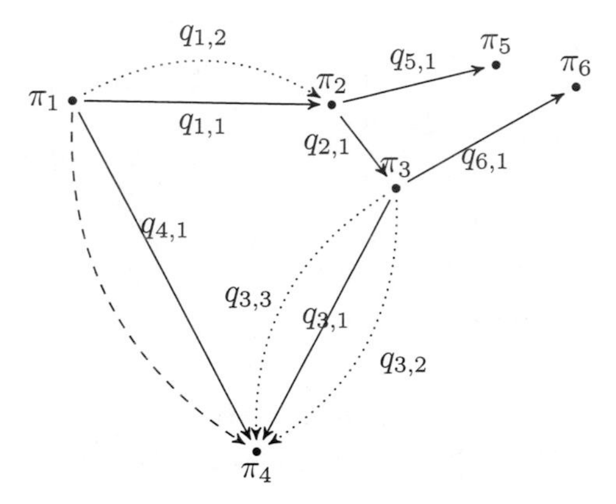

(b) Arc flow and node potential variables q and π.

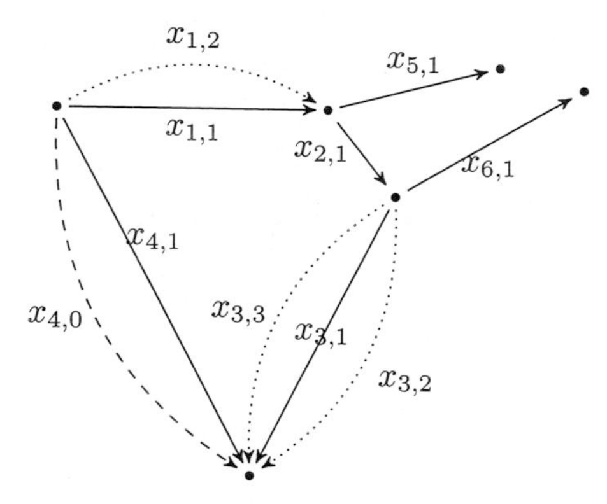

(c) Discrete (switching) variables x. The variables $x_{2,1}$ and $x_{5,1}$ and $x_{6,1}$ are fixed to 1.

Figure 3.3: Example of a network with binary and continuous variables of the associated topology optimization problem (3.2.1). The dashed arcs correspond to the valve in closed mode. The dotted arcs represent loops. Loops are available for the original arcs a_1 and a_3.

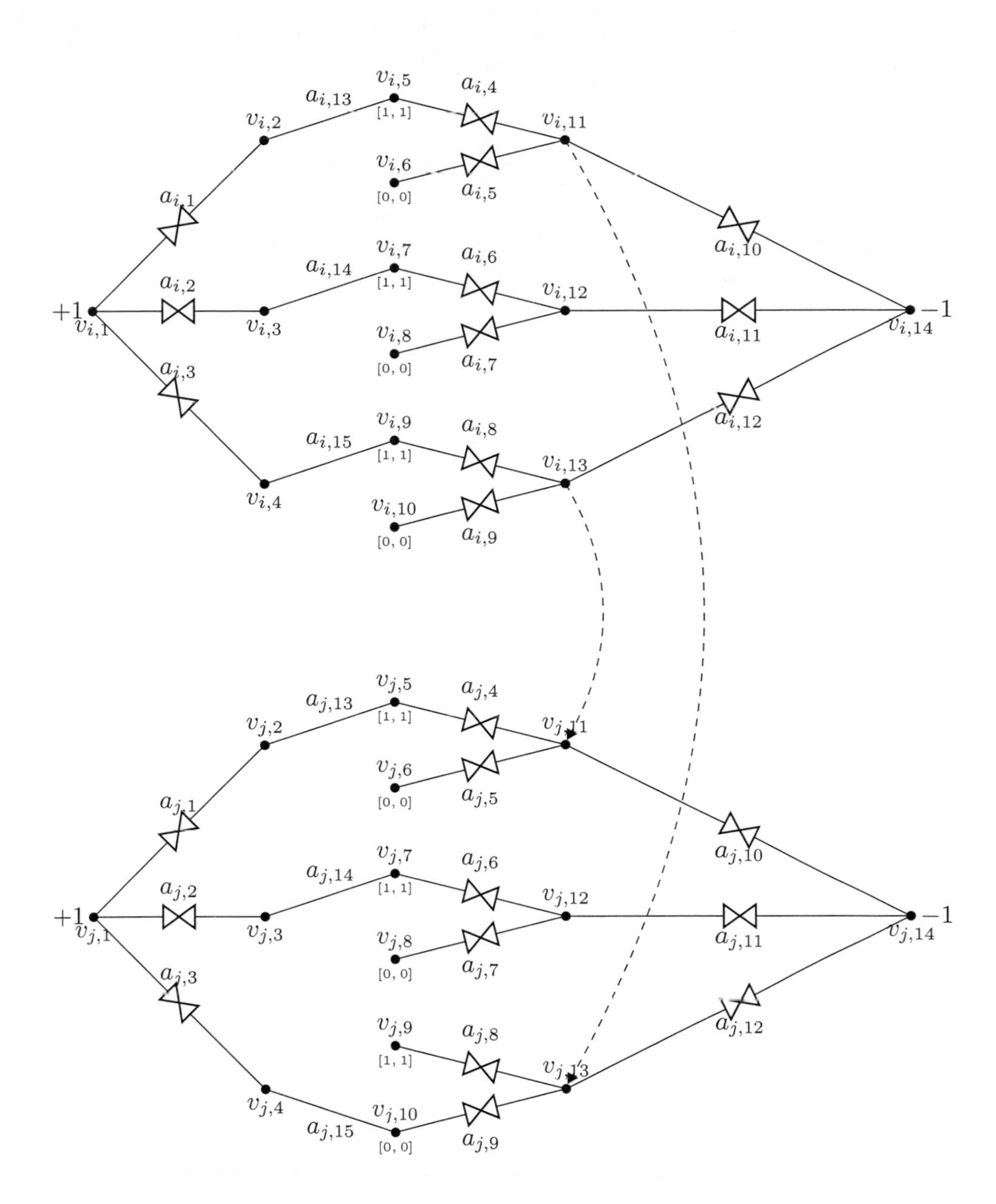

Figure 3.4: Two different clauses Z_i, Z_j with $i < j$ of an instance of 3SAT are associated with an instance of the topology optimization problem. The nodes and arcs associated with clause Z_i are indexed by i. Every arc is either a pipe or a valve restricted to nonnegative arc flows. The flow on the dashed arcs is fixed to zero. The intervals that are shown give lower and upper bounds for the node potential values. All other node potential values are unbounded. This figure exactly matches the instance $\mathcal{Z} = \{Z_i, Z_j\}$ with $Z_i = \{x_1, x_2, x_3\}$ and $Z_j = \{x_3, x_4, \overline{x}_1\}$, where $x_1, \ldots, x_4$ are the boolean variables.

allow nonnegative arc flow only. Let us denote the set of pipes by $A_p \subseteq A$ and the set of valves by $A_{va} \subseteq A$. Then the MINLP model (3.2.1) for this instance writes as

$$
\begin{aligned}
& \exists\, q, \pi && && (3.3.1)\\
\text{s.t.} \quad & x_{a,1} = 1 \Rightarrow \pi_v - \pi_w = 0 && \forall\, a = (v,w) \in A, \\
& x_{a,0} = 1 \Rightarrow \qquad q_a = 0 && \forall\, a \in A, \\
& \sum_{a \in \delta_A^+(v)} q_a - \sum_{a \in \delta_A^-(v)} q_a = d_v && \forall\, v \in V, \\
& x_{a,0} + x_{a,1} = 1 && \forall\, a \in A_{va}, \\
& x_{a,1} = 1 && \forall\, a \in A_p, \\
& \underline{\pi}_v \le \pi_v \le \overline{\pi}_v && \forall\, v \in V, \\
& 0 \le q_a \le \overline{q}_a && \forall\, a \in A, \\
& \pi_v \in \mathbb{R} && \forall\, v \in V, \\
& q_a \in \mathbb{R} && \forall\, a \in A, \\
& x_{a,0}, x_{a,1} \in \{0,1\} && \forall\, a \in A.
\end{aligned}
$$

Each clause Z_i is associated with a subnetwork of (V, A). The nodes and arcs of this subnetwork are indicated with index i. It consists of nodes $v_{i,1}, \ldots, v_{i,14}$, 12 valves and 3 pipes. The inflow at node $v_{i,1}$ equals one ($d_{v_{i,1}} = 1$) which is led out at node $v_{i,14}$ ($d_{v_{i,14}} = -1$). All other nodes of the subnetwork are transshipment nodes ($d_{v_{i,2}} = d_{v_{i,3}} = \ldots = 0$). The potential value at the nodes $v_{i,5}, \ldots, v_{i,10}$ is fixed by $\pi_{v_{i,5}} = \pi_{v_{i,7}} = \pi_{v_{i,9}} = 1$ and $\pi_{v_{i,6}} = \pi_{v_{i,8}} = \pi_{v_{i,10}} = 0$. For the other nodes of the subnetwork there are no restrictions on the node potential values. The topology of two subnetworks differs only in the arcs between node sets $\{v_{i,2}, v_{i,5}, v_{i,6}\}$ and $\{v_{i,3}, v_{i,7}, v_{i,8}\}$ and $\{v_{i,4}, v_{i,9}, v_{i,10}\}$. Here connecting arcs exist as follows:

$$
\begin{aligned}
(v_{i,2}, v_{i,5}) \in A &:\Leftrightarrow l_{i,1} = x_{i,1}, \\
(v_{i,2}, v_{i,6}) \in A &:\Leftrightarrow l_{i,1} = \overline{x}_{i,1}, \\
(v_{i,3}, v_{i,7}) \in A &:\Leftrightarrow l_{i,2} = x_{i,2}, \\
(v_{i,3}, v_{i,8}) \in A &:\Leftrightarrow l_{i,2} = \overline{x}_{i,2}, \\
(v_{i,4}, v_{i,9}) \in A &:\Leftrightarrow l_{i,3} = x_{i,3},
\end{aligned}
$$

$$(v_{i,4}, v_{i,10}) \in A :\Leftrightarrow l_{i,3} = \overline{x}_{i,3}.$$

For each pair of clauses Z_i, Z_j with index $i < j$ there is an arc between the subnetworks associated with Z_i and Z_j connecting $v_{i,10+k}$ and $v_{j,10+k'}$ if and only if $x_{i,k} = x_{j,k'}$ for each $k, k' \in \{1,2,3\}$. The flow on these arcs is fixed to zero, i.e., upper and lower flow bounds equal zero.

Theorem 3.3.1:
The topology optimization problem (3.2.1) *is NP-hard.*

Proof. We prove that 3SAT polynomially transforms to the topology optimization problem (3.3.1). We consider a feasible solution of 3SAT and define a feasible solution for (3.3.1). Initially we set a node potential of value zero for all nodes and flow value zero for all arcs. In the discussion below we will change some of these variables to different values. For each clause Z_i we open and close valves of the corresponding subnetwork by the following instructions:

Case $x_{i,1} = 1$**:** We open valve $a_{i,4}$ and close $a_{i,5}$.

Case $x_{i,1} = 0$**:** We open valve $a_{i,5}$ and close $a_{i,4}$.

Case $x_{i,2} = 1$**:** We open valve $a_{i,6}$ and close $a_{i,7}$.

Case $x_{i,2} = 0$**:** We open valve $a_{i,7}$ and close $a_{i,6}$.

Case $x_{i,3} = 1$**:** We open valve $a_{i,8}$ and close $a_{i,9}$.

Case $x_{i,3} = 0$**:** We open valve $a_{i,9}$ and close $a_{i,8}$.

Then we set the π values as follows for each clause Z_i. For those $\pi_{v_{i,\cdot}}$ which are fixed by its bounds we set $\pi_{v_{i,5}} := \pi_{v_{i,7}} := \pi_{v_{i,9}} := 1$ and $\pi_{v_{i,6}} := \pi_{v_{i,8}} := \pi_{v_{i,10}} := 0$. The other values $\pi_{v_{i,\cdot}}$ which are not fixed by its bounds are set as described below:

$$\begin{aligned} \pi_{v_{i,2}} &:= \pi_{v_{i,11}} := x_{i,1}, \\ \pi_{v_{i,3}} &:= \pi_{v_{i,12}} := x_{i,2}, \\ \pi_{v_{i,4}} &:= \pi_{v_{i,13}} := x_{i,3}. \end{aligned} \tag{3.3.2}$$

For the remaining valves we proceed by applying one of the following cases for open and closed status:

Case 1, $l_{i,1} = 1$**:** We open the valves $a_{i,1}$ and $a_{i,10}$ and close the remaining valves of the subnetwork. We send one unit of flow from $v_{i,1}$ to $v_{i,14}$ along the unique path containing $a_{i,1}$ and $a_{i,10}$. We set $\pi_{v_{i,1}} := \pi_{v_{i,2}}$ and $\pi_{v_{i,14}} := \pi_{v_{i,11}}$.

Case 2, $l_{i,1} = 0, l_{i,2} = 1$: We open the valves $a_{i,2}$ and $a_{i,11}$ and close the remaining valves of the subnetwork. We send one unit of flow from $v_{i,1}$ to $v_{i,14}$ along the unique path containing $a_{i,2}$ and $a_{i,11}$. We set $\pi_{v_{i,1}} := \pi_{v_{i,3}}$ and $\pi_{v_{i,14}} := \pi_{v_{i,12}}$.

Case 3, $l_{i,1} = 0, l_{i,2} = 0, l_{i,3} = 1$: We open the valves $a_{i,3}$ and $a_{i,12}$ and close the remaining valves of the subnetwork. We send one unit of flow from $v_{i,1}$ to $v_{i,14}$ along the unique path containing $a_{i,3}$ and $a_{i,12}$. We set $\pi_{v_{i,1}} := \pi_{v_{i,4}}$ and $\pi_{v_{i,14}} := \pi_{v_{i,13}}$.

We note that at least one literal is "true" which implies that exactly one of the previous cases applies. This in turn implies that the nomination for each subnetwork associated with Z_i is transported through the subnetwork. The arcs between two subnetworks associated with Z_i and Z_j for $i, j \leq r$ and $i \neq j$ ensure $\pi_{v_{i,10+k}} = \pi_{v_{j,10+k'}}$ if $x_{i,k} = x_{j,k'}$ for $k, k' \in \{1, 2, 3\}$. These conditions are fulfilled because of (3.3.2). Thus we obtain a feasible solution for the topology optimization problem (3.3.1).

Now we consider a feasible solution of the topology optimization problem (3.3.1) and derive a feasible solution for 3SAT. We proceed by assigning values to the x variables as follows:

Case 1 (valve $a_{i,4}$ or $a_{i,5}$ open): If $a_{i,4}$ is open, then we set $x_{i,1} := 1$. Otherwise we set $x_{i,1} := 0$. We note that at most one of the valves $a_{i,4}$ and $a_{i,5}$ is open because otherwise $1 = \pi_{v_{1,5}} = \pi_{v_{1,11}} = \pi_{v_{1,6}} = 0$ leads to a contradiction.

Case 2 (valve $a_{i,6}$ or $a_{i,7}$ open): If $a_{i,6}$ is open, then we set $x_{i,2} := 1$. Otherwise we set $x_{i,2} := 0$. We note that at most one of the valves $a_{i,6}$ and $a_{i,7}$ is open because otherwise $1 = \pi_{v_{1,7}} = \pi_{v_{1,12}} = \pi_{v_{1,8}} = 0$ leads to a contradiction.

Case 3 (valve $a_{i,8}$ or $a_{i,9}$ open): If $a_{i,8}$ is open, then we set $x_{i,3} := 1$. Otherwise we set $x_{i,3} := 0$. We note that at most one of the valves $a_{i,8}$ and $a_{i,9}$ is open because otherwise $1 = \pi_{v_{1,9}} = \pi_{v_{1,13}} = \pi_{v_{1,10}} = 0$ leads to a contradiction.

At least one of these cases applies because of the construction of the network and the nomination. The definition above is well-defined because of the arcs connecting two different subnetworks are associated with two different clauses. From these settings we obtain:

$$
\begin{aligned}
q_{a_{i,1}} > 0 \quad &\Rightarrow \quad q_{a_{i,4}} + q_{a_{i,5}} > 0 \quad \overset{\text{(case 1 applies)}}{\Rightarrow} \quad l_{i,1} = 1, \\
q_{a_{i,2}} > 0 \quad &\Rightarrow \quad q_{a_{i,6}} + q_{a_{i,7}} > 0 \quad \overset{\text{(case 2 applies)}}{\Rightarrow} \quad l_{i,2} = 1, \\
q_{a_{i,3}} > 0 \quad &\Rightarrow \quad q_{a_{i,8}} + q_{a_{i,9}} > 0 \quad \overset{\text{(case 3 applies)}}{\Rightarrow} \quad l_{i,3} = 1.
\end{aligned}
\tag{3.3.3}
$$

We note that the cases apply because no positive flow is valid on a valve in closed state. By construction of the nomination, we have $q_{a_{i,1}} + q_{a_{i,2}} + q_{a_{i,3}} = 1$ for each $i = 1, \dots, r$. Thus we obtain from (3.3.3) that $l_{i,1} \vee l_{i,2} \vee l_{i,3}$ is "true". Thus clause Z_i is fulfilled for every $i = 1, \dots, r$. We proceed by assigning 1 to all variables x_k which are not set by the previous definitions. This does not change the status of a clause and thus, the instance of 3SAT is feasible. □

3.4 The Passive and Active Transmission Problem

The topology optimization problem (3.2.1) has two types of subproblems. We call them *passive* and *active transmission problems.* Both problems state the problem of transmitting gas between entries and exits through a network where the operation modes of valves, control valves and compressors are fixed. In Chapter 4 and 6 we present efficient solution methods for solving them. The passive transmission problem is obtained by fixing the variables x and y in (3.2.1). We note that, if the network contains only valves and pipes or in the more general case $\underline{y} = \overline{y}$, this problem is obtained after fixing all binary decisions. We define the arc set A' so that it contains all the arcs that are allowed to carry flow, i.e., $A' := \{(a,i) \in A_X : x_{a,i} = 1, i > 0\}$. For the ease of notation we always write $a \in A'$ as abbreviation for $(a,i) \in A'$. Accordingly we set $\gamma_a := \gamma_{\tilde{a}}$ and $A_a := A_{\tilde{a}}$ and $b_a := b_{\tilde{a}}$ for $a = (\tilde{a}, i) \in A'$. Further we define $\tilde{\beta}_a := \beta_a y_a$ for the current value y_a for each arc $a \in A'$. Now the passive transmission problem writes as

$$\exists\, q, \pi, p \tag{3.4.1a}$$

$$\text{s.t.} \quad \alpha_a\, q_a |q_a|^{k_a} - \tilde{\beta}_a - (\pi_v - \gamma_a \pi_w) = 0 \quad \forall a = (v,w) \in A', \tag{3.4.1b}$$

$$A_a\, (q_a, p_v, p_w)^T \le b_a \quad \forall a = (v,w) \in A', \tag{3.4.1c}$$

$$\sum_{a \in \delta^+_{A'}(v)} q_a - \sum_{a \in \delta^-_{A'}(v)} q_a = d_v \quad \forall v \in V, \tag{3.4.1d}$$

$$p_v |p_v| - \pi_v = 0 \quad \forall v \in V, \tag{3.4.1e}$$

$$\pi_v \le \overline{\pi}_v \quad \forall v \in V, \tag{3.4.1f}$$

$$\pi_v \ge \underline{\pi}_v \quad \forall v \in V, \tag{3.4.1g}$$

$$q_a \le \overline{q}_a \quad \forall a \in A', \tag{3.4.1h}$$

$$q_a \ge \underline{q}_a \quad \forall a \in A', \tag{3.4.1i}$$

$$p_v, \pi_v \in \mathbb{R} \quad \forall\, v \in V, \tag{3.4.1j}$$

$$q_a \in \mathbb{R} \quad \forall\, a \in A'. \tag{3.4.1k}$$

This problem is called passive transmission problem because the flow on every arc is directly related to the node potential difference at the end nodes of the arc, except those arcs $a \in A'$ where $\alpha_a = 0$. Hence each arc $a \in A'$ can be regarded as a pipeline within the context of this passive transmission problem. In turn, pipelines are passive network elements.

The active transmission problem is obtained by fixing all binary variables x in (3.2.1). Again we denote by $A' := \{(a, i) \in A_X : x_{a,i} = 1, i > 0\}$ the set of arcs that are allowed to carry flow. Now the active transmission problem writes as follows:

$$\exists\, q, \pi, p, y \tag{3.4.2a}$$

$$\text{s. t.} \quad \alpha_a\, q_a |q_a|^{k_a} - \beta_a y_a - (\pi_v - \gamma_a \pi_w) = 0 \quad \forall\, a = (v, w) \in A', \tag{3.4.2b}$$

$$\sum_{a \in \delta^+_{A'}(v)} q_a - \sum_{a \in \delta^-_{A'}(v)} q_a = d_v \quad \forall\, v \in V, \tag{3.4.2c}$$

$$A_a\, (q_a, p_v, p_w)^T \le b_a \quad \forall\, a = (v, w) \in A', \tag{3.4.2d}$$

$$p_v |p_v| - \pi_v = 0 \quad \forall\, v \in V, \tag{3.4.2e}$$

$$\pi_v \le \overline{\pi}_v \quad \forall\, v \in V, \tag{3.4.2f}$$

$$\pi_v \ge \underline{\pi}_v \quad \forall\, v \in V, \tag{3.4.2g}$$

$$q_a \le \overline{q}_a \quad \forall\, a \in A', \tag{3.4.2h}$$

$$q_a \ge \underline{q}_a \quad \forall\, a \in A', \tag{3.4.2i}$$

$$y_a \le \overline{y}_a \quad \forall\, a \in A', \tag{3.4.2j}$$

$$y_a \ge \underline{y}_a \quad \forall\, a \in A', \tag{3.4.2k}$$

$$p_v, \pi_v \in \mathbb{R} \quad \forall\, v \in V, \tag{3.4.2l}$$

$$q_a, y_a \in \mathbb{R} \quad \forall\, a \in A'. \tag{3.4.2m}$$

This problem is called active transmission problem because it might contain active elements like compressors which are in active mode. In this case the flow is not directly related to the node potential values at the end nodes of the arc associated

instance	nodes	pipes	valves	other active elements
net1	20	29	0	0
net2	40	39	0	6
net3	85	80	44	5
net4	143	142	103	6
net5	396	402	261	29
net6	661	614	33	42
net7	4165	3983	308	133

Table 3.1: Information about the dimension of the networks. We use seven networks with different topologies for our computational studies. A visualization of these networks is shown in Figure 3.5 to Figure 3.11.

with this compressor. This is the crucial difference to the passive transmission problem.

Taking the definition of the active transmission problem we give a different view on the topology optimization problem: The task is to compute a cost-optimal choice of the binary variables x such that the corresponding active transmission problem is feasible.

3.5 Computational Setup

For our computational studies seven different networks net1 – net7 with different nominations are considered, see Figure 3.5 to Figure 3.11. These networks contain all types of active elements for pressure regulation, i.e., valves, control valves, and compressors. The dimensions of the underlying graphs are summarized in Table 3.1. The networks are obtained from literature and from our cooperation partner OGE.

Recall that we distinguish between three types of networks in this thesis as described in Section 1.5. The networks net4 and net5 with contracted compressors and control valves form the first type of network, namely those which contain only pipes and valves. The networks net1 and net2 with contracted active elements are considered with additional loop arcs in our computations and form the second type of network. The networks net3 – net7 contain compressors and control valves and form the third type of network.

For our computations in the following chapters we imposed a time limit of 11 h for the topology expansion instances and a time limit of 4 h for the nomination validation instances. In the experiments our main criteria for measuring performance are the number of solved instances, the running time, the number of branch-and-bound nodes needed to prove optimality and the gap. To average values over all instances of the test set, we use a *shifted geometric mean*. The shifted geometric mean of values $t_1, \dots, t_n$ with shift s is defined as $\sqrt[n]{\prod(t_i + s)} - s$. We use a shift of $s = 10$

for time, $s = 100$ for nodes and $s = 0$ for gap in order to reduce the effect of very easy instances in the mean values. Further, using a geometric mean prevents hard instances at or close to the time limit from overly dominating the measures. Thus the shifted geometric mean has the advantage that it reduces the influence of outliers in both directions.

As hardware for our computations we use a cluster of 64bit Intel Xeon X5672 CPUs at 3.20 GHz with 12 MByte cache and 48 GB main memory, running an OpenSuse 12.1 Linux with a gcc 4.6.2 compiler. As MINLP solver we use the mixed-integer nonlinear branch-and-bound framework SCIP 3.0.1 as already suggested in Chapter 2. Further we use the following software packages: CPLEX 12.1 (CPLEX) as linear programming solver, and IPOPT 3.10 (Wächter and Biegler 2006) as nonlinear solver within SCIP. SCIP reports an optimality gap which is defined as the ratio between the best upper bound u (i.e., the objective function value of best feasible solution) and the best lower bound ℓ. That is, gap $= (\frac{u-\ell}{\ell}) \cdot 100\,\%$. The framework (Lamatto++) is used as framework integrating the different software and handling the input data. Hyperthreading and Turboboost, special features of the Intel CPUs, are disabled. In all computational experiments we run only one job per node to reduce random noise in the measured running time that might be caused by delays if multiple processes share common resources, in particular the memory bus.

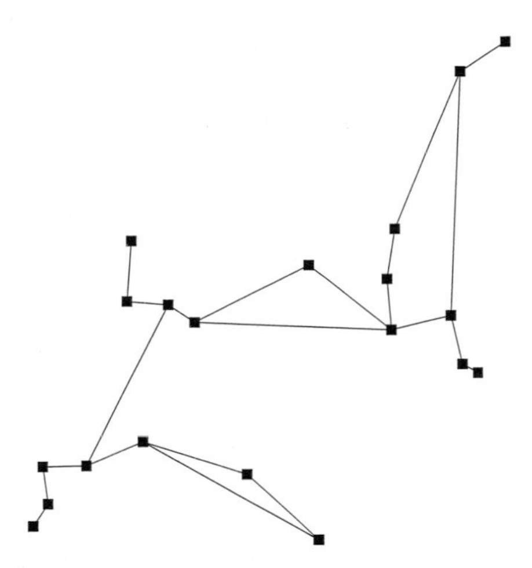

Figure 3.5: The test network net1. The data for this network can be found in (GAMS Model Library), and computational results for this network are given by De Wolf and Smeers (2000) and De Wolf and Bakhouya (2008). The network represents an approximation of the backbone network of the Belgium natural gas network. For our computational experiments we will consider a single nomination which we upscale by increasing values $2.0, 2.1, \ldots$ in order to model an increase of the transported gas.

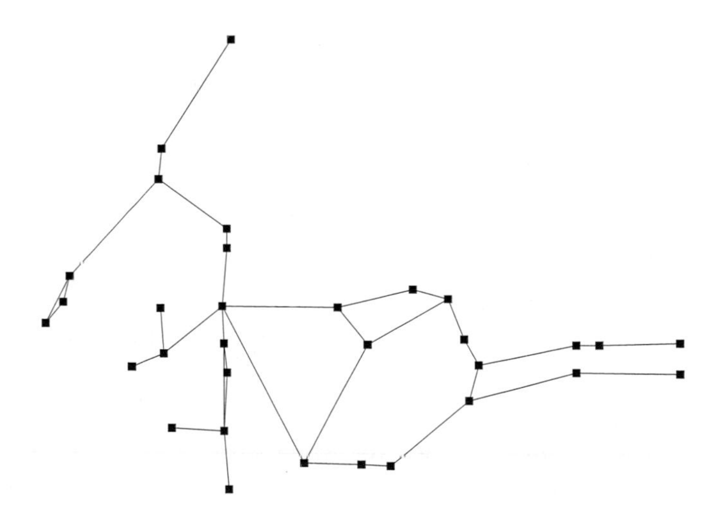

Figure 3.6: The test network net2. It is an approximation of parts of the German gas network in the Rhine-Main-Ruhr area. More precisely the length and the diameters of the pipelines are real-world data while other parameters like roughness or compressor data are set to realistic mean values. Altered data of network net2 with similar characteristic is publicly available at URL http://gaslib.zib.de under the name gaslib-40. For our computational experiments we will consider a single nomination which we upscale by increasing values $2.0, 2.1, \ldots$ in order to model an increase of the transported gas.

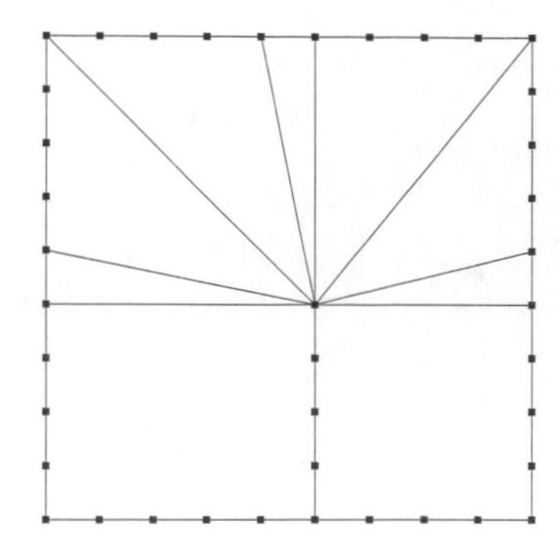

Figure 3.7: The test network net3. For almost every arc on the outer square there exists an additional parallel arc in series with a valve which is not visible in this picture. This test network was created by request of the cooperation partner. The aim was to analyze the optimal solution of the topology optimization problem for several different nominations. Especially the difference between a selection of loops and diagonal pipelines in an optimal solution was of particular interest. We consider three different nominations which we upscale by increasing values $2.5, 3, 4, 5$ in order to model an increase of the transported gas.

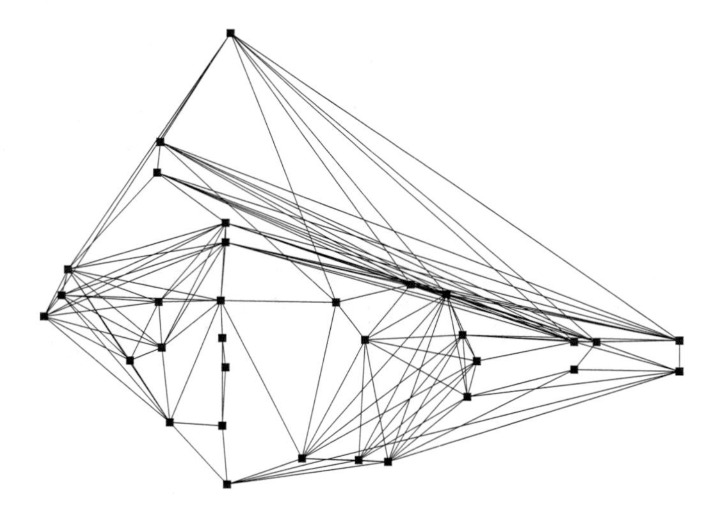

Figure 3.8: The test network net4. It is an extension of the network net2. The additional arcs were obtained manually. They represent each a pipeline in series with a valve. The length of these pipelines is set to the geographical distance between the end nodes. Cost associated with these pipelines reflect the building cost. For our computational experiments we will consider a single nomination which we upscale by increasing values $1.0, 1.1, 1.2, \ldots$ in order to model an increase of the transported gas.

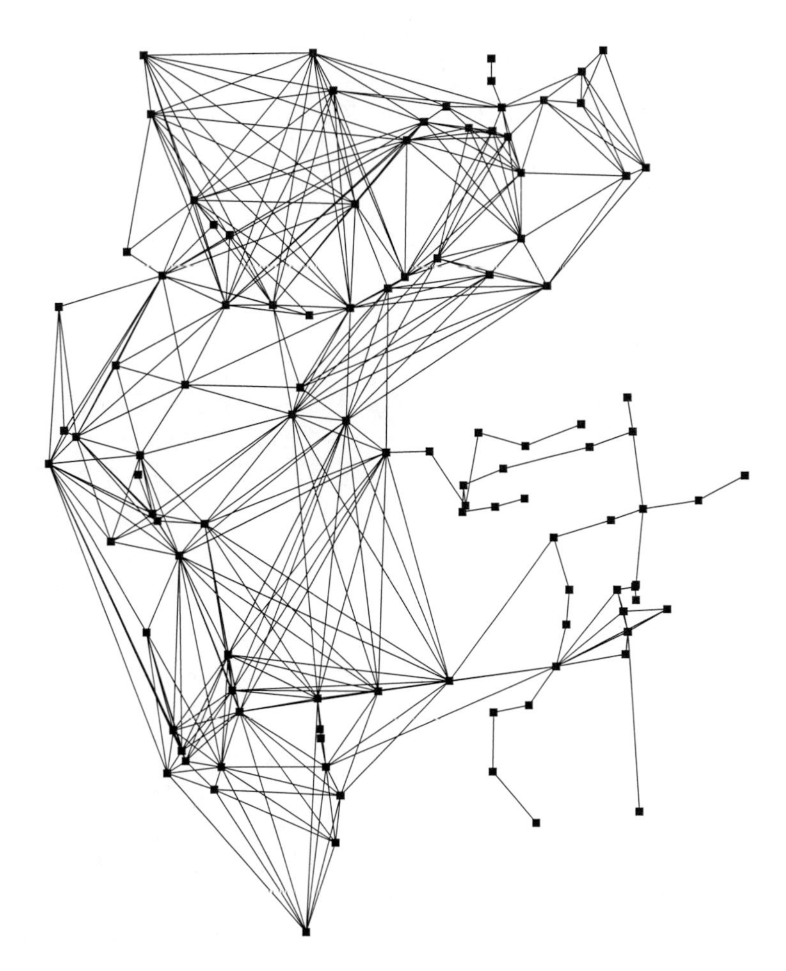

Figure 3.9: The test network net5. This network is an approximation of the German gas network for the high calorific gas operated by the cooperation partner. More precisely the length and the diameters of the pipelines are real-world data while other parameters like roughness or compressor data are set to realistic mean values. The additional arcs were obtained manually. They represent each a pipeline in series with a valve. The length of these additional pipelines is set to the geographical distance between the end nodes. Cost associated with these pipelines reflect the building cost. Altered data of the underlying network, which contains no extensions, is publicly available at URL `http://gaslib.zib.de` under the name gaslib-135. For our computational experiments we will consider a single nomination which we upscale by increasing values $1.0, 1.1, 1.2, \ldots$ in order to model an increase of the transported gas.

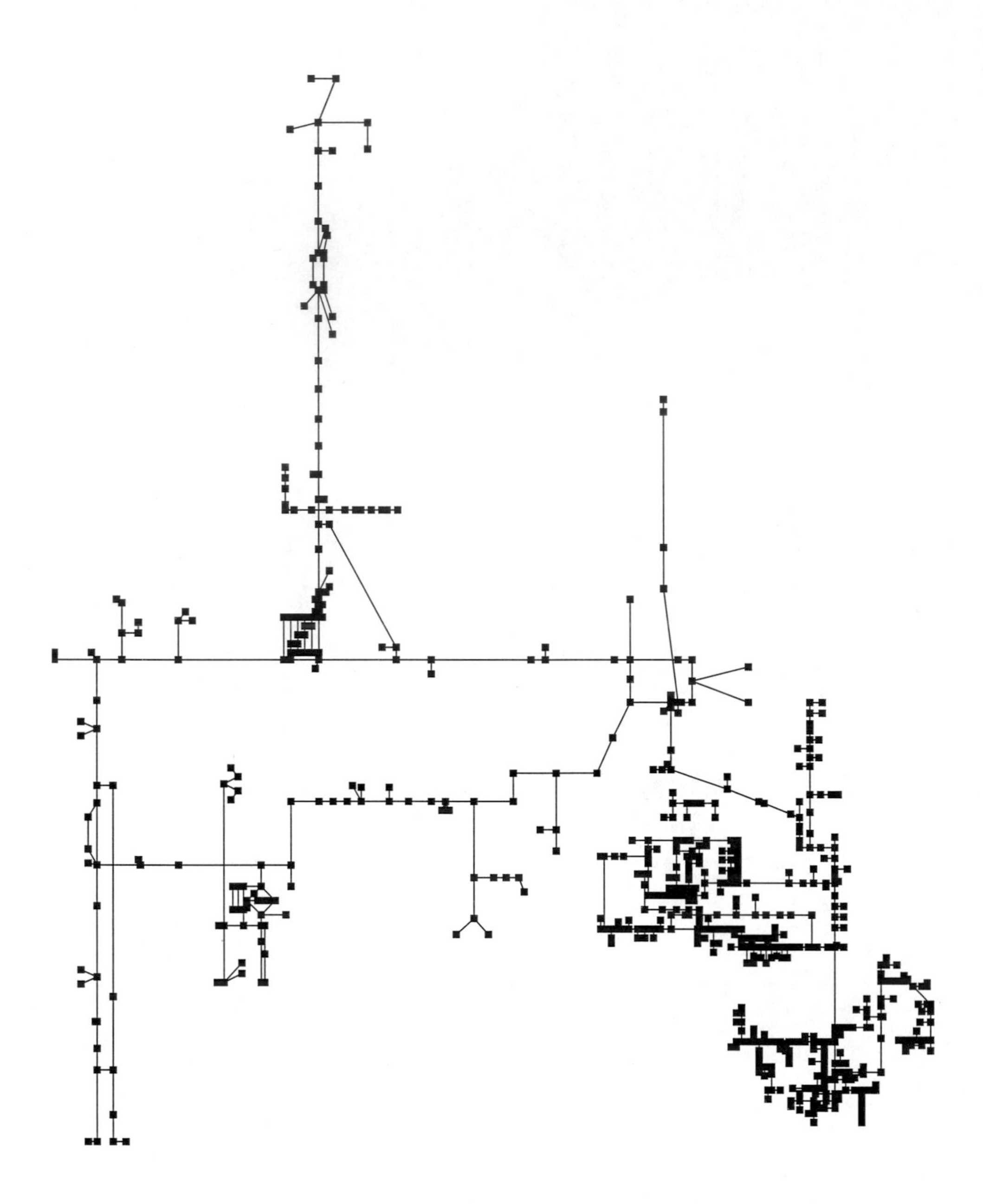

Figure 3.10: The real-world network net6. Data and nominations are provided by the cooperation partner OGE. The network is located in the northern part of Germany.

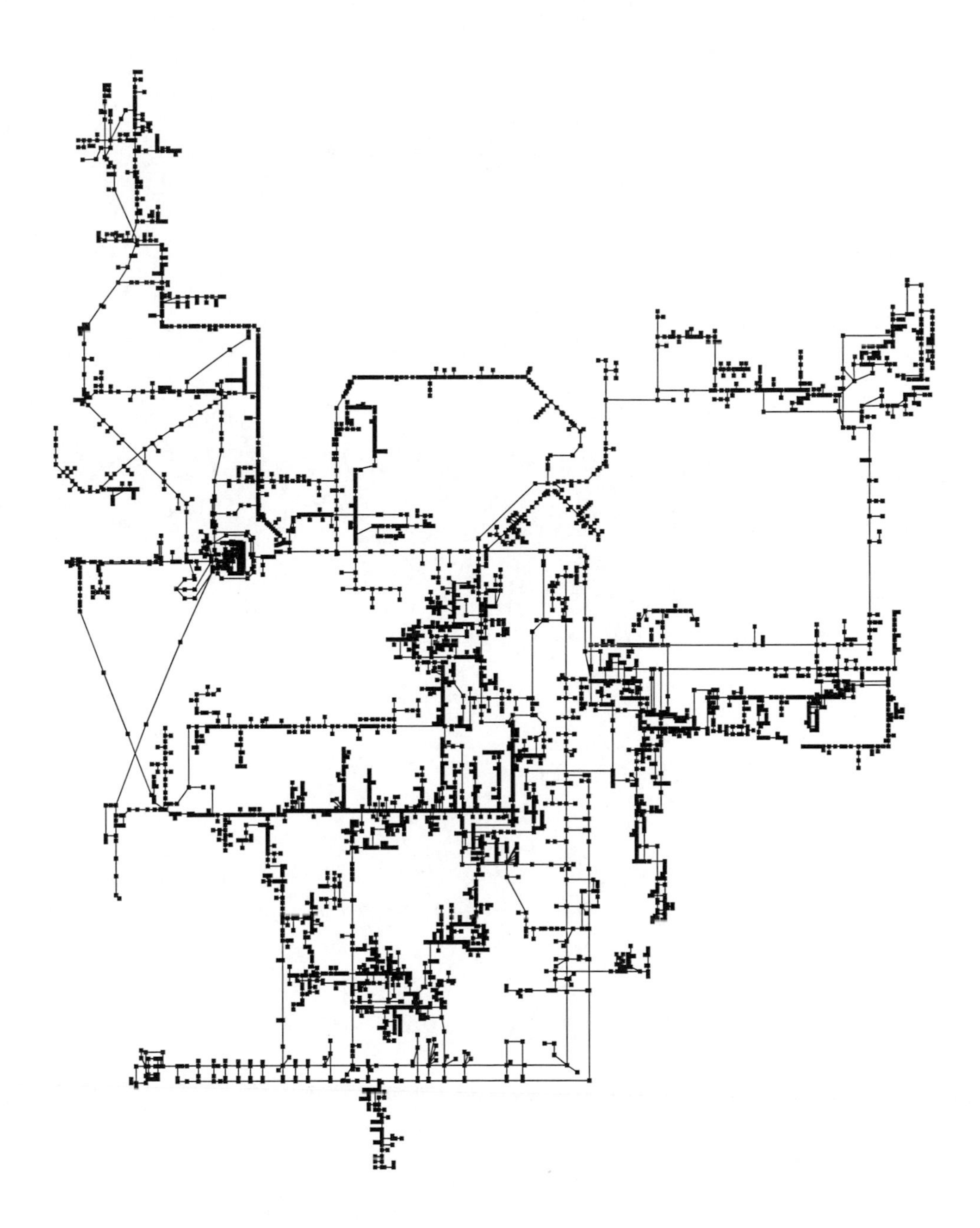

Figure 3.11: The real-world network net7. Data and nominations are provided by the cooperation partner OGE. The network is located in the Rhine-Main-Ruhr area of Germany.

Chapter 4

Efficiently Solving the Passive Transmission Problem

In this chapter we focus on the topology optimization problem (3.2.1) in the case $\underline{y} = \overline{y}$. This case includes the first type of network we consider in this thesis. Recall that these networks consist of pipes and valves only. We present a method for solving the passive transmission problem (3.4.1) to global optimality efficiently. As previously discussed this problem results from fixing the variables x and y in (3.2.1) while the fixation of y already follows from its bounds. Therefore (3.4.1) is a continuous nonlinear feasibility problem.

We use the solution method to speed up the solution process of the topology optimization problem (3.2.1) as follows: Recall that (3.2.1) is solved within the branch-and-bound framework implemented by SCIP while the nonlinearity is handled by spatial branching. We adapt this framework and solve all nodes of the branch-and-bound tree which correspond to passive transmission problems by the efficient solution method which we present in this chapter. Thus there is no need to continue with spatial branching in these nodes. A similar approach is followed by Raghunathan (2013). He considers an MINLP arising in the context of designing water networks and globally solves subproblems of this MINLP separately. Gentilini et al. (2013) consider the traveling salesman problem with neighborhoods as a non-convex MINLP. Certain subproblems turn out to be convex. These are solved to global optimality in a separate step within a branch-and-bound approach for the overall problem.

As the passive transmission problem (3.4.1) is continuous and nonlinear we want to apply a nonlinear solver for computing an optimal solution. A problem that can occur is that the passive transmission problem might be infeasible, which cannot be detected efficiently by nonlinear solvers. Our solution approach is to consider different relaxations of the passive transmission problem, which are either feasible or their infeasibility can be detected in a preprocessing step. All relaxations neglect the constraints (3.4.1c) and (3.4.1e). Further they successively relax all other constraints: The first relaxation is obtained by relaxing the variable bounds (3.4.1f)–(3.4.1i), the second one by relaxing the flow conservation constraints (3.4.1d) and the third one by relaxing the potential-flow-coupling constraints (3.4.1b). It turns out that under some assumptions two of these relaxations are convex and hence can be solved efficiently by a nonlinear solver.

As highlighted earlier we solve the topology optimization problem by a special tailored version of SCIP. It turns out that this approach enables around 29 % more instances of the first type of network to be solved to global optimality within our given time frame compared to solving the overall topology optimization problem by branch-and-bound, separation, and spatial branching. On average the run time is reduced by 72 % on those instances which are globally solvable by SCIP. In the case that a passive transmission problem is infeasible, the dual solutions of the convex relaxations provide an explanation for the infeasibility.

The outline of this chapter is as follows: In Section 4.1 we state the passive transmission problem again to improve readability. In Sections 4.2 - 4.4 we present three different relaxations of the passive transmission problem where we assume $\alpha_a > 0$ for all arcs $a \in A'$. In addition all relaxations neglect constraints (3.4.1c) and (3.4.1e). In Section 4.5 we show how to handle the general case without these assumptions. Finally computational results are given in Section 4.6.

4.1 Notation

Let us recall the passive transmission problem (3.4.1). It is a subproblem of the topology optimization problem (3.2.1) obtained by fixing all binary variables x and all slack variables y of compressors and control valves. We write the passive transmission problem as

$$\exists\, q, \pi, p \tag{4.1.1a}$$

$$\text{s.t.} \quad \alpha_a\, q_a |q_a|^{k_a} - \tilde{\beta}_a - (\pi_v - \gamma_a \pi_w) = 0 \quad \forall\, a = (v,w) \in A', \tag{4.1.1b}$$

$$A_a\, (q_a, p_v, p_w)^T \leq b_a \quad \forall\, a = (v,w) \in A', \tag{4.1.1c}$$

$$\sum_{a \in \delta^+_{A'}(v)} q_a - \sum_{a \in \delta^-_{A'}(v)} q_a = d_v \quad \forall\, v \in V, \tag{4.1.1d}$$

$$p_v |p_v| - \pi_v = 0 \quad \forall\, v \in V, \tag{4.1.1e}$$

$$\pi_v \leq \overline{\pi}_v \quad \forall\, v \in V, \tag{4.1.1f}$$

$$\pi_v \geq \underline{\pi}_v \quad \forall\, v \in V, \tag{4.1.1g}$$

$$q_a \leq \overline{q}_a \quad \forall\, a \in A', \tag{4.1.1h}$$

$$q_a \geq \underline{q}_a \quad \forall\, a \in A', \tag{4.1.1i}$$

$$p_v, \pi_v \in \mathbb{R} \quad \forall\, v \in V, \tag{4.1.1j}$$

$$q_a \in \mathbb{R} \quad \forall\, a \in A'. \tag{4.1.1k}$$

Here $\tilde{\beta}_a := \beta_a y_a$ for the current value y_a. Recall that the arc set $A' := \{(a,i) \in A_X : x_{a,i} = 1, i > 0\}$ contains all arcs such that the flow is not fixed to zero. Throughout this chapter we assume w.l.o.g. that (V, A') is a connected graph. If this is not the case, we split the problem such that we obtain a passive transmission problem for each connected component.

4.2 Relaxation of Domains

Let us consider the following *domain relaxation* of the passive transmission problem (4.1.1) with $\alpha_a > 0$ for each arc $a \in A'$. We introduce a slack variable $\Delta_v \in \mathbb{R}_{\geq 0}$ in order to relax the potential value of node $v \in V$ and another slack variable $\Delta_a \in \mathbb{R}_{\geq 0}$ for relaxing the flow of arc $a \in A'$. Then the domain relaxation writes as

$$\min \sum_{v \in V} \Delta_v + \sum_{a \in A'} \Delta_a \tag{4.2.1a}$$

$$\text{s.t.} \quad \alpha_a\, q_a |q_a|^{k_a} - \tilde{\beta}_a - (\pi_v - \gamma_a \pi_w) = 0 \quad \forall\, a = (v,w) \in A', \tag{4.2.1b}$$

$$\sum_{a \in \delta^+_{A'}(v)} q_a - \sum_{a \in \delta^-_{A'}(v)} q_a = d_v \quad \forall\, v \in V, \tag{4.2.1c}$$

$$\pi_v - \Delta_v \leq \overline{\pi}_v \quad \forall\, v \in V, \tag{4.2.1d}$$

$$\pi_v + \Delta_v \geq \underline{\pi}_v \quad \forall\, v \in V, \tag{4.2.1e}$$

$$q_a - \Delta_a \leq \overline{q}_a \quad \forall\, a \in A', \tag{4.2.1f}$$

$$q_a + \Delta_a \geq \underline{q}_a \quad \forall\, a \in A', \tag{4.2.1g}$$

$$\pi_v \in \mathbb{R} \quad \forall\, v \in V, \tag{4.2.1h}$$

$$q_a \in \mathbb{R} \quad \forall\, a \in A', \tag{4.2.1i}$$

$$\Delta_v \in \mathbb{R}_{\geq 0} \quad \forall\, v \in V, \tag{4.2.1j}$$

$$\Delta_a \in \mathbb{R}_{\geq 0} \quad \forall\, a \in A'. \tag{4.2.1k}$$

In the following we show that this nonlinear optimization problem is feasible and convex, see Section 4.2.1 and Section 4.2.2. Hence it can be solved very efficiently to global optimality. Note that the convexity here is not given by the constraints but by the feasible solution space of the relaxation. Finally in Section 4.2.3 we will give an interpretation of the dual solution of a KKT point of (4.2.1). This dual solution forms a network flow that is induced by dual node potentials. It fulfills constraints that are similar to (4.2.1b) and (4.2.1c). In the case that the passive transmission problem (4.1.1) is infeasible, this dual solution allows to filter those parts of the network (V, A') which imply the infeasibility.

4.2.1 Existence of a Solution

The existence of a primal solution for (4.2.1) was shown by Collins et al. (1978) and Maugis (1977). In the following we review their method which basically consists of solving the nonlinear optimization problems (4.2.2) and (4.2.3). Note that their approach only works for constant heights, i.e., $\gamma_a = 1$ for all $a \in A'$. In the subsequent part of this section we extend their method and show how the case of inhomogeneous heights, i.e., $\gamma_a \neq 1$ for some $a \in A'$, can be treated.

Collins et al. considered the convex nonlinear optimization problem

$$\begin{aligned} & \min \sum_{a\in A'} \int_{q_a^0}^{q_a} \Phi_a(t)\ dt \\ \text{s.t.} \quad & \sum_{a\in\delta^-_{A'}(v)} q_a - \sum_{a\in\delta^+_{A'}(v)} q_a = -d_v \qquad \forall\, v \in V, \\ & q_a \in \mathbb{R} \qquad \forall\, a \in A', \end{aligned} \tag{4.2.2}$$

where $\Phi_a(\cdot)$ is a continuous strictly monotone function. Further q_a^0 is a root of $\Phi_a(\cdot)$ which implies that the objective is convex. In the context of our study we set $\Phi_a(q_a) := \alpha_a\, q_a |q_a|^{k_a} - \tilde{\beta}_a$. Then $\Phi_a(q_a)$ is strictly monotone increasing because $\alpha_a > 0$ by our assumption.

Lemma 4.2.1 (Collins et al. (1978) and Maugis (1977)):
The nonlinear optimization problem (4.2.2) *is convex. Its optimal solution yields a feasible solution for* (4.2.1) *in the case* $\gamma_a = 1$ *for each arc* $a \in A'$.

Proof. The constraints of (4.2.2) are linear and the objective is a sum of convex functions. Hence (4.2.2) is a convex optimization problem.

Furthermore the objective and all constraints of (4.2.2) are continuously differentiable. From Theorem 2.4.2 and (2.4.2a) of the KKT conditions (2.4.2) for (4.2.2) we obtain for optimal primal values q^* that there exist dual values μ^* such that

$$\mu_v^* - \mu_w^* = \Phi_a(q_a^*)$$

for each arc $a = (v, w) \in A'$. By setting the node potential $\pi^* := \mu^*$, and $\Delta_v^* := \max\{0, \pi_v^* - \overline{\pi}_v, \underline{\pi}_v - \pi_v^*\}$ for each node $v \in V$, and $\Delta_a^* := \max\{0, q_a^* - \overline{q}_a, \underline{q}_a - q_a^*\}$ for each arc $a \in A'$ we obtain a primal feasible solution (q^*, π^*, Δ^*) of (4.2.1) with $\gamma_a = 1$ (for all $a \in A'$). □

We note that the convex optimization problem (4.2.2) is not only useful for theoretical purpose. Raghunathan (2013) uses (4.2.2) for computing a solution of the passive transmission problem (4.1.1) on a real-world application.

We further remark that Collins et al. (1978) provide a different proof for the existence of a solution for (4.2.1) (with $\gamma_a = 1$) by considering the following nonlinear program:

$$\begin{aligned} \min \quad & \sum_{a=(v,w)\in A'} \int_{\Delta_a^0}^{\pi_v - \pi_w} \Phi_a^{-1}(t)\,\mathrm{d}t - \sum_{v \in V} \int_0^{\pi_v} d_v \,\mathrm{d}t \\ \text{s.t.} \quad & \pi_v \in \mathbb{R} \quad \forall v \in V. \end{aligned} \tag{4.2.3}$$

Here Φ_a^{-1} is the inverse of Φ_a and Δ_a^0 is a root of Φ_a^{-1}.

Lemma 4.2.2 (Collins et al. (1978) and Maugis (1977)):
The nonlinear optimization problem (4.2.3) *is convex. Its optimal solution yields a feasible solution for* (4.2.1) *in the case* $\gamma_a = 1$ *for each arc* $a \in A'$.

Note that problem (4.2.3) might be unbounded. In this case there exists no optimal solution and hence we cannot ensure the feasibility of (4.2.1) by making use of Lemma 4.2.2.

Proof. The objective is a sum of convex functions. Hence (4.2.3) is convex. For an optimal solution π^* of (4.2.3) we define q^* by

$$q_a^* := \Phi_a^{-1}(\pi_v^* - \pi_w^*) \quad \Leftrightarrow \quad \Phi_a(q_a^*) = \pi_v^* - \pi_w^*$$

for each arc $a = (v, w) \in A'$ while Φ_a^{-1} is the inverse function of Φ_a. Furthermore the objective of (4.2.3) is continuously differentiable. From Theorem 2.4.2 and (2.4.2a) of the KKT conditions (2.4.2) for (4.2.3) we obtain that the following flow conservation constraints are fulfilled:

$$\sum_{a \in \delta^+_{A'}(v)} q^*_a - \sum_{a \in \delta^-_{A'}(v)} q^*_a - d_v = 0 \qquad \forall v \in V.$$

By setting $\Delta^*_v := \max\{0, \pi^*_v - \overline{\pi}_v, \underline{\pi}_v - \pi^*_v\}$ for each node $v \in V$ and $\Delta^*_a := \max\{0, q^*_a - \overline{q}_a, \underline{q}_a - q^*_a\}$ for each arc $a \in A'$ we obtain a primal feasible solution (q^*, π^*, Δ^*) of (4.2.1) with $\gamma_a = 1$ (for all $a \in A'$). □

After this brief literary review we now turn to the general case $\gamma_a \neq 1$ for some arc $a \in A'$. We show how to obtain a feasible solution for domain relaxation (4.2.1) by using the convex optimization problem (4.2.2).

Definition 4.2.3:

Let (V, E) be the undirected version of (V, A') obtained by removing the orientation of the arcs $a \in A'$. This way each arc $a \in A'$ uniquely corresponds to an edge in $e \in E$ and vice versa. Let r be any node in V. For a node $v \in V$ denote by $P_r(v)$ an undirected path from r to v. Let $v_1, \dots, v_n$ be the nodes of this path, $e_1, \dots, e_{n-1}$ the edges and $a_1, \dots, a_{n-1}$ the corresponding arcs. Recall our assumption for this chapter that (V, A') is connected. We define

$$\gamma_{r,v} := \left(\prod_{i : a_i = (v_i, v_{i+1})} \gamma_{a_i} \right) \left(\prod_{i : a_i = (v_{i+1}, v_i)} \gamma_{a_i}^{-1} \right).$$

We have $\gamma_{r,v} > 0$ as $\gamma_a > 0$ for every arc $a \in A'$. The definition is actually independent of the path $P_r(v)$. To see this, let P' be a different r-v-path in (V, E). Consider the cycle C from r to v on path P, and back from v to r on path P' in reverse order. Denote the reverse path of P' by Q'. Denote the nodes of this path by $\tilde{v}_1, \dots, \tilde{v}_m$ with $\tilde{v}_1 = v$ and $\tilde{v}_m = r$, the edges by $\tilde{e}_1, \dots, \tilde{e}_{m-1}$ and the arcs by $\tilde{a}_1, \dots, \tilde{a}_{m-1}$. According to our assumption in Section 3.1 we have

$$1 = \left(\prod_{i : a_i = (v_i, v_{i+1})} \gamma_a \right) \left(\prod_{i : a_i = (v_{i+1}, v_i)} \gamma_a^{-1} \right) \left(\prod_{i : \tilde{a}_i = (\tilde{v}_{i+1}, \tilde{v}_i)} \gamma_a \right) \left(\prod_{i : \tilde{a}_i = (\tilde{v}_i, \tilde{v}_{i+1})} \gamma_a^{-1} \right).$$

Hence $\gamma_{r,v}$ is uniquely defined.

Using this value $\gamma_{r,v}$ we define the function π'_v by

$$\pi'_v(\pi) := \gamma_{r,v}\,\pi_v \tag{4.2.4}$$

for every node $v \in V$. As a consequence of (4.2.4) we obtain lower and upper bounds of $\pi'_v(\pi)$ from $\underline{\pi}'_v := \pi'_v(\underline{\pi})$ and $\overline{\pi}'_v := \pi'_v(\overline{\pi})$, respectively, for each node $v \in V$.

It follows from elementary calculations that

$$\pi'_v(\pi) - \pi'_w(\pi) = \gamma_{r,v}\pi_v - \underbrace{\gamma_{r,w}}_{=\gamma_{r,v}\gamma_a}\,\pi_w = \gamma_{r,v}\,(\pi_v - \gamma_a\pi_w) \tag{4.2.5}$$

holds for each arc $a = (v, w) \in A'$.

We use the previous definitions and show how to obtain a feasible solution for the domain relaxation (4.2.1) in the general case $\gamma_a \neq 1$ for some arc $a \in A'$.

Lemma 4.2.4:
The optimization problem (4.2.1) is feasible.

Proof. We use $\gamma_{r,v}$ from Definition 4.2.3 and equation (4.2.5). Now we compute a local optimum q^* of the problem

$$\begin{aligned} \min \sum_{a\in A'} \int_{q_a^0}^{q_a} \Phi_a(t)\,dt & \\ \text{s.t.} \quad \sum_{a\in\delta^-_{A'}(v)} q_a - \sum_{a\in\delta^+_{A'}(v)} q_a = -d_v & \quad \forall v \in V, \\ q_a \in \mathbb{R} & \quad \forall a \in A', \end{aligned}$$

where $\Phi_a(q_a) := \gamma_{r,v}\alpha_a\, q_a|q_a|^{k_a} - \gamma_{r,v}\tilde{\beta}_a$ and q_a^0 is a root of this function for each arc $a = (v, w) \in A'$.

By Lemma 4.2.1 this optimization problem yields a feasible solution (q^*, π^*) for a modified version of the domain relaxation (4.2.1) which is obtained by replacing the constraints (4.2.1b), (4.2.1d), (4.2.1e) by

$$\begin{aligned} \gamma_{r,v}\alpha_a\, q_a|q_a|^{k_a} - \gamma_{r,v}\tilde{\beta}_a - (\pi_v - \pi_w) = 0 & \quad \forall a = (v, w) \in A', \\ \pi_v - \Delta_v \leq \pi'_v(\overline{\pi}) & \quad \forall v \in V, \\ \pi_v + \Delta_v \geq \pi'_v(\underline{\pi}) & \quad \forall v \in V, \end{aligned}$$

where we used (4.2.4).

Using (q^*, π^*) we are going to show how to obtain a feasible solution for the domain relaxation (4.2.1) where the previous modifications do not take place. We define

$$\hat{\pi}_v := \pi_v'^{-1}(\pi^*)$$

for each node $v \in V$. Recall that $\gamma_{r,v} > 0$ because $\gamma_a > 0$ for each arc $a \in A$. This implies that π'^{-1} is well defined. In combination with (4.2.5) we obtain

$$\alpha_a\, q_a^* |q_a^*|^{k_a} - \tilde{\beta}_a = \gamma_{r,v}^{-1}\left(\pi_v^* - \pi_w^*\right) = \hat{\pi}_v - \gamma_a \hat{\pi}_w \qquad \forall\, a = (v, w) \in A'.$$

Thus $(q^*, \hat{\pi})$ is a vector which is feasible for the constraints (4.2.1b) and (4.2.1c) of the original domain relaxation (4.2.1). Setting $\Delta_v^* := \max\{0, \hat{\pi}_v - \overline{\pi}_v, \underline{\pi}_v - \hat{\pi}_v\}$ for each node $v \in V$ and $\Delta_a^* := \max\{0, q_a^* - \overline{q}_a, \underline{q}_a - q_a^*\}$ for each arc $a \in A'$ we obtain a primal feasible solution $(q^*, \hat{\pi}, \Delta^*)$ of (4.2.1). □

4.2.2 Characterization of the Feasible Region

In the previous section we proved that the domain relaxation (4.2.1) is feasible. In the following we are going to prove that (4.2.1) is a convex optimization problem. First we show that the solution flow q is unique and the feasible node potentials π form a straight line. From this we conclude that (4.2.1) is convex.

By Maugis (1977) the uniqueness of the flow vector q follows from the strict convexity of the objectives of (4.2.2) and (4.2.3) in the case $\gamma_a = 1$ for each arc $a \in A'$. Nevertheless, we give a proof for the more general case that there exists an arc $a \in A'$ with $\gamma_a \neq 1$. For this proof we make use of the following theorem and the lemma below:

Theorem 4.2.5:
Let $q \in \mathbb{R}_{\geq 0}^{A'}$ be a network flow in the (V, A'). There exists a family $\mathcal{P}$ of paths and a family $\mathcal{C}$ of circuits in (V, A') along with values $f_P > 0, P \in \mathcal{P}$ and $f_C > 0, C \in \mathcal{C}$ such that

$$q_a = \sum_{\substack{P \in \mathcal{P}:\\ a \in A'(P)}} f_P' + \sum_{\substack{C \in \mathcal{C}:\\ a \in A'(C)}} f_C' \qquad \forall\, a \in A'.$$

If q is a circulation then it holds $\mathcal{P} = \varnothing$.

Proof. Let q be a network flow between two nodes s and t, i.e., an s-t-flow. In this case the theorem follows from Korte and Vygen (2007), Theorem 8.8.

The more general case of multiple sources and sinks is obtained by adding a super node s for the sources and another super node t for the sinks. We set

$$\tilde{q}_{(s,v)} := \sum_{a\in\delta^+_{A'}(v)} q_a - \sum_{a\in\delta^-_{A'}(v)} q_a \qquad \tilde{q}_{(w,t)} := \sum_{a\in\delta^-_{A'}(w)} q_a - \sum_{a\in\delta^+_{A'}(w)} q_a$$

for the sources v and sinks w and $\tilde{q}_a := q_a$ for all arcs $a \in A'$. For each source $v \in V$ we add the arc (s, v) and for each sink $w \in V$ we add the arc (w, t) to the graph (V, A'). Then we apply Theorem 8.8 to the extended graph (V, A') and the network flow $\tilde{q}$ therein. □

Lemma 4.2.6:
Let (q', π', Δ') and (q'', π'', Δ'') be two feasible solutions of the domain relaxation (4.2.1). W.l.o.g. it holds $q' \geq q''$.

Proof. Let (q', π', Δ') and (q'', π'', Δ'') be two solutions of (4.2.1). Recall that the network (V, A') is connected. If there exists an arc a with $q'_a < q''_a$ then we reorient this arc $a = (v, w)$ as follows: We remove it from the graph and add the backward (or antiparallel) arc $\overleftarrow{a} = (w, v)$ to A'. From the constraint

$$\alpha_a\, q_a |q_a|^{k_a} - \beta_a y_a - (\pi_v - \gamma_a \pi_w) = 0$$

we obtain with

$$\begin{aligned} &\alpha_{\overleftarrow{a}} := \alpha_a \gamma_a^{-1} \in \mathbb{R}_{\geq 0}, \quad \beta_{\overleftarrow{a}} := \beta_a \gamma_a^{-1} \in \mathbb{R}_{\geq 0}, \quad \gamma_{\overleftarrow{a}} := \gamma_a^{-1} \in \mathbb{R}_{\geq 0} \setminus \{0\}, \\ &\overline{y}_{\overleftarrow{a}} := -\underline{y}_a \in \mathbb{R}, \quad \underline{y}_{\overleftarrow{a}} := -\overline{y}_a \in \mathbb{R}, \quad \overline{q}_{\overleftarrow{a}} := -\underline{q}_a \in \mathbb{R}, \quad \underline{q}_{\overleftarrow{a}} := -\overline{q}_a \in \mathbb{R} \end{aligned}$$

the equivalent equation for the backward arc $\overleftarrow{a}$

$$\alpha_{\overleftarrow{a}}\, q_{\overleftarrow{a}} |q_{\overleftarrow{a}}|^{k_a} - \beta_{\overleftarrow{a}} y_{\overleftarrow{a}} - (\pi_w - \gamma_{\overleftarrow{a}} \pi_v) = 0. \tag{4.2.6}$$

Note that we used the relation $\gamma_{\overleftarrow{a}} \gamma_a = 1$. Hence constraint (4.2.1b) for arc a is replaced by constraint (4.2.6) for arc $\overleftarrow{a}$. Note that this constraint fulfills our assumption for the constants $\alpha_{\overleftarrow{a}}, \beta_{\overleftarrow{a}}$ and $\gamma_{\overleftarrow{a}}$ as described in Section 3.2 where the constraints are explained. Note that replacing a by $\overleftarrow{a}$ goes along with the relation $q_{\overleftarrow{a}} = -q_a$ for the arc flow variables. For a solution of (q^*, π^*, Δ^*) of the original domain relaxation we define $\tilde{q}$ by $\tilde{q}_a := q^*_a$ and $\tilde{\Delta}^{\pm}_a := \Delta^{\pm *}_a$ if we did not change the orientation of a and $\tilde{q}_a := -q^*_a$, $\tilde{\Delta}^+_{\overleftarrow{a}} := \Delta^{-*}_a$ and $\tilde{\Delta}^-_{\overleftarrow{a}} := \Delta^{+*}_a$ if we adapted the arc orientation. Further we set $\tilde{\Delta}^{\pm}_v := \Delta^{\pm *}_v$ for every node $v \in V$. Then it holds that

(q^*, π^*, Δ^*) is feasible for the domain relaxation on the original graph (V, A') if and only if $(\tilde{q}, \pi^*, \tilde{\Delta})$ is feasible for the domain relaxation on the modified graph.

We adapt the direction of all arcs $a \in A'$ with $q'_a < q''_a$ and end up with a modified version of the passive transmission problem of the same type as the original passive transmission problem. Further we have two feasible solutions $(\tilde{q}', \pi', \tilde{\Delta}')$ and $(\tilde{q}'', \pi'', \tilde{\Delta}'')$. □

Lemma 4.2.7:
There exist vectors $\tilde{q} \in \mathbb{R}^{A'}$, $\tilde{\pi} \in \mathbb{R}^V$ and $\theta \in \mathbb{R}^V_{\geq 0}$, such that the following holds: (q^*, π^*, Δ^*) is feasible for problem (4.2.1) if and only if there exists $t \in \mathbb{R}$ with $\pi^* = \tilde{\pi} + t\theta$ and $q^* = \tilde{q}$.

Proof. It follows from Lemma 4.2.4 that (4.2.1) is feasible. Let us assume the existence of two different solutions (q', π', Δ') and (q'', π'', Δ'') of (4.2.1) with $q' \neq q''$ and $\pi' \neq \pi''$. First we prove by contradiction that the primal solution flow of (4.2.1) is unique. In a second step we analyze the difference $\pi' - \pi''$.

The difference $q' - q''$ is a network flow in (V, A') consisting of circulations only. First we focus on the case $q' \geq q''$. By Theorem 4.2.5 the difference $q' - q''$ can be split into flow along circuits $C_1, \ldots, C_n$. Thus we obtain from $q' \geq q''$, that there exist flow values $q_{C_i} > 0, i = 1, \ldots, n$ such that

$$q'_a - q''_a = \sum_{\substack{i=1,\ldots,n: \\ a \in A'(C_i)}} q'_{C_i} \qquad \forall\, a \in A'.$$

Note that the sum can be empty for an arc $a \in A'$. We consider a single circuit C out of $C_1, \ldots, C_n$. For each arc $a \in A'(C)$ we have $q'_a > q''_a$. As abbreviation we set $\Phi_a(q) := \alpha_a\, q|q|^{k_a} - \tilde{\beta}_a$. Then Φ_a is a continuous strictly monotone function. We use equation (4.2.1b) to obtain for all arcs $a = (v, w) \in A'(C)$:

$$q'_a > q''_a \;\Leftrightarrow\; \Phi_a(q'_a) > \Phi_a(q''_a) \;\Leftrightarrow\; \pi'_v - \gamma_a \pi'_w > \pi''_v - \gamma_a \pi''_w.$$

Let the nodes of the circuit C be ordered such that $V(C) = \{v_1, \ldots, v_\ell\}$ and (as abbreviation) $v_{\ell+1} := v_1$, and the arcs be ordered such that $a_i = (v_i, v_{i+1})$ holds. We consider all arcs of the circuit C and rewrite the following telescopic sums as

$$\sum_{i=1}^{\ell} \left(\prod_{j=1}^{i-1} \gamma_{a_j} \right) \left(\pi'_{v_i} - \gamma_{a_i} \pi'_{v_{i+1}} \right) > \sum_{i=1}^{\ell} \left(\prod_{j=1}^{i-1} \gamma_{a_j} \right) \left(\pi''_{v_i} - \gamma_{a_i} \pi''_{v_{i+1}} \right)$$

$$\Leftrightarrow \qquad \pi'_{v_1}\left(1-\prod_{a\in A'(C)}\gamma_a\right) > \pi''_{v_1}\left(1-\prod_{a\in A'(C)}\gamma_a\right).$$

Note that $\prod_{a\in A'(C)}\gamma_a = 1$ (see Section 3.2). Hence we have

$$q'_a > q''_a, a \in A'(C) \quad \Leftrightarrow \quad \pi'_{v_1}\cdot 0 > \pi''_{v_1}\cdot 0,$$

which implies that $q' \neq q''$ is infeasible. This contradicts our assumption $q' \neq q''$. So the solution flow $q' = q''$ of (4.2.1) is unique.

So far we considered the case $q' \geq q''$. If there exists an arc $a \in A'$ with $q'_a < q''_a$ then we apply Lemma 4.2.6 and change the orientation of this arc. When reorienting we obtain two flow vectors $\tilde{q}'$ and $\tilde{q}''$ which differ from q' and q'' only in the signs of the reoriented arcs. For these new flow vectors we apply the result from above and obtain $\tilde{q}' = \tilde{q}''$. Coming back to the original orientation we change back the signs and obtain $q' = q''$.

Now we prove that the feasible node potentials π of (4.2.1) form a straight line. Therefor let (q', π', Δ') and (q', π'', Δ'') be two different feasible solutions of (4.2.1). Let r be some node $r \in V$. Let $w \in V$ be any other node in V. Consider the graph (V, E) which is obtained from (V, A') by removing the orientation. This way each arc $a \in A'$ uniquely corresponds to an edge $e \in E$. Consider an undirected r-w-path P in (V, E) with nodes $r = v_1, \dots, v_k = w$, edges $e_1, \dots, e_{k-1}$ and corresponding arcs $a_1, \dots, a_{k-1}$. For an arc $a_j = (v_{j+1}, v_j)$ we rewrite equation (4.2.1b) as

$$-\gamma_{a_j}^{-1}\Phi_{a_j}(q_{a_j}) = \pi_{v_j} - \gamma_{a_j}^{-1}\pi_{v_{j+1}}.$$

For all other corresponding arcs we do not apply this reformulation. As abbreviation we set

$$\tau_\ell := \prod_{\substack{j=1\\ a_j=(v_j,v_{j+1})}}^{\ell} \gamma_{a_j} \prod_{\substack{j=1\\ a_j=(v_{j+1},v_j)}}^{\ell} \gamma_{a_j}^{-1}.$$

Then we obtain the equality

$$\begin{aligned}\pi'_r - \pi'_w\tau_{k-1} = &\sum_{\substack{j=1:\\ a_j=(v_j,v_{j+1})}}^{k-1}\left(\prod_{i=1}^{j-1}\tau_i\right)\Phi_{a_j}(q_{a_j})\\ &- \sum_{\substack{j=1:\\ a_j=(v_{j+1},v_j)}}^{k-1}\left(\prod_{i=1}^{j-1}\tau_i\right)\gamma_{a_j}^{-1}\Phi_{a_j}(q_{a_j}) = \pi''_r - \pi''_w\tau_{k-1}.\end{aligned}$$

This is equivalent to

$$\pi'_r - \pi''_r = (\pi'_w - \pi''_w)\,\tau_{k-1}.$$

We define $\theta_w := \tau_{k-1}^{-1} \neq 0$. This setting is well-defined, i.e., independent of the actual path P from r to w as discussed in Definition 4.2.3. We set $t := \pi''_r - \pi'_r$. Then the solution π'' can be expressed as $\pi''_w = \pi'_w + t\theta_w$ for all $w \in V$. This proves the lemma. □

Corollary 4.2.8:
The feasible solution vectors (q, π) of the domain relaxation (4.2.1) form a convex space: The feasible flow q is unique, while the feasible node potential values π form a straight line.

Proof. This follows from Lemma 4.2.7. □

Let us briefly discuss the result of this corollary. Figure 4.1 shows the node potential values for a test network having 34 nodes. The node potential bounds are shown as straight lines (lower bound of 500 and upper bound of 6000). Solutions for four different passive transmission problems are shown in different colors: Each dot represents the node potential value at the respective node. By Lemma 4.2.7 the feasible node potentials for the passive transmission problem are on a straight line. This corresponds to shifting all dots in Figure 4.1, which correspond to the same passive transmission problem, up or down at the same time. For three of the four problems it is not possible to move all node potential values inside the bounds, hence these solutions are infeasible. Only for the solution corresponding to the circles, all values are inside the bounds and this solution is feasible.

Now, the combination of Lemma 4.2.4 and Corollary 4.2.8 allows us to characterize the domain relaxation (4.2.1).

Theorem 4.2.9:
The domain relaxation (4.2.1) is a feasible and convex relaxation of the passive transmission problem (4.1.1) in the case $\alpha_a > 0$ for all arcs $a \in A'$.

Proof. A solution (q^*, π^*, p^*) of the passive transmission problem (4.1.1) is feasible only if $(q^*, \pi^*, 0)$ is a feasible solution of (4.2.1). Hence (4.2.1) is a relaxation of the passive transmission problem (4.1.1).

It follows from Lemma 4.2.4 that the domain relaxation (4.2.1) is feasible. The convexity of the feasible solution space for (q, π) follows from Corollary 4.2.8, which states that the solution space is an affine subspace. The flow is unique, while the feasible vectors π form a straight line. The additional constraints (4.2.1d)-(4.2.1g)

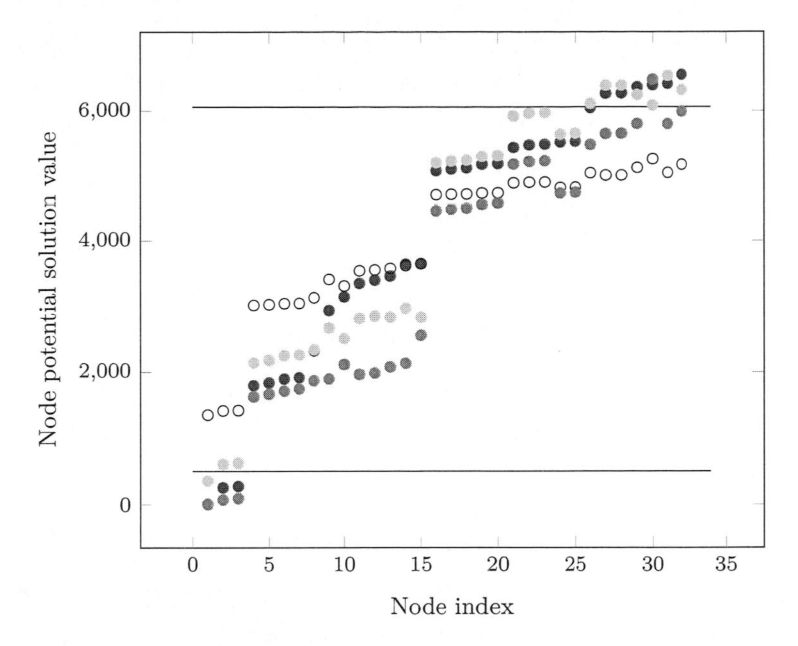

Figure 4.1: Feasible node potential values π for the domain relaxation (4.2.1) of four different passive transmission problems (4.1.1) on a gas network with 32 nodes. The node potential values are feasible for the passive transmission problems within the marked bounds, other values are infeasible. By Corollary 4.2.8 the node potentials π form a straight line. This corresponds to shifting all dots, which correspond to the same passive transmission problem, up or down at the same time.

are linear and do not affect the feasibility (i.e., rendering the problem infeasible) and the affine subspace. In total, problem (4.2.1) is a convex optimization problem over a non-empty set of feasible solutions. □

In the following Lemma we turn to the more general case where we do not assume $\alpha_a > 0$ for the arcs $a \in A'$. This implies that constraint (4.2.1b) can be written as $\pi_v = \pi_w$ for an arc $(v, w) \in A'$.

Lemma 4.2.10:
The domain relaxation (4.2.1) *is a feasible and convex optimization problem in the general case that there exist arcs* $(v, w) \in A'$ *where constraint* (4.2.1b) *writes as* $\pi_v = \pi_w$.

Proof. The feasibility in the more general case follows from Lemma 4.2.2 because the proof does not depend on the assumption $\alpha_a > 0$.

The convexity is seen as follows: Let A'' denote the set of arcs $a = (v, w) \in A'$ which are not modeled by $\pi_v = \pi_w$, i.e., $A'' = \{a \in A' \mid \alpha_a > 0\}$. For a solution (q^*, π^*, Δ^*) of the domain relaxation (4.2.1) it holds that q_a^* is unique for all arcs $a \in A''$. Otherwise, if there exist two solution flows q', q'' of the domain relaxation with $q'_{a'} \neq q''_{a'}$ for an arc $a' \in A''$, then $(q'_{a'})_{a' \in A''}$ and $(q''_{a'})_{a' \in A''}$ would be two different feasible solution vectors for the domain relaxation of the preprocessed passive transmission problem where all arcs $a \in A' \setminus A''$ are contracted. This is a contradiction to Lemma 4.2.7. It follows that the domain relaxation is a linear program, when fixing the unique arc flow q_a on all arcs $a = (v, w) \in A''$. This proves the convexity. □

4.2.3 Interpretation of Lagrange Multipliers

Assume that the passive transmission problem (4.1.1) is infeasible and the domain relaxation (4.2.1) has a positive optimal objective value. It turns out (see Lemma 4.2.11) that there exist Lagrange multipliers for the optimal solution such that the optimal solution and the multipliers form a KKT point of (4.2.1). These multipliers have a practical interpretation. They form a generalized network flow in (V, A') which is coupled with node potentials, similar to a primal solution (q^*, π^*) of the passive transmission problem. This network flow may characterize the infeasibility as discussed in Example 4.2.13, and hence allows a visualization of the conflicting parts of the network yielding the infeasibility.

Lemma 4.2.11:

Let (q^*, π^*, Δ^*) be an optimal solution of the domain relaxation (4.2.1). There exist Lagrange multipliers (μ^*, λ^*) which consecutively correspond to the equality and inequality constraints of (4.2.1), respectively, such that $(q^*, \pi^*, \Delta^*, \mu^*, \lambda^*)$ is a KKT point of (4.2.1). These multipliers are characterized as follows: $(\mu^*_a)_{a \in A'}$ is a general network flow in (V, A') which is induced by dual node potentials $(\mu^*_v)_{v \in V}$. More precisely the multipliers (μ^*, λ^*) are a feasible solution for

$$\mu_a \frac{d\Phi_a}{dq_a}(q^*_a) + \lambda^+_a - \lambda^-_a = \mu_v - \mu_w \qquad \forall\, a = (v, w) \in A',$$

$$\sum_{a \in \delta^+_{A'}(v)} \mu_a - \sum_{a \in \delta^-_{A'}(v)} \gamma_a \mu_a = \lambda^+_v - \lambda^-_v \qquad \forall\, v \in V.$$

Hereby the dual node flow $\lambda^+_v - \lambda^-_v$ is restricted by

	exit		*innode*		*entry*	
$\lambda^+_v \in$				$\{0\}$	$[0,1]$	$\{1\}$
$\lambda^-_v \in$	$\{1\}$	$[0,1]$	$\{0\}$			
node potential	$\pi^*_v < \underline{\pi}_v$	$\pi^*_v = \underline{\pi}_v$	$\underline{\pi}_v < \pi^*_v$	$\pi^*_v < \overline{\pi}_v$	$\pi^*_v = \overline{\pi}_v$	$\pi^*_v > \overline{\pi}_v$

for each node $v \in V$, and the dual compression $\lambda_a^+ - \lambda_a^-$ is constrained by

	compression		*neutral*		*regulation*	
$\lambda_a^+ \in$				$\{0\}$	$[0,1]$	$\{1\}$
$\lambda_a^- \in$	$\{1\}$	$[0,1]$	$\{0\}$			
flow	$q_a^* < \underline{q}_a$	$q_a^* = \underline{q}_a$	$q_a^* > \underline{q}_a$	$q_a^* < \overline{q}_a$	$q_a^* = \overline{q}_a$	$q_a^* > \overline{q}_a$

for each arc $a \in A'$.

Proof. Let (q^*, π^*, Δ^*) be an optimal solution of (4.2.1). Recall that every locally optimal solution is globally optimal because of the convexity of (4.2.1) by Theorem 4.2.9. The objective and the constraints of (4.2.1) are continuously differentiable. Assume there exist dual values (μ^*, λ^*) such that $(q^*, \pi^*, \Delta^*, \mu^*, \lambda^*)$ is a KKT point of domain relaxation (4.2.1). A proof of existence follows from the next Lemma 4.2.12. Let us write the conditions (2.4.2a) of the KKT conditions (2.4.2). We denote the Lagrange multipliers by $(\mu, \lambda) = (\mu_v, \mu_a, \lambda_v^+, \lambda_v^-, \lambda_a^+, \lambda_a^-)_{v \in V, a \in A'}$, such that $\mu_v, \mu_a \in \mathbb{R}$ and $\lambda_v^+, \lambda_v^-, \lambda_a^+, \lambda_a^- \in \mathbb{R}_{\geq 0}$. We set $\Phi_a(q_a) := \alpha_a \, q_a |q_a|^{k_a} - \tilde{\beta}_a$. Then the Lagrange function of problem (4.2.1) has the form

$$
\begin{aligned}
L(q, \pi, \Delta, \mu, \lambda) = & \sum_{v \in V} \Delta_v + \sum_{a \in A'} \Delta_a \\
& + \sum_{\substack{a \in A' \\ a=(v,w)}} \mu_a \left(\Phi_a(q_a) - (\pi_v - \gamma_a \pi_w) \right) \\
& + \sum_{v \in V} \mu_v \left(d_v - \sum_{a \in \delta^+_{A'}(v)} q_a + \sum_{a \in \delta^-_{A'}(v)} q_a \right) \\
& + \sum_{v \in V} \left(\lambda_v^+ \left(\pi_v - \Delta_v - \overline{\pi}_v \right) + \lambda_v^- \left(\underline{\pi}_v - \pi_v - \Delta_v \right) \right) \\
& + \sum_{a \in A'} \left(\lambda_a^+ \left(q_a - \Delta_a - \overline{q}_a \right) + \lambda_a^- \left(\underline{q}_a - q_a - \Delta_a \right) \right) \\
& - \sum_{v \in V} \lambda_v \Delta_v - \sum_{a \in A'} \lambda_a \Delta_a .
\end{aligned}
$$

From (2.4.2a) we obtain that the KKT point $(q^*, \pi^*, \Delta^*, \mu^*, \lambda^*)$ is feasible for

$$
\frac{\partial L}{\partial q_a} = 0 \Rightarrow \quad \mu_a \frac{d\Phi_a}{dq_a}(q_a^*) + \lambda_a^+ - \lambda_a^- = \mu_v - \mu_w \quad \forall \, a = (v, w) \in A', \tag{4.2.7a}
$$

$$
\frac{\partial L}{\partial \pi_v} = 0 \Rightarrow \sum_{a \in \delta^+_{A'}(v)} \mu_a - \sum_{a \in \delta^-_{A'}(v)} \mu_a \gamma_a = \lambda_v^+ - \lambda_v^- \quad \forall \, v \in V, \tag{4.2.7b}
$$

$$\frac{\partial L}{\partial \Delta_v} = 0 \Rightarrow \qquad \lambda_v^+ + \lambda_v^- + \lambda_v = 1 \qquad \forall\, v \in V, \tag{4.2.7c}$$

$$\frac{\partial L}{\partial \Delta_a} = 0 \Rightarrow \qquad \lambda_a^+ + \lambda_a^- + \lambda_a = 1 \qquad \forall\, a \in A'. \tag{4.2.7d}$$

We conclude from (4.2.7c), (4.2.7d), and the complementarity condition (2.4.2e) that $(q^*, \pi^*, \Delta^*, \mu^*, \lambda^*)$ fulfills:

$$\begin{aligned} &\pi_v > \underline{\pi}_v \Rightarrow \lambda_v^- = 0, && \pi_v < \overline{\pi}_v \Rightarrow \lambda_v^+ = 0, \\ &\pi_v = \underline{\pi}_v \Rightarrow \lambda_v^- \in [0,1], && \pi_v = \overline{\pi}_v \Rightarrow \lambda_v^+ \in [0,1], \qquad \forall\, v \in V \\ &\pi_v < \underline{\pi}_v \Rightarrow \lambda_v^- = 1, && \pi_v > \overline{\pi}_v \Rightarrow \lambda_v^+ = 1, \end{aligned} \tag{4.2.8a}$$

$$\begin{aligned} &q_a > \underline{q}_a \Rightarrow \lambda_a^- = 0, && q_a < \overline{q}_a \Rightarrow \lambda_a^+ = 0, \\ &q_a = \underline{q}_a \Rightarrow \lambda_a^- \in [0,1], && q_a = \overline{q}_a \Rightarrow \lambda_a^+ \in [0,1], \qquad \forall\, a \in A' \\ &q_a < \underline{q}_a \Rightarrow \lambda_a^- = 1, && q_a > \overline{q}_a \Rightarrow \lambda_a^+ = 1. \end{aligned} \tag{4.2.8b}$$

The tables of the Lemma follow from (4.2.8a) and (4.2.8b), respectively, and the constraints follow from (4.2.7a) and (4.2.7b).

The interpretation of these conditions is as follows: Equality (4.2.7b) indicates, that $(\mu_a)_{a \in A'}$ represents a network flow in (V, A') where each arc $a \in A'$ has μ_a as its flow variable. The in- and out-flows at sources and sinks are given by $\lambda_v^+ - \lambda_v^-$, and the relation of these values with the arc flows is given by the weighted flow conservation (4.2.7b) (also called generalized flow conservation, see Oldham (1999) and the references therein). As this flow conservation is weighted, the node flow must not necessarily be balanced, i.e., $\sum_{v \in V}(\lambda_v^+ - \lambda_v^-) \neq 0$ might hold. The implications (4.2.8a) ensure that a nonzero entry flow is only allowed, if $\pi_v \geq \overline{\pi}_v$. Furthermore, a nonzero exit flow can only occur at a node fulfilling $\pi_v \leq \underline{\pi}_v$. Looking at equation (4.2.7a), the dual value μ_v can be interpreted as a dual node potential at node v. The values λ_a^+, λ_a^- enforce a dual decrease or increase of the potential values and so react like a dual active element (compressor or control valve) restricted by (4.2.8b). □

Lemma 4.2.12:
Let (q^*, π^*, Δ^*) be an optimal solution of the domain relaxation (4.2.1)*. There exist Lagrange multipliers (μ^*, λ^*) which consecutively correspond to the equality and inequality constraints of* (4.2.1)*, respectively, such that $(q^*, \pi^*, \Delta^*, \mu^*, \lambda^*)$ is a KKT point of* (4.2.1)*.*

Proof. We prove that (μ^*, λ^*) exist such that the conditions (4.2.7) are fulfilled. It follows from the optimality of (q^*, π^*, Δ^*) (especially of the Δ variables) that there exist $\lambda^*_v, \lambda^{*\pm}_v, \lambda^*_a, \lambda^{*\pm}_a \in \mathbb{R}_{\geq 0}$ such that (4.2.7c), (4.2.7d) and the complementarity conditions (2.4.2e) are fulfilled. As the dual flow conservation (4.2.7b) must be fulfilled by (μ^*, λ^*) we observe an additional constraint for $\lambda^{*\pm}$. It is obtained as follows: Summing up the dual flow conservation (4.2.7b) using $\gamma_{r,v}$ from Definition 4.2.3 yields

$$\begin{aligned}\sum_{v\in V} \gamma_{r,v}^{-1}(\lambda_v^{+*} - \lambda_v^{-*}) &= \sum_{v\in V} \gamma_{r,v}^{-1} \left(\sum_{a\in\delta^+_{A'}(v)} \mu^*_a - \sum_{a\in\delta^-_{A'}(v)} \gamma_a \mu^*_a \right) \\ &= \sum_{a\in A'} \mu^*_a(\gamma_{r,v}^{-1} - \gamma_a\gamma_{r,w}^{-1}) = \sum_{a\in A'} \mu^*_a(\gamma_{r,v}^{-1} - \gamma_{r,v}^{-1}) = 0.\end{aligned}$$

We conclude from the connectivity of the graph (V, A') that $\lambda^{*\pm}_v, v \in V$ has to be chosen such that the balancing constraint

$$\sum_{v\in V} \gamma_{r,v}^{-1}(\lambda_v^{*+} - \lambda_v^{*-}) = 0 \tag{4.2.9}$$

is fulfilled in addition to the previously mentioned restrictions. Otherwise there does not exist a dual arc flow μ fulfilling (4.2.7b). In order to show that (4.2.9) is an additional feasible restriction for the definition of $\lambda^{*\pm}_v, v \in V$ we apply Lemma 4.2.7. Thereby the solution space of π of the domain relaxation (4.2.1) is a straight line, i.e., $\pi^* + \theta t$ is feasible for every $t \in \mathbb{R}$. We rewrite the objective function of (4.2.1) by

$$f(t) := \sum_v \left(\max\{0, (\pi^*_v + \theta_v t) - \overline{\pi}_v\} + \max\{0, \underline{\pi}_v - (\pi^*_v + \theta_v t)\}\right).$$

This function $f(t)$ is convex and optimal for $t = 0$ because of the local optimality of (q^*, π^*, Δ^*). This implies the following conditions:

$$\forall t' > 0 : f(t') \geq f(0) \Rightarrow \sum_{v\in V:\pi^*_v \geq \overline{\pi}_v} \theta_v \geq \sum_{v\in V:\pi^*_v < \underline{\pi}_v} \theta_v, \tag{4.2.10a}$$

$$\forall t' < 0 : f(t') \geq f(0) \Rightarrow \sum_{v\in V:\pi^*_v > \overline{\pi}_v} \theta_v \leq \sum_{v\in V:\pi^*_v \leq \underline{\pi}_v} \theta_v. \tag{4.2.10b}$$

Recall that $\lambda^{*\pm}_v$ is fixed for those nodes $v \in V$ with $\pi_v > \overline{\pi}_v$ or $\pi_v < \underline{\pi}_v$ by (4.2.7c) and the complementarity conditions (2.4.2e). For those nodes with $\pi_v > \overline{\pi}_v$ we have $\lambda^{*+}_v - \lambda^{*-}_v = 1$ and for the other ones with $\pi_v < \underline{\pi}_v$ we have $\lambda^{*+}_v - \lambda^{*-}_v = -1$. Now we distinguish between three cases:

Case 1:

$$\sum_{v \in V : \pi_v^* > \overline{\pi}_v} \theta_v = \sum_{v \in V : \pi_v^* < \underline{\pi}_v} \theta_v$$

Setting $\lambda_v^{*+} = \lambda_v^{*-} = 0$ for every node $v \in V$ with $\underline{\pi}_v \leq \pi_v \leq \overline{\pi}_v$ yields (4.2.9). Recall the relation $\theta_v := \gamma_{r,v}^{-1}, v \in V$ from the proof of Lemma 4.2.7.

Case 2:

$$\sum_{v \in V : \pi_v^* > \overline{\pi}_v} \theta_v < \sum_{v \in V : \pi_v^* < \underline{\pi}_v} \theta_v$$

We set $\lambda_v^{*+} = \lambda_v^{*-} = 0$ for every node $v \in V$ with $\underline{\pi}_v \leq \pi_v < \overline{\pi}_v$ and $\lambda_v^{*+} \in [0,1]$, $\lambda_v^{*-} = 0$ for every node $v \in V$ with $\pi_v = \overline{\pi}_v$ such that (4.2.9) is fulfilled. This setting is possible due to (4.2.10b).

Case 3:

$$\sum_{v \in V : \pi_v^* > \overline{\pi}_v} \theta_v > \sum_{v \in V : \pi_v^* < \underline{\pi}_v} \theta_v$$

We set $\lambda_v^{*+} = \lambda_v^{*-} = 0$ for every node $v \in V$ with $\underline{\pi}_v < \pi_v \leq \overline{\pi}_v$ and $\lambda_v^{*+} = 0$, $\lambda_v^{*-} \in [0,1]$ for every node $v \in V$ with $\pi_v = \underline{\pi}_v$ such that (4.2.9) is fulfilled. This setting is possible due to (4.2.10b).

We note that exactly one of the above cases applies which means that the previous discussion yields a feasible definition for $\lambda_v^{*\pm}, v \in V$. Recall that the values of $\lambda_v^*, v \in V$ have to be set according to (4.2.7c).

Now we compute a local optimum $(\mu_a^*)_{a \in A'}$ of the problem

$$\begin{aligned}
\min \quad & \sum_{a \in A'} \int_{\mu_a^0}^{\mu_a} \gamma_{r,v}^{-1} \, \tilde{\Phi}_a(t) \; \mathrm{d}t \\
\text{s.t.} \quad & \sum_{a \in \delta_{A'}^-(v)} \gamma_a \mu_a - \sum_{a \in \delta_{A'}^+(v)} \mu_a = -\lambda_v^{*+} + \lambda_v^{*-} && \forall\, v \in V, \\
& \mu_a \in \mathbb{R} && \forall\, a \in A',
\end{aligned}$$

where $\tilde{\Phi}_a(\mu) := \mu_a \frac{d\Phi_a}{dq_a}(q_a^*) + \lambda_a^{*+} - \lambda_a^{*-}$ and μ_a^0 is a root of this function for each arc $a = (v,w) \in A'$. The nonlinear optimization problem is feasible due to the previous discussion. Similar as in the proof of Lemma 4.2.1 we conclude that there exists $(\mu_v')_{v \in V}$ such that

$$\mu_v' - \gamma_a \mu_w' = \gamma_{r,v}^{-1} \, \tilde{\Phi}_a(\mu_a^*)$$

holds for every arc $a = (v, w) \in A'$. We define $\mu_v^* := \gamma_{r,v}\mu_v'$ and make use of (4.2.5) to obtain

$$\mu_v^* - \mu_w^* = \tilde{\Phi}_a(\mu_a^*)$$

for every arc $a = (v, w) \in A'$. This way we obtain a solution (μ^*, λ^*) which is feasible for (4.2.7a) and (4.2.7b). Hence $(q^*, \pi^*, \Delta^*, \mu^*, \lambda^*)$ is a KKT point of (4.2.1). □

Example 4.2.13:

Figure 4.2 shows a visualization of primal and dual flow of a KKT point of the domain relaxation (4.2.1) *in a network of practical dimension (*`net6`*). The color of a node corresponds to the primal / dual node potential. The arc width represents the primal / dual flow value (the thicker the more flow), while its color depicts the mean value of the node potentials at both end nodes. The figures show an infeasible primal flow, where the node potentials at some entries are above their respective upper limit and some exit node potentials are below their respective lower limit. (Note that there are other primal entries and exits, too.) Those nodes exceeding their respective upper limit are dual entries, those which exceed their respective lower limit are dual exits.*

For this example the arc set A'' which contains all arcs having a nonzero dual flow, i.e., $A'' = \{a \in A' \mid \mu_a \neq 0\}$, characterizes the infeasibility of the nomination d in the following sense: Let $(\tilde{q}, \tilde{\pi}, \tilde{\Delta})$ be a solution of the domain relaxation (4.2.1). *We observe from Figure 4.2 that there exists a subset $A''' \subseteq A''$ such that from*

$$\alpha_a \tilde{q}_a |\tilde{q}_a|^{k_a} - \tilde{\beta}_a - (\tilde{\pi}_v - \gamma_a \tilde{\pi}_w) = 0$$

for all arcs $a = (v, w) \in A'''$ the existence of nodes $s, t \in V$ such that $\tilde{\pi}_s - \tilde{\pi}_t > \overline{\pi}_s - \underline{\pi}_t$ follows.

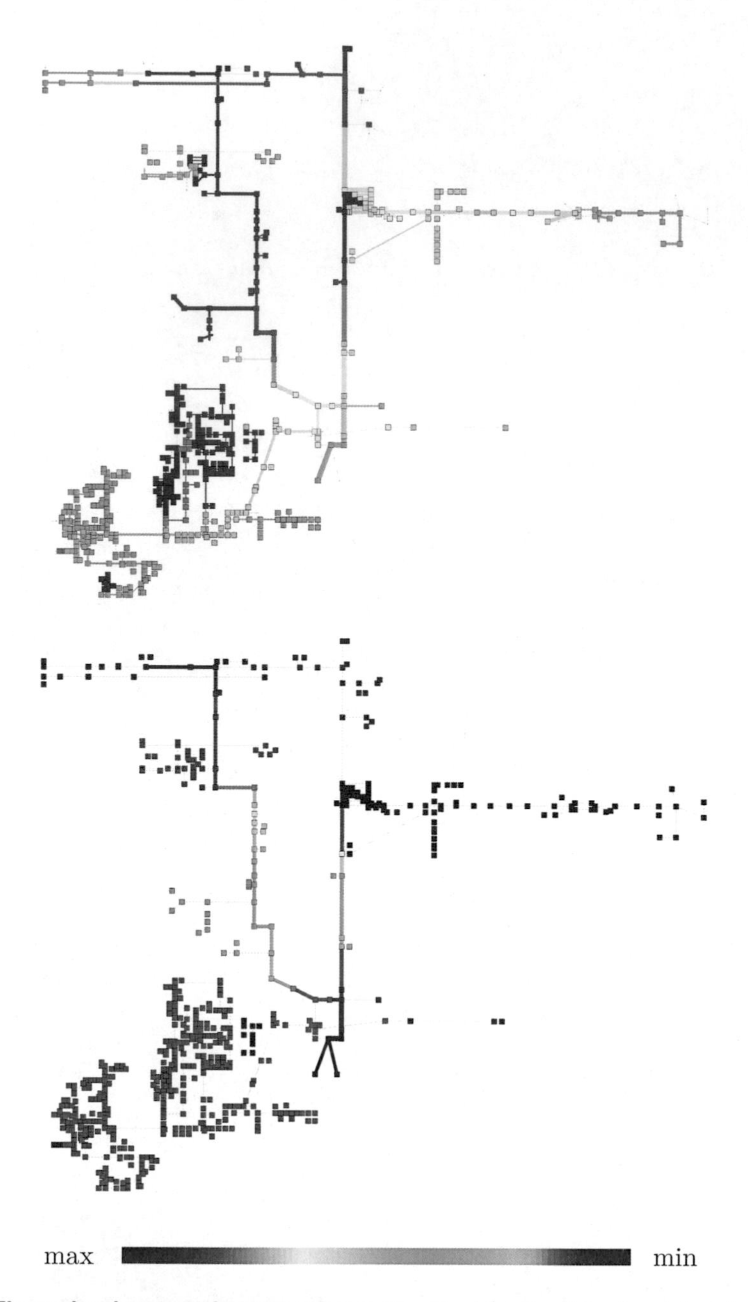

Figure 4.2: Flow and node potential corresponding to the primal (upper picture) and dual (lower picture) parts of a KKT point of domain relaxation (4.2.1) for network `net6`. The line width represents the flow value (the thicker the more flow), while its color depicts the mean value of the node pressures at both end nodes. The nodes are depicted by squares and the node colors represent the node pressures. All active elements which are not closed are in bypass mode and the corresponding arcs are contracted. It follows from Corollary 4.2.8 that the primal solution flow is unique.

4.3 Relaxation of Flow Conservation Constraints

In this section we consider the *flow conservation relaxation* of the passive transmission problem with $\alpha_a > 0$ for each arc $a \in A'$. This relaxation is obtained by relaxing the flow conservation constraints (4.1.1d). It is as follows:

$$\min \sum_{v\in V} \left(\Delta_v^+ + \Delta_v^-\right) + \sum_{a\in A'} \left(\Delta_a^+ + \Delta_a^-\right) \tag{4.3.1a}$$

$$\text{s.t.} \quad \alpha_a\, q_a|q_a|^{k_a} - \tilde{\beta}_a - (\pi_v - \gamma_a \pi_w) = 0 \quad \forall\, a \in A',\ a = (v,w), \tag{4.3.1b}$$

$$\sum_{a\in\delta^+_{A'}(v)} (q_a - (\Delta_a^+ - \Delta_a^-)) - \sum_{a\in\delta^-_{A'}(v)} (q_a - (\Delta_a^+ - \Delta_a^-)) - (\Delta_v^+ - \Delta_v^-) = d_v \quad \forall\, v \in V, \tag{4.3.1c}$$

$$\pi_v \le \overline{\pi}_v \quad \forall\, v \in V, \tag{4.3.1d}$$

$$\pi_v \ge \underline{\pi}_v \quad \forall\, v \in V, \tag{4.3.1e}$$

$$q_a \le \overline{q}_a \quad \forall\, a \in A', \tag{4.3.1f}$$

$$q_a \ge \underline{q}_a \quad \forall\, a \in A', \tag{4.3.1g}$$

$$\Delta_v^- \,(\overline{\pi}_v - \pi_v) = 0 \quad \forall\, v \in V, \tag{4.3.1h}$$

$$\Delta_v^+ \,(\pi_v - \underline{\pi}_v) = 0 \quad \forall\, v \in V, \tag{4.3.1i}$$

$$\Delta_a^- \,(\overline{q}_a - q_a) = 0 \quad \forall\, a \in A', \tag{4.3.1j}$$

$$\Delta_a^+ \,(q_a - \underline{q}_a) = 0 \quad \forall\, a \in A', \tag{4.3.1k}$$

$$\pi_v \in \mathbb{R} \quad \forall\, v \in V, \tag{4.3.1l}$$

$$q_a \in \mathbb{R} \quad \forall\, a \in A', \tag{4.3.1m}$$

$$\Delta_v^+, \Delta_v^- \in \mathbb{R}_{\ge 0} \ \forall\, v \in V, \tag{4.3.1n}$$

$$\Delta_a^+, \Delta_a^- \in \mathbb{R}_{\ge 0} \ \forall\, a \in A'. \tag{4.3.1o}$$

On an arc $a \in A'$ a positive slack value $\Delta_a^+ > 0$ or $\Delta_a^- > 0$ is feasible only if the flow variable q_a reaches its bounds $\underline{q}_a$ or $\overline{q}_a$, respectively. Accordingly, a positive slack value $\Delta_v^+ > 0$ or $\Delta_v^- > 0$ at a node $v \in V$ is feasible only if the potential value π_v attains a boundary value $\underline{\pi}_v$ or $\overline{\pi}_v$, respectively.

We will show how to solve this nonlinear optimization problem (4.3.1) efficiently. It can be infeasible, but this infeasibility can be detected in a preprocessing step as described in Section 4.3.1. Otherwise, if this preprocessing does not detect infeasibility, then the problem is feasible and convex, see Section 4.3.2 and Section 4.3.3. Hence it can be solved very efficiently to global optimality. Note that the convexity here is not given by the constraints but by the feasible solution space of the relaxation. Finally in Section 4.3.4 we will give an interpretation of the dual solution of a KKT point of (4.3.1). Similar to the discussion in Section 4.2.3 this dual solution forms a general network flow that is induced by a dual node potential.

4.3.1 Preprocessing

Problem (4.3.1) can be infeasible. This happens, if the flow bounds (4.3.1f) and (4.3.1g) enforce such a high amount of flow on an arc, that the potential loss (as deduced by equation (4.3.1b)) is in conflict with the bounds on the node potentials on both end nodes of the arc which are given by (4.3.1d) and (4.3.1e). The conflicting constraints in this case are

$$\begin{aligned} \alpha_a \, q_a |q_a|^{k_a} - \tilde{\beta}_a &= \pi_v - \gamma_a \pi_w, \\ \underline{q}_a \leq q_a &\leq \overline{q}_a, \\ \underline{\pi}_v \leq \pi_v &\leq \overline{\pi}_v, \\ \underline{\pi}_w \leq \pi_w &\leq \overline{\pi}_w, \end{aligned}$$

for arc $a = (v, w) \in A'$. This conflict situation can easily be detected by solving the linear program

$$\begin{aligned} &\exists \pi \\ \text{s.t.} \quad &\Phi_a(\underline{q}_a) \leq \pi_v - \gamma_a \pi_w \leq \Phi_a(\overline{q}_a) && \forall a = (v, w) \in A', \\ &\underline{\pi}_v \leq \pi_v \leq \overline{\pi}_v && \forall v \in V, \\ &\pi_v \in \mathbb{R} && \forall v \in V, \end{aligned} \tag{4.3.2}$$

with $\Phi_a(q) := \alpha_a \, q|q|^{k_a} - \tilde{\beta}_a$. If this LP turns out to be infeasible, then the passive transmission problem is infeasible, because the constraints that induce the infeasibility are part of the passive transmission problem. Otherwise it follows from Lemma 4.3.3 (stated and proven in the following section) that the flow conservation relaxation (4.3.1) is feasible. In the following discussion we assume that we applied this preprocessing technique which means to solve LP (4.3.2) and did not detect infeasibility.

4.3.2 Existence of a Solution

We consider the preprocessed problem (4.3.1) where the preprocessing is applied as described in Section 4.3.1. Recall that (4.2.2) and (4.2.3), which were first presented by Collins et al. (1978) and Maugis (1977), allowed to compute a feasible solution for the domain relaxation (4.2.1). We extend these convex optimization problems in order to compute a feasible solution for the flow conservation relaxation (4.3.1). This approach only works for constant heights, i.e., $\gamma_a = 1$ for all $a \in A'$. In the subsequent part of this section we show how to treat the general case of inhomogeneous heights, i.e., $\gamma_a \neq 1$ for some $a \in A'$.

First we describe the extension of (4.2.2). We introduce slack variables and add further terms to the objective function. Again we set $\Phi_a(q_a) := \alpha_a\, q_a |q_a|^{k_a} - \tilde{\beta}_a$. Then this extension is of the following form:

$$\begin{aligned} \min \quad & \sum_{a \in A'} \int_{q_a^0}^{q_a} \Phi_a(t)\, dt \\ & + \sum_{v \in V} \left(\overline{\pi}_v\, \Delta_v^- - \underline{\pi}_v\, \Delta_v^+ \right) + \sum_{a \in A'} \left(\Phi_a(\overline{q}_a)\, \Delta_a^- - \Phi_a(\underline{q}_a)\, \Delta_a^+ \right) \end{aligned} \tag{4.3.3a}$$

$$\begin{aligned} \text{s.t.} \quad & \sum_{a \in \delta_{A'}^-(v)} \left(q_a - (\Delta_a^+ - \Delta_a^-) \right) \\ & - \sum_{a \in \delta_{A'}^+(v)} \left(q_a - (\Delta_a^+ - \Delta_a^-) \right) + (\Delta_v^+ - \Delta_v^-) = -d_v \quad \forall v \in V, \end{aligned} \tag{4.3.3b}$$

$$\Delta_v^\pm \geq 0 \quad \forall v \in V, \tag{4.3.3c}$$

$$\Delta_a^\pm \geq 0 \quad \forall a \in A', \tag{4.3.3d}$$

$$\Delta_v^\pm \in \mathbb{R} \quad \forall v \in V, \tag{4.3.3e}$$

$$q_a, \Delta_a^\pm \in \mathbb{R} \quad \forall a \in A'. \tag{4.3.3f}$$

Here q_a^0 is the root of $\Phi_a(\cdot)$. In the next lemma we characterize this nonlinear optimization problem by analyzing the KKT conditions for this constraint system.

Lemma 4.3.1:

The nonlinear optimization problem (4.3.3) *is convex. Every optimal solution for*

(4.3.3) *can be transformed into a feasible solution for* (4.3.1)*, if* $\gamma_a = 1$ *for all arcs* $a \in A'$.

Proof. The optimization problem (4.3.3) is convex, as Φ_a is a monotone increasing function which implies that the objective consists of a sum of convex functions. Furthermore the constraints are of linear type.

Let (q^*, Δ^*) be a local minimum of (4.3.3). Because of the convexity every local minimum is global. The objective and the constraints of (4.3.3) are continuously differentiable. Hence, by Theorem 2.4.2, there exist dual values (π^*, λ^*), which consecutively correspond to the equality and inequality constraints of (4.3.3), such that $(q^*, \Delta^*, \pi^*, \lambda^*)$ is a KKT point of (4.3.3). The Lagrange function with the Lagrange multipliers $(\pi, \lambda) = (\pi_v, \lambda_v^+, \lambda_v^-, \lambda_a^+, \lambda_a^-)_{v \in V, a \in A'}$ with $\pi_v \in \mathbb{R}, v \in V$ and $\lambda_v^+, \lambda_v^- \in \mathbb{R}_{\geq 0}, v \in V$ and $\lambda_a^+, \lambda_a^- \in \mathbb{R}_{\geq 0}, a \in A'$ is as follows:

$$
\begin{aligned}
&L(q, \Delta, \pi, \lambda) \\
&= \sum_{a \in A'} \int_{q_a^0}^{q_a} \Phi_a(t)\, \mathrm{d}t + \sum_{v \in V} \left(\overline{\pi}_v\, \Delta_v^- - \underline{\pi}_v\, \Delta_v^+ \right) + \sum_{a \in A'} \left(\Phi_a(\overline{q}_a)\, \Delta_a^- - \Phi_a(\underline{q}_a)\, \Delta_a^+ \right) \\
&+ \sum_{v \in V} \pi_v \left(d_v - \sum_{a \in \delta_{A'}^+(v)} (q_a - (\Delta_a^+ - \Delta_a^-)) + \sum_{a \in \delta_{A'}^-(v)} (q_a - (\Delta_a^+ - \Delta_a^-)) \right) \\
&- \sum_{v \in V} \pi_v \left(\sum_{v \in V} (\Delta_v^- - \Delta_v^+) \right) \\
&- \sum_{v \in V} \left(\lambda_v^+ \Delta_v^+ + \lambda_v^- \Delta_v^- \right) - \sum_{a \in A'} \left(\lambda_a^+ \Delta_a^+ + \lambda_a^- \Delta_a^- \right).
\end{aligned}
$$

We obtain from (2.4.2a) of the KKT conditions (2.4.2) that $(q^*, \Delta^*, \pi^*, \lambda^*)$ is feasible for

$$\frac{\partial L}{\partial q_a} = 0 \Rightarrow \quad \Phi_a(q_a) = \pi_v - \pi_w \quad \forall\, a = (v, w) \in A', \tag{4.3.4a}$$

$$\frac{\partial L}{\partial \Delta_v^-} = 0 \Rightarrow \quad \overline{\pi}_v - \lambda_v^- = \pi_v \quad \forall\, v \in V, \tag{4.3.4b}$$

$$\frac{\partial L}{\partial \Delta_v^+} = 0 \Rightarrow \quad \underline{\pi}_v + \lambda_v^+ = \pi_v \quad \forall\, v \in V, \tag{4.3.4c}$$

$$\frac{\partial L}{\partial \Delta_a^-} = 0 \Rightarrow \quad \Phi_a(\overline{q}_a) - \lambda_a^- = \pi_v - \pi_w \quad \forall\, a = (v, w) \in A', \tag{4.3.4d}$$

$$\frac{\partial L}{\partial \Delta_a^+} = 0 \Rightarrow \quad \Phi_a(\underline{q}_a) + \lambda_a^+ = \pi_v - \pi_w \quad \forall\, a = (v, w) \in A'. \tag{4.3.4e}$$

It follows from (4.3.4a) that the vector (q^*, π^*) satisfies constraint (4.3.1b). Using (4.3.4b)–(4.3.4c) in combination with the complementarity conditions (2.4.2e) which write here as

$$\lambda_v^+ \Delta_v^+ = 0 \quad \text{and} \quad \lambda_v^- \Delta_v^- = 0, \tag{4.3.5a}$$

and

$$\lambda_a^+ \Delta_a^+ = 0 \quad \text{and} \quad \lambda_a^- \Delta_a^- = 0, \tag{4.3.5b}$$

we observe that the following constraints are fulfilled by (Δ^*, π^*):

$$\begin{aligned} \Delta_v^- (\overline{\pi}_v - \pi_v) &= 0, \\ \Delta_v^+ (\pi_v - \underline{\pi}_v) &= 0. \end{aligned}$$

Using the strict monotonicity of Φ_a and (4.3.4a), (4.3.4d), (4.3.4e), (4.3.5b), we obtain that (q^*, Δ^*) fulfills

$$\begin{aligned} \Delta_a^- (\Phi_a(\overline{q}_a) - \Phi_a(q_a)) = 0 \quad &\Rightarrow \quad \Delta_a^- (\overline{q}_a - q_a) = 0, \\ \Delta_a^+ (\Phi_a(q_a) - \Phi_a(\underline{q}_a)) = 0 \quad &\Rightarrow \quad \Delta_a^+ (q_a - \underline{q}_a) = 0. \end{aligned}$$

So constraints (4.3.1h)–(4.3.1k) are fulfilled by (q^*, Δ^*, π^*). Note that $\lambda \geq 0$. Therefore (4.3.1d)–(4.3.1g) are also fulfilled because of (4.3.4a)–(4.3.4e). Furthermore the flow conservation constraints (4.3.1c) are fulfilled due to constraints (4.3.3b). Altogether, (q^*, π^*, Δ^*) is a feasible solution for (4.3.1), if $\gamma_a = 1, a \in A'$. □

Note that problem (4.3.3) might be unbounded. In this case there exists no optimal solution and that is why we cannot ensure the feasibility of (4.3.1) by Lemma 4.3.1. Thus we consider another optimization problem which extends (4.2.3) as follows:

$$\min \sum_{a=(v,w)\in A'} \int_{\Delta_a^0}^{\pi_v - \pi_w} \Phi_a^{-1}(t)\, \mathrm{d}t - \sum_{v\in V} \int_0^{\pi_v} d_v \, \mathrm{d}t \tag{4.3.6a}$$

$$\text{s.t.} \quad \pi_v \leq \overline{\pi}_v \quad \forall v \in V, \tag{4.3.6b}$$

$$\pi_v \geq \underline{\pi}_v \quad \forall v \in V, \tag{4.3.6c}$$

$$\pi_v - \pi_w \leq \Phi_a(\overline{q}_a) \quad \forall a = (v,w) \in A', \tag{4.3.6d}$$

$$\pi_v - \pi_w \geq \Phi_a(\underline{q}_a) \quad \forall a = (v,w) \in A', \tag{4.3.6e}$$

$$\pi_v \in \mathbb{R} \quad \forall v \in V. \tag{4.3.6f}$$

Here Δ_a^0 is the root of the function $\Phi_a^{-1}(\cdot)$, the inverse of $\Phi_a(\cdot)$.

Lemma 4.3.2:
The nonlinear optimization problem (4.3.6) *is convex and bounded. Its optimal solution yields a feasible solution for* (4.3.1)*, if* $\gamma_a = 1$ *for all arcs* $a \in A'$.

Note that problem (4.3.6) is bounded and hence allows to compute a feasible solution for the flow conservation relaxation (4.3.1) which is not guaranteed for the previous problem (4.3.3). However, we consider (4.3.3) and (4.3.6) for theoretical purpose only.

Proof. We note that the constraints of (4.3.6) are of linear type. The objective function is convex, because of the definition of Δ_a^0. Hence (4.3.6) is convex.

The problem (4.3.6) might be infeasible because of the linear constraints, but this situation is excluded because of the preprocessing described in Section 4.3.1. The linear constraints ensure that the optimization problem is bounded. Hence there exists a local optimum which we denote by π^*. This is a global minimum because of the convexity.

The objective and the constraints of (4.3.6) are continuously differentiable. Hence, by Theorem 2.4.2, there exist dual values Δ^* consecutively corresponding to the inequality constraints of (4.3.6) such that (π^*, Δ^*) is a KKT point of (4.3.6). We denote the Lagrange function with Lagrange multipliers $\Delta_v^+, \Delta_v^- \in \mathbb{R}_{\geq 0}$ for each node $v \in V$ and $\Delta_a^+, \Delta_a^- \in \mathbb{R}_{\geq 0}$ for each arc $a \in A'$ as follows:

$$\begin{aligned} L(\pi, \Delta) = & \sum_{a=(v,w)\in A'} \int_{\Delta_a^0}^{\pi_v - \pi_w} \Phi_a^{-1}(t)\,\mathrm{d}t - \sum_{v\in V} \int_0^{\pi_v} d_v\,\mathrm{d}t \\ & + \sum_{v\in V} (\Delta_v^-(\pi_v - \overline{\pi}_v) + \Delta_v^+(\underline{\pi}_v - \pi_v)) \\ & + \sum_{a=(v,w)\in A'} \Delta_a^-(\pi_v - \pi_w - \Phi_a(\overline{q}_a)) \\ & + \sum_{a=(v,w)\in A'} \Delta_a^+(\Phi_a(\underline{q}_a) - (\pi_v - \pi_w)). \end{aligned}$$

We obtain from (2.4.2a) of the KKT conditions (2.4.2) that (π^*, Δ^*) is feasible for

$$\frac{\partial L}{\partial \pi_v} = 0 \Rightarrow \sum_{a=(v,w)\in\delta_{A'}^+(v)} (\Phi_a^{-1}(\pi_v - \pi_w) - (\Delta_a^+ - \Delta_a^-))$$

$$- \sum_{a=(v,w)\in\delta^-_{A'}(v)} (\Phi_a^{-1}(\pi_v - \pi_w) - (\Delta_a^+ - \Delta_a^-))$$
$$-(\Delta_v^+ - \Delta_v^-) = d_v \qquad \forall\, v \in V.$$

Setting q^* by

$$q_a^* := \Phi_a^{-1}(\pi_v^* - \pi_w^*) \quad \Leftrightarrow \quad \Phi_a(q_a^*) = \pi_v^* - \pi_w^* \qquad \forall\, a = (v,w) \in A' \tag{4.3.7}$$

we derive that (q^*, π^*, Δ^*) fulfills (4.3.1b) and the flow conservation constraints (4.3.1c).

Constraints (4.3.1d) and (4.3.1e) are fulfilled by (q^*, π^*, Δ^*) because of constraints (4.3.6b) and (4.3.6c). Constraints (4.3.1f) and (4.3.1g) are fulfilled by (q^*, π^*, Δ^*) because of constraints (4.3.7), (4.3.6d) and (4.3.6e) and the strictly monotonicity (hence bijectivity) of Φ_a and Φ_a^{-1}. From the complementary slackness conditions we observe that the following constraints are fulfilled by (q^*, π^*, Δ^*):

$$\Delta_v^-(\overline{\pi}_v - \pi_v) = 0, \qquad \Delta_v^+(\pi_v - \underline{\pi}_v) = 0 \qquad \forall\, v \in V,$$
$$\Delta_a^-(\overline{q}_a - q_a) = 0, \qquad \Delta_a^+(q_a - \underline{q}_a) = 0 \qquad \forall\, a \in A'.$$

This gives (4.3.1h)–(4.3.1k). Therefore, if $\gamma_a = 1, a \in A'$, then (q^*, π^*, Δ^*) is a solution for (4.3.1). □

Making use of (4.3.6) we are able to prove the following characterization of the feasibility of the flow conservation relaxation (4.3.1) in the general case that there exists an arc $a \in A'$ such that $\gamma_a \neq 1$.

Lemma 4.3.3:
If the preprocessing described in Section 4.3.1 does not detect infeasibility, then the flow conservation relaxation (4.3.1) *is feasible.*

Proof. If the preprocessing described in Section 4.3.1 detects infeasibility, then the passive transmission problem is infeasible as discussed in that section. Otherwise, we proceed as follows. We use $\gamma_{r,v}$ from Definition 4.2.3, the function $\pi'_v(\pi)$ from (4.2.4) and equation (4.2.5). Now we compute a local optimum π^* for the following problem

$$\min \sum_{a=(v,w)\in A'} \int_{\Delta_a^0}^{\pi_v - \pi_w} \Phi_a'^{-1}(t)\,\mathrm{d}t - \sum_{v\in V} \int_0^{\pi_v} d_v\,\mathrm{d}t$$
$$\text{s.t.} \qquad \pi_v \le \pi'_v(\overline{\pi}) \qquad \forall\, v \in V,$$

$$
\begin{aligned}
\pi_v &\geq \pi'_v(\underline{\pi}) && \forall v \in V, \\
\pi_v - \pi_w &\leq \Phi'_a(\overline{q}_a) && \forall a = (v,w) \in A', \\
\pi_v - \pi_w &\geq \Phi'_a(\underline{q}_a) && \forall a = (v,w) \in A', \\
\pi_v &\in \mathbb{R} && \forall v \in V,
\end{aligned}
$$

where $\Phi'^{-1}_a(\cdot)$ is the inverse of $\Phi'_a(q_a) := \gamma_{r,v}\alpha_a\, q_a|q_a|^{k_a} - \gamma_{r,v}\tilde{\beta}_a$ and Δ^0_a is a root of this inverse function for each arc $a = (v,w) \in A'$.

This problem is feasible because of the assumption that we applied the preprocessing described in Section 4.3.1. By Lemma 4.3.2 this optimization problem yields a feasible solution (q^*, π^*, Δ^*) for a modified version of the flow conservation relaxation (4.3.1) which is obtained by replacing the constraints (4.3.1b), (4.3.1d), (4.3.1e), (4.3.1h), (4.3.1i) by

$$
\begin{aligned}
\gamma_{r,v}\alpha_a\, q_a|q_a|^{k_a} - \gamma_{r,v}\tilde{\beta}_a - (\pi_v - \pi_w) &= 0 && \forall a = (v,w) \in A', \\
\pi_v &\leq \pi'_v(\overline{\pi}) && \forall v \in V, \\
\pi_v &\geq \pi'_v(\underline{\pi}) && \forall v \in V, \\
\Delta^-_v\ (\pi'_v(\overline{\pi}) - \pi_v) &= 0 && \forall v \in V, \\
\Delta^+_v\ (\pi_v - \pi'_v(\underline{\pi})) &= 0 && \forall v \in V.
\end{aligned}
$$

Using (q^*, π^*, Δ^*) we are going to show how to obtain a feasible solution for the flow conservation relaxation (4.3.1) which is not modified. We define

$$
\hat{\pi}_v := \pi'^{\,-1}_v\,(\pi^*).
$$

for each node $v \in V$. We obtain $\underline{\pi} \leq \hat{\pi} \leq \overline{\pi}$. Furthermore we obtain $\Delta^-_v\ (\overline{\pi}_v - \hat{\pi}_v) = 0$ and $\Delta^+_v\ (\hat{\pi}_v - \underline{\pi}_v) = 0$. In combination with (4.2.5) we obtain

$$
\alpha_a\, q^*_a|q^*_a|^{k_a} - \tilde{\beta}_a = \gamma^{-1}_{r,v}\,(\pi^*_v - \pi^*_w) = \hat{\pi}_v - \gamma_a\hat{\pi}_w \qquad \forall a = (v,w) \in A'.
$$

Thus $(q^*, \hat{\pi}, \Delta^*)$ is a feasible solution for the flow conservation relaxation (4.3.1). □

4.3.3 Characterization of the Feasible Region

The next lemmata characterize the feasible region of the flow conservation relaxation (4.3.1). They state that all local optimal solutions differ only in the π values, which lie together on a straight line segment. This then allows to prove that the flow conservation relaxation is either infeasible, which can be detected by the preprocessing described in Section 4.3.1, or a feasible convex optimization problem.

We start with proving a simple result which is needed in the following part of this section.

Lemma 4.3.4:
Let $(q', \pi'), (q'', \pi'') \in \mathbb{R}^{A'} \times \mathbb{R}^V$ *with* $q' \leq q''$ *be two vectors fulfilling*

$$\alpha_a \, q_a |q_a|^{k_a} - \tilde{\beta}_a = \pi_v - \pi_w \quad \forall \, a = (v, w) \in A'$$

with $\alpha_a > 0$ *for all arcs* $a \in A'$*. Let* P *be a* v*-*w*-path in* (V, A')*. The following implications hold for the node potential values in* v *and* w*:*

$$\begin{aligned} \pi'_v \geq \pi''_v \quad &\Rightarrow \quad \pi'_w \geq \pi''_w, \\ \pi'_w \leq \pi''_w \quad &\Rightarrow \quad \pi'_v \leq \pi''_v, \end{aligned}$$

and

$$\begin{aligned} \pi'_v > \pi''_v \quad &\Rightarrow \quad \pi'_w > \pi''_w, \\ \pi'_w < \pi''_w \quad &\Rightarrow \quad \pi'_v < \pi''_v. \end{aligned}$$

If there exists an arc a *in this path with* $q'_a < q''_a$ *then it holds:*

$$\begin{aligned} \pi'_v \geq \pi''_v \quad &\Rightarrow \quad \pi'_w > \pi''_w, \\ \pi'_w \leq \pi''_w \quad &\Rightarrow \quad \pi'_v < \pi''_v. \end{aligned}$$

Proof. Let P be a v-w-path in (V, A'). Let the nodes of P be given by $v_1, \ldots, v_{n+1}$ where $v_1 = v$ and $v_{n+1} = w$ and connecting arcs by $a_1, \ldots, a_n$. We obtain from $q'_{a_i} \leq q''_{a_i}$:

$$\begin{aligned} \pi'_{v_i} - \gamma_{a_i} \pi'_{v_{i+1}} &= \alpha_{a_i} \, q'_{a_i} |q'_{a_i}|^{k_{a_i}} - \tilde{\beta}_{a_i} \\ &\leq \alpha_{a_i} \, q''_{a_i} |q''_{a_i}|^{k_{a_i}} - \tilde{\beta}_{a_i} = \pi''_{v_i} - \gamma_{a_i} \pi''_{v_{i+1}} \end{aligned} \tag{4.3.8}$$

From this we obtain, again using $q'_{a_i} \leq q''_{a_i}$ for every $i = 1, \ldots, n$:

$$\begin{aligned} \pi'_{v_1} \geq \pi''_{v_1} \quad &\Rightarrow \pi'_{v_2} \geq \pi''_{v_2} \Rightarrow \ldots \Rightarrow \pi'_{v_{n+1}} \geq \pi''_{v_{n+1}}, \\ \pi'_{v_{n+1}} \leq \pi''_{v_{n+1}} &\Rightarrow \pi'_{v_n} \leq \pi''_{v_n} \Rightarrow \ldots \Rightarrow \quad \pi'_{v_1} \leq \pi''_{v_1}. \end{aligned}$$

If there exists an arc a_{i_0} with $q'_{a_{i_0}} < q''_{a_{i_0}}$, then inequality (4.3.8) is strict. We obtain:

$$\begin{aligned} \pi'_{v_1} \geq \pi''_{v_1} &\Rightarrow \pi'_{v_2} \geq \pi''_{v_2} \Rightarrow \ldots \Rightarrow \pi'_{v_{i_0+1}} > \pi''_{v_{i_0+1}} \Rightarrow \ldots \Rightarrow \pi'_{v_{n+1}} > \pi''_{v_{n+1}}, \\ \pi'_{v_{n+1}} \leq \pi''_{v_{n+1}} &\Rightarrow \pi'_{v_n} \leq \pi''_{v_n} \Rightarrow \ldots \Rightarrow \pi'_{v_{i_0}} < \pi''_{v_{i_0}} \Rightarrow \ldots \Rightarrow \pi'_{v_1} < \pi''_{v_1}. \end{aligned}$$

□

Lemma 4.3.5:
There exists a vector $\tilde{q} \in \mathbb{R}^{A'}$ such that the following holds: (q^*, π^*, Δ^*) is a feasible solution of (4.3.1) if and only if $q^* = \tilde{q}$.

Proof. Assume that there exist two solutions (q', π', Δ') and (q'', π'', Δ'') of problem (4.3.1). Note that there exists at least one by Lemma 4.3.3. We are going to prove that $q' = q''$. We proceed as follows:

1. At first we show that there exist $\Delta'''^{+}, \Delta'''^{-} \in \mathbb{R}^{A'}$ such that the vector $(q'', \pi'', (\Delta'^{\pm}_v, \Delta'''^{\pm}_a)_{v \in V, a \in A'})$ is feasible for (4.3.1).

2. Afterwards we conclude $q' = q''$.

In order to prove the first item we have to show the following because of (4.3.1h) and (4.3.1i): If there exists a node $v \in V$ with one of the values $\Delta'^{+}_v, \Delta'^{-}_v, \Delta''^{+}_v, \Delta''^{-}_v$ not equal to zero, then the node potentials π'_v and π''_v are equal, i.e., $\pi'_v = \pi''_v$.

Let $v^* \in V$ be a node such that one of the values $\Delta'^{+}_{v^*}, \Delta'^{-}_{v^*}, \Delta''^{\ +}_{v^*}, \Delta''^{\ -}_{v^*}$ is not equal to zero. In the case that $\Delta'^{+}_{v^*} - \Delta'^{-}_{v^*} = \Delta''^{\ +}_{v^*} - \Delta''^{\ -}_{v^*} \neq 0$ we conclude from (4.3.1h) and (4.3.1i) that $\pi'_{v^*} = \pi''_{v^*}$. In the case $\Delta'^{+}_{v^*} - \Delta'^{-}_{v^*} = \Delta''^{\ +}_{v^*} - \Delta''^{\ -}_{v^*} = 0$ and some value $\Delta'^{+}_{v^*}, \Delta'^{-}_{v^*}, \Delta''^{\ +}_{v^*}, \Delta''^{\ -}_{v^*} \neq 0$ we conclude that $\underline{\pi}_{v^*} = \overline{\pi}_{v^*}$ and hence $\pi'_{v^*} = \pi''_{v^*}$. Next we consider the remaining case $\Delta'^{+}_{v^*} - \Delta'^{-}_{v^*} \neq \Delta''^{\ +}_{v^*} - \Delta''^{\ -}_{v^*}$.

We define $\hat{q}_a$ for all arcs $a \in A'$ by

$$\hat{q}_a := (q''_a - (\Delta''^{+}_a - \Delta''^{-}_a)) - (q'_a - (\Delta'^{+}_a - \Delta'^{-}_a)).$$

W.l.o.g. we assume that $\Delta'^{+}_{v^*} - \Delta'^{-}_{v^*} < \Delta''^{\ +}_{v^*} - \Delta''^{\ -}_{v^*}$ (otherwise, exchange the two solutions). Hence v^* acts as a source for $\hat{q}$.

W.l.o.g. we assume $\hat{q} \geq 0$. If this is not the case then we change the orientation of every arc $a \in A'$ with $\hat{q}_a < 0$ in the same way as described in the proof of Lemma 4.2.6. This implies $q''_a - (\Delta''^{+}_a - \Delta''^{-}_a) \geq q'_a - (\Delta'^{+}_a - \Delta'^{-}_a)$ for every arc $a \in A'$. This in combination with (4.3.1j) and (4.3.1k) implies

$$q''_a \geq q'_a \quad \forall\, a \in A'. \tag{4.3.9}$$

Using Theorem 4.2.5 we split the flow $\hat{q}$ into a flows along paths $P_1, \dots, P_m$ and flows along cycles $C_1, \dots, C_n$. We denote the path flow values by $\hat{q}_{P_i} > 0$, $i = 1, \dots, m$ and the cycle flow values by $\hat{q}_{C_i} > 0, i = 1, \dots, n$. Then

$$\hat{q}_a = \sum_{\substack{i=1,\dots,m:\\ a \in A'(P_i)}} \hat{q}_{P_i} + \sum_{\substack{i=1,\dots,n:\\ a \in A'(C_i)}} \hat{q}_{C_i} \qquad \forall\, a \in A'.$$

Let $\hat{q}_{P_\ell}$ be a path flow that starts at source node v^* and ends at some other sink node w^*. Then we have

$$\Delta'^{+}_{v^*} - \Delta'^{-}_{v^*} < \Delta''^{+}_{v^*} - \Delta''^{-}_{v^*} \quad \text{and} \quad \Delta'^{+}_{w^*} - \Delta'^{-}_{w^*} > \Delta''^{+}_{w^*} - \Delta''^{-}_{w^*}. \tag{4.3.10}$$

We analyze the node potential differences in the two end nodes of the path P_ℓ, that is, $\pi''_{v^*} - \pi''_{w^*}$ versus $\pi'_{v^*} - \pi'_{w^*}$, and show that $\pi'_{v^*} = \pi''_{v^*}$ and $\pi'_{w^*} = \pi''_{w^*}$. Therefor we distinguish three cases. Note that the signs of $\Delta'^{+}_{v^*} - \Delta'^{-}_{v^*}$ and $\Delta''^{+}_{w^*} - \Delta''^{-}_{w^*}$ are unknown.

Case 1, $\Delta'^{+}_{v^*} - \Delta'^{-}_{v^*} > 0$. From constraint (4.3.1i) for v^* it follows that $\pi'_{v^*} = \underline{\pi}_{v^*}$ and (4.3.10) implies $\pi''_{v^*} = \underline{\pi}_{v^*}$. We distinguish three subcases.

Case 1.1, $\Delta'^{+}_{w^*} - \Delta'^{-}_{w^*} > 0$. Then again it follows that $\pi'_{w^*} = \underline{\pi}_{w^*}$. From Lemma 4.3.4 we obtain using $\pi''_{v^*} = \pi'_{v^*}$ and (4.3.9) that $\pi''_{w^*} \leq \pi'_{w^*}$ holds. From $\underline{\pi}_{w^*} \leq \pi''_{w^*} \leq \pi'_{w^*} = \underline{\pi}_{w^*}$ we obtain $\pi''_{w^*} = \underline{\pi}_{w^*}$. Hence $\pi'_{v^*} = \pi''_{v^*}$ and $\pi'_{w^*} = \pi''_{w^*}$. From (4.3.9) we conclude $q''_a = q'_a$ for every arc a of path P_ℓ because otherwise $q''_a > q'_a$ for an arc of path P_ℓ in combination with $\pi''_{v^*} = \pi'_{v^*}$ means $\pi''_{w^*} < \pi'_{w^*}$ by Lemma 4.3.4.

Case 1.2, $\Delta'^{+}_{w^*} - \Delta'^{-}_{w^*} = 0$. By (4.3.10) we have $\Delta''^{+}_{w^*} - \Delta''^{-}_{w^*} < 0$. According to constraint (4.3.1h) we obtain that $\pi''_{w^*} = \pi_{w^*}$. Now the node potentials π'' are at the lower boundary on the source and at the upper boundary on the sink side ($\pi''_{v^*} = \underline{\pi}_{v^*}, \pi''_{w^*} = \overline{\pi}_{w^*}$). From Lemma 4.3.4 we obtain using $\pi'_{v^*} = \pi''_{v^*}$ and (4.3.9) that $\pi'_{w^*} \geq \pi''_{w^*}$ holds. From $\overline{\pi}_{w^*} \geq \pi'_{w^*} \geq \pi''_{w^*} = \overline{\pi}_{w^*}$ we obtain $\pi''_{w^*} = \pi'_{w^*}$. Hence $\pi'_{v^*} = \pi''_{v^*}$ and $\pi'_{w^*} = \pi''_{w^*}$. From (4.3.9) we conclude $q''_a = q'_a$ for every arc a of path P_ℓ because otherwise $q''_a > q'_a$ for an arc of path P_ℓ in combination with $\pi''_{v^*} = \pi'_{v^*}$ means $\pi''_{w^*} < \pi'_{w^*}$ by Lemma 4.3.4.

Case 1.3, $\Delta'^{+}_{w^*} - \Delta'^{-}_{w^*} < 0$. Then from constraint (4.3.1h) we have $\pi''_{w^*} = \overline{\pi}_{w^*}$, and we are in the same situation as in Case 1.2.

Case 2, $\Delta'^{+}_{v^*} - \Delta'^{-}_{v^*} = 0$. From (4.3.1i) in combination with $0 = \Delta'^{+}_{v^*} - \Delta'^{-}_{v^*} < \Delta''^{+}_{v^*} - \Delta''^{-}_{v^*}$ by (4.3.10) we get that $\pi''_{v^*} = \underline{\pi}_{v^*}$.

Case 2.1, $\Delta'^{+}_{w*} - \Delta'^{-}_{w*} > 0$. Then again it follows that $\pi'_{w*} = \underline{\pi}_{w*}$. From Lemma 4.3.4 we obtain using $\underline{\pi}_{v*} = \pi''_{v*} \leq \pi'_{v*}$ and (4.3.9) that $\pi''_{w*} \leq \pi'_{w*}$ holds. From $\underline{\pi}_{w*} \leq \pi''_{w*} \leq \pi'_{w*} = \underline{\pi}_{w*}$ we obtain $\pi''_{w*} = \pi'_{w*}$. Again we obtain from Lemma 4.3.4 using $\pi'_{w*} \leq \pi''_{w*}$ and (4.3.9) that $\pi'_{v*} \leq \pi''_{v*}$ holds. Hence $\pi'_{v*} = \pi''_{v*}$ and $\pi'_{w*} = \pi''_{w*}$. From (4.3.9) we conclude $q''_a = q'_a$ for every arc a of path P_ℓ because otherwise $q''_a > q'_a$ for an arc of path P_ℓ in combination with $\pi''_{v*} = \pi'_{v*}$ means $\pi''_{w*} < \pi'_{w*}$ by Lemma 4.3.4.

Case 2.2, $\Delta'^{+}_{w*} - \Delta'^{-}_{w*} = 0$. By (4.3.10) we have $\Delta''^{\ +}_{w*} - \Delta''^{\ -}_{w*} < 0$. According to constraint (4.3.1h) we obtain that $\pi''_{w*} = \overline{\pi}_{w*}$. Now the node potentials π'' are at the lower boundary on the source and at the upper boundary on the sink side ($\pi''_{v*} = \underline{\pi}_{v*}, \pi''_{w*} = \overline{\pi}_{w*}$). From Lemma 4.3.4 we obtain using $\pi'_{v*} \geq \pi''_{v*} = \underline{\pi}_{v*}$ and (4.3.9) that $\pi'_{w*} \geq \pi''_{w*}$ holds. From $\overline{\pi}_{w*} \geq \pi'_{w*} \geq \pi''_{w*} = \overline{\pi}_{w*}$ we conclude $\pi''_{w*} = \pi'_{w*}$. Again we obtain from Lemma 4.3.4 using $\pi'_{w*} \leq \pi''_{w*}$ and (4.3.9) that $\pi'_{v*} \leq \pi''_{v*}$ holds. Hence $\pi'_{v*} = \pi''_{v*}$ and $\pi'_{w*} = \pi''_{w*}$. From (4.3.9) we conclude $q''_a = q'_a$ for every arc a of path P_ℓ because otherwise $q''_a > q'_a$ for an arc of path P_ℓ in combination with $\pi''_{v*} = \pi'_{v*}$ means $\pi''_{w*} < \pi'_{w*}$ by Lemma 4.3.4.

Case 2.3, $\Delta'^{+}_{w*} - \Delta'^{-}_{w*} < 0$. Then from constraint (4.3.1h) we have $\pi''_{w*} = \overline{\pi}_{w*}$, and we are in the same situation as in Case 2.2.

Case 3, $\Delta'^{+}_{v*} - \Delta'^{-}_{v*} < 0$. Then it follows from (4.3.1h) that $\pi'_{v*} = \overline{\pi}_{v*}$.

Case 3.1, $\Delta'^{+}_{w*} - \Delta'^{-}_{w*} > 0$. Then it follows from (4.3.1i) that $\pi'_{w*} = \underline{\pi}_{w*}$. Now $\pi'_{v*} = \overline{\pi}_{v*}$ and $\pi'_{w*} = \underline{\pi}_{w*}$ holds. From Lemma 4.3.4 we obtain using $\pi''_{v*} \leq \overline{\pi}_{v*} = \pi'_{v*}$ and (4.3.9) that $\pi'_{w*} \geq \pi''_{w*}$ holds. From $\underline{\pi}_{w*} = \pi'_{w*} \geq \pi''_{w*} \geq \underline{\pi}_{w*}$ we conclude $\pi''_{w*} = \pi'_{w*}$. Again we obtain from Lemma 4.3.4 using $\pi'_{w*} \leq \pi''_{w*}$ and (4.3.9) that $\pi'_{v*} \leq \pi''_{v*}$ holds. Hence $\pi'_{v*} = \pi''_{v*}$ and $\pi'_{w*} = \pi''_{w*}$. Again from (4.3.9) we conclude $q''_a = q'_a$ for every arc a of path P_ℓ because otherwise $q''_a > q'_a$ for an arc of path P_ℓ in combination with $\pi''_{v*} = \pi'_{v*}$ means $\pi''_{w*} < \pi'_{w*}$ by Lemma 4.3.4.

Case 3.2, $\Delta'^{+}_{w*} - \Delta'^{-}_{w*} = 0$. We derive $0 = \Delta'^{\ +}_{w*} - \Delta'^{\ -}_{w*} > \Delta''^{\ +}_{w*} - \Delta''^{\ -}_{w*}$ from (4.3.10). In combination with (4.3.1i) we obtain $\pi''_{w*} = \overline{\pi}_{w*}$. Furthermore we have $\pi'_{w*} \leq \overline{\pi}_{w*} = \pi''_{w*}$. From Lemma 4.3.4 and (4.3.9) we conclude $\pi''_{v*} \geq \pi'_{v*}$. From $\overline{\pi}_{v*} \geq \pi''_{v*} \geq \pi'_{v*} = \overline{\pi}_{v*}$ we obtain $\pi'_{v*} = \pi''_{v*}$. Again from Lemma 4.3.4 and (4.3.9) we conclude $\pi'_{w*} \geq \pi''_{w*}$ which then in combination with the previous observation $\pi'_{w*} \leq \pi''_{w*}$ implies $\pi'_{w*} = \pi''_{w*}$. We derive $q''_a = q'_a$ for every arc a of path P_ℓ because otherwise $q''_a > q'_a$ for an arc of path P_ℓ in combination with $\pi''_{v*} = \pi'_{v*}$ means $\pi''_{w*} < \pi'_{w*}$ by Lemma 4.3.4.

Case 3.3, $\Delta'^{+}_{w^*} - \Delta'^{-}_{w^*} < 0$. Then we get from (4.3.1h) that $\pi''_{w^*} = \overline{\pi}_{w^*}$. So we are in the same situation as in Case 3.2, hence the same conclusions remain valid.

From these cases we conclude that $\pi'_{v^*} = \pi''_{v^*}$ for any node $v^* \in V$ such that at least one value $\Delta'^{+}_{v^*}, \Delta'^{-}_{v^*}, \Delta''^{+}_{v^*}, \Delta''^{-}_{v^*}$ is not equal to zero. This implies that $(\pi''_v, \Delta'^{\pm}_v)_{v\in V}$ is feasible for constraints (4.3.1h) and (4.3.1i).

In order to complete the proof of the first item we turn to the definition of $(\Delta'''^{\pm}_a)_{a\in A'}$. The flow $\hat{q}$ is a network flow with node flow $(\Delta''^{+}_v - \Delta''^{-}_v) - (\Delta'^{+}_v - \Delta'^{-}_v)$ for every node $v \in V$. Hence we obtain from the previous 3×3 cases that $q''_a = q'_a$ for every arc $a \in A'(P_i), i = 1, \dots, m$. This implies for every arc $a \in A'(P_i), i = 1, \dots, m$:

$$(\Delta''^{-}_a - \Delta''^{+}_a) - (\Delta'^{-}_a - \Delta'^{+}_a) = \hat{q}_a = \sum_{\substack{i=1,\dots,m:\\ a\in A'(P_i)}} \hat{q}_{P_i} + \sum_{\substack{i=1,\dots,n:\\ a\in A'(C_i)}} \hat{q}_{C_i} > 0. \qquad (4.3.11)$$

We define

$$(\Delta'''^{-}_a - \Delta'''^{+}_a) := (\Delta''^{-}_a - \Delta''^{+}_a) - \sum_{\substack{i=1,\dots,m:\\ a\in A'(P_i)}} \hat{q}_{P_i} \qquad (4.3.12)$$

for every arc $a \in A'$ where at least one value Δ'''^{-}_a or Δ'''^{+}_a equals zero. Now we consider an arc $a \in A'(P_i), i = 1, \dots, m$. From $\Delta'''^{-}_a > 0$ it follows by $\hat{q}_{P_i} > 0$, $i = 1, \dots, m$ and (4.3.12) that $\Delta''^{-}_a > 0$. This implies $q''_a = \overline{q}_a$. From $\Delta'''^{+}_a > 0$ it follows from (4.3.11) that $\Delta'^{+}_a > 0$ holds. This implies $q''_a = q'_a = \underline{q}_a$. For all other arcs $a \in A'$ which are not considered before it holds that they are not part of any path $P_i, i = 1 \dots, m$. Hence $\Delta'''^{-}_a - \Delta'''^{+}_a = \Delta''^{-}_a - \Delta''^{+}_a$ holds for them and (4.3.1j) and (4.3.1k) are fulfilled. We conclude that $(q''_a, \Delta'''^{\pm}_a)_{a\in A'}$ is feasible for constraints (4.3.1j) and (4.3.1k).

It follows from (4.3.11) and (4.3.12) that the flow conservation constraint (4.3.1c) is fulfilled by $(q'', (\Delta'^{\pm}_v, \Delta'''^{\pm}_a)_{v\subset V, a\subset A'})$. Hence a feasible vector for (4.3.1) is given by $(q'', \pi'', (\Delta'^{\pm}_v, \Delta'''^{\pm}_a)_{v\in V, a\in A'})$. This proves the first item.

Turning to the second item we define $\hat{q}'_a := (q''_a - (\Delta'''^{+}_a - \Delta'''^{-}_a)) - (q'_a - (\Delta'^{+}_a - \Delta'^{-}_a))$ for each arc $a \in A'$. Then $\hat{q}'$ consists of circulations only because of

$$\sum_{a\in\delta^{+}_{A'}(v)} \hat{q}'_a - \sum_{a\in\delta^{-}_{A'}(v)} \hat{q}'_a = d_v + (\Delta'^{+}_v - \Delta'^{-}_v) - (d_v + (\Delta'^{+}_v - \Delta'^{-}_v)) = 0 \quad \forall v \in V.$$

But a circulation can only take place in the $(\Delta^{\pm}_a)_{a\in A'}$ variables because of the following reason: By (4.3.9) it holds $q'' \geq q'$. If there exists an arc $a \in A'$ with $q''_a > q'_a$, then, using $\alpha_a > 0, a \in A'$ and (4.3.1b), we derive the contradiction $0 > 0$ similar as done in the end of the proof of Lemma 4.2.7. Thus we conclude $q' = q''$, which proves the Lemma. □

Theorem 4.3.6:
The flow conservation relaxation (4.3.1) *is a relaxation of the passive transmission problem* (4.1.1). *The relaxation is either infeasible, which can be detected by the preprocessing technique described in Section 4.3.1, or a feasible convex optimization problem.*

Proof. Every feasible solution (q^*, π^*, p^*) of the passive transmission problem (4.1.1) yields a feasible solution $(q^*, \pi^*, 0)$ for (4.3.1). Hence (4.3.1) is a relaxation of the passive transmission problem (4.1.1).

If infeasibility is not detected during the preprocessing described in Section 4.3.1, then (4.3.1) is feasible by Lemma 4.3.3. In this case we consider two feasible solutions (q', π', Δ') and (q'', π'', Δ'') of (4.3.1). From Lemma 4.3.5 we know that q' and q'' are equal, i.e., $q' = q''$. By the same argumentation as in the last part of the proof of Lemma 4.2.7 we obtain the existence of $\theta \in \mathbb{R}^V_{\geq 0}$ such that there exists $t \in \mathbb{R}$ such that $\pi'' = \pi' + t\theta$. The definition of θ is independent of the two solutions. Now we distinguish two cases:

1. If there exist two nodes v and w with $\pi'_v = \overline{\pi}_v$ and $\pi'_w = \underline{\pi}_w$ then $\theta \geq 0$ implies either $t = 0$ or $\theta = 0$ and hence $\pi' = \pi''$. This means that the two solutions (q', π', Δ') and (q'', π'', Δ'') differ only in the variables $\Delta = (\Delta^{\pm}_a, \Delta^{\pm}_v)_{a \in A', v \in V}$. Fixing the node potential variables $\pi = \pi'$ and the flow variables $q = q'$ in (4.3.1) results in a linear program. This means that $(q', \pi', \epsilon\Delta' + (1 - \epsilon)\Delta'')$ is a feasible solution for (4.3.1) for every $\epsilon \in [0, 1]$. We conclude that the feasible solution space of (4.3.1) is convex in this case.

2. It holds either $\pi'_v > \underline{\pi}_v$ for every node $v \in V$ or $\pi'_v < \overline{\pi}_v$ for every node $v \in V$. This means either ${\Delta'_v}^+ = 0$ for all nodes $v \in V$ or ${\Delta'_v}^- = 0$ for all nodes $v \in V$. Summing (4.3.1c) over all nodes $v \in V$ yields the condition $\sum_{v \in V}({\Delta'_v}^+ - {\Delta'_v}^-) = 0$. We conclude ${\Delta'_v}^{\pm} = 0$ for all nodes $v \in V$. From $\theta \geq 0$ we obtain that there does not exist a solution (q', π''', Δ''') of (4.3.1) with $\pi'''_v = \overline{\pi}_v$ and $\pi'''_w = \underline{\pi}_w$ for two nodes $v, w \in V$. Hence the previous discussion also applies for π'' and ${\Delta''_v}^{\pm}, v \in V$ and we obtain ${\Delta''_v}^{\pm} = 0, v \in V$. Fixing the node slack variables ${\Delta_v}^{\pm} = 0$ and the flow variables $q = q'$ in (4.3.1) results in a linear program. This means that $(q', \epsilon_1\pi' + (1 - \epsilon_1)\pi'', \epsilon_2\Delta' + (1 - \epsilon_2)\Delta'')$ is a feasible solution for (4.3.1) for every $\epsilon_1, \epsilon_2 \in [0, 1]$. We conclude that the feasible solution space of (4.3.1) is convex in this case.

Exactly one of the previous cases applies. This proves that the feasible solution space of (4.3.1) is convex and hence (4.3.1) is a convex optimization problem. □

4.3.4 Interpretation of Lagrange Multipliers

Assume that the passive transmission problem (4.1.1) is infeasible and the flow conservation relaxation (4.3.1) has a positive optimal objective value. It turns out (see Lemma 4.3.7) that there exist Lagrange multipliers for the optimal solution such that the optimal solution and the multipliers form a KKT point of (4.3.1). These multipliers have a practical interpretation. They form a generalized network flow which is coupled with node potentials, similar to a primal solution (q^*, π^*) of the passive transmission problem. This result is comparable to the interpretation in Section 4.2.3.

Lemma 4.3.7:

Let (q^*, π^*, Δ^*) be an optimal solution of the flow conservation relaxation (4.3.1). *Let (μ^*, λ^*) be Lagrange multipliers which consecutively correspond to the equality and inequality constraints of* (4.3.1), *respectively, such that $(q^*, \pi^*, \Delta^*, \mu^*, \lambda^*)$ is a KKT point of* (4.3.1). *These multipliers are characterized as follows: $(\mu_a^*)_{a \in A'}$ is a general network flow in (V, A') which is induced by dual node potentials $(\mu_v^*)_{v \in V}$. More precisely the multipliers (μ^*, λ^*) are a feasible solution for*

$$\mu_a \frac{d\Phi_a}{dq_a}(q_a^*) + \lambda_a^+ - \lambda_a^- = \mu_v - \mu_w + \mu_a^+ \Delta_a^{-*} - \mu_a^- \Delta_a^{+*} \quad \forall\, a - (v, w) \in A',$$

$$\sum_{a \in \delta_{A'}^+(v)} \mu_a - \sum_{a \in \delta_{A'}^-(v)} \gamma_a \mu_a = \lambda_v^+ - \lambda_v^- + \mu_v^+ \Delta_v^{-*} - \mu_v^- \Delta_v^{+*} \quad \forall\, v \in V.$$

Hereby the dual node potential μ_v is restricted by

$\mu_v \in$	$\{1\}$	$[-1, 1]$	$\{-1\}$
node slack	$\Delta_v^{-*} > 0$	$\Delta_v^{-*} = 0 = \Delta_v^{+*}$	$\Delta_v^{+*} > 0$

for each node $v \in V$, and the dual node potential difference $\mu_v - \mu_w$ is constrained by

$\mu_v - \mu_w \in$	$\{1\}$	$[-1, 1]$	$\{-1\}$
arc slack	$\Delta_a^{-*} > 0$	$\Delta_a^{-*} = 0 = \Delta_a^{+*}$	$\Delta_a^{+*} > 0$

for each arc $a = (v, w) \in A'$.

Proof. Let (q^*, π^*, Δ^*) be an optimal solution of (4.3.1). Recall that a local optimal solution is globally optimal because of the convexity of (4.3.1) by Theorem 4.3.6. Let (μ^*, λ^*) be dual values such that $(q^*, \pi^*, \Delta^*, \mu^*, \lambda^*)$ is a KKT point of flow conservation relaxation (4.3.1). Let us write the conditions (2.4.2a) of the KKT conditions (2.4.2). We denote the Lagrange multipliers by $(\mu, \lambda) =$

$(\mu_v, \mu_v^+, \mu_v^-, \mu_a, \mu_a^+, \mu_a^-, \lambda_v^+, \lambda_v^-, \tilde{\lambda}_v^+, \tilde{\lambda}_v^-, \lambda_a^+, \lambda_a^-, \tilde{\lambda}_a^+, \tilde{\lambda}_a^-)_{v\in V, a\in A'}$, such that for the domains $\mu_v, \mu_v^+, \mu_v^-, \mu_a, \mu_a^+, \mu_a^- \in \mathbb{R}$ and $\lambda_v^+, \lambda_v^-, \tilde{\lambda}_v^+, \tilde{\lambda}_v^-, \lambda_a^+, \lambda_a^-, \tilde{\lambda}_a^+, \tilde{\lambda}_a^- \in \mathbb{R}_{\geq 0}$ holds. Furthermore we define the function $\Phi_a(q_a) := \alpha_a\, q_a |q_a|^{k_a} - \tilde{\beta}_a$. Then the Lagrange function of problem (4.3.1) has the form

$$
\begin{aligned}
&L(q, \pi, \Delta, \mu, \lambda) \\
&= \sum_{v\in V} \left(\Delta_v^+ + \Delta_v^-\right) + \sum_{a\in A'} \left(\Delta_a^+ + \Delta_a^-\right) \\
&+ \sum_{a=(v,w)\in A'} \mu_a \left(\Phi_a(q_a) - (\pi_v - \gamma_a \pi_w)\right) \\
&+ \sum_{v\in V} \mu_v \left(d_v - \sum_{a\in\delta_{A'}^+(v)} (q_a - \Delta_a^+ + \Delta_a^-) + \sum_{a\in\delta_{A'}^-(v)} (q_a - \Delta_a^+ + \Delta_a^-) \right) \\
&+ \sum_{v\in V} \mu_v (\Delta_v^+ - \Delta_v^-) \\
&+ \sum_{v\in V} \left(\lambda_v^+ \left(\pi_v - \overline{\pi}_v\right) + \lambda_v^- \left(\underline{\pi}_v - \pi_v\right)\right) \\
&+ \sum_{a\in A'} \left(\lambda_a^+ \left(q_a - \overline{q}_a\right) + \lambda_a^- \left(\underline{q}_a - q_a\right)\right) \\
&+ \sum_{v\in V} \left(\mu_v^+ \left(\overline{\pi}_v - \pi_v\right) \Delta_v^- + \mu_v^- \left(\pi_v - \underline{\pi}_v\right) \Delta_v^+ \right) \\
&+ \sum_{a\in A'} \left(\mu_a^+ \left(\overline{q}_a - q_a\right) \Delta_a^- + \mu_a^- \left(q_a - \underline{q}_a\right) \Delta_a^+ \right) \\
&- \sum_{v\in V} \left(\tilde{\lambda}_v^+ \Delta_v^+ + \tilde{\lambda}_v^- \Delta_v^- \right) \\
&- \sum_{a\in A'} \left(\tilde{\lambda}_a^+ \Delta_a^+ + \tilde{\lambda}_a^- \Delta_a^- \right).
\end{aligned}
$$

From (2.4.2a) we obtain that the KKT point $(q^*, \pi^*, \Delta^*, \mu^*, \lambda^*)$ is feasible for

$$
\begin{aligned}
\frac{\partial L}{\partial q_a} = 0 \Rightarrow \quad & -\mu_a^+ \Delta_a^- + \mu_a^- \Delta_a^+ \\
& + \mu_a \frac{d\Phi_a}{dq_a}(q_a) + \lambda_a^+ - \lambda_a^- = \mu_v - \mu_w \qquad \forall\, a \in A',\ a = (v,w), \qquad (4.3.13a) \\
\frac{\partial L}{\partial \pi_v} = 0 \Rightarrow \quad & \mu_v^+ \Delta_v^- - \mu_v^- \Delta_v^+ \\
& + \sum_{a\in\delta_{A'}^+(v)} \mu_a - \sum_{a\in\delta_{A'}^-(v)} \gamma_a \mu_a = \lambda_v^+ - \lambda_v^- \quad \forall\, v \in V, \qquad (4.3.13b) \\
\frac{\partial L}{\partial \Delta_v^+} = 0 \Rightarrow \quad & \mu_v + \mu_v^- \left(\pi_v - \underline{\pi}_v\right) - \tilde{\lambda}_v^+ = -1 \qquad \forall\, v \in V, \qquad (4.3.13c)
\end{aligned}
$$

$$\frac{\partial L}{\partial \Delta_v^-} = 0 \Rightarrow \quad -\mu_v + \mu_v^+ (\overline{\pi}_v - \pi_v) - \tilde{\lambda}_v^- = -1 \qquad \forall\, v \in V, \tag{4.3.13d}$$

$$\frac{\partial L}{\partial \Delta_a^+} = 0 \Rightarrow \mu_v - \mu_w + \mu_a^- (q_a - \underline{q}_a) - \tilde{\lambda}_a^+ = -1 \qquad \forall\, a \in A',\ a = (v,w), \tag{4.3.13e}$$

$$\frac{\partial L}{\partial \Delta_a^-} = 0 \Rightarrow \mu_w - \mu_v + \mu_a^+ (\overline{q}_a - q_a) - \tilde{\lambda}_a^- = -1 \qquad \forall\, a \in A',\ a = (v,w). \tag{4.3.13f}$$

We conclude from (4.3.13c)-(4.3.13f), and the complementarity condition (2.4.2e) that $(q^*, \pi^*, \Delta^*, \mu^*, \lambda^*)$ fulfills:

$$\Delta_v^- > 0 \Rightarrow \mu_v = 1, \qquad \Delta_v^+ > 0 \Rightarrow \mu_v = -1, \qquad \forall\, v \in V, \tag{4.3.14a}$$

$$\Delta_a^- > 0 \Rightarrow \mu_v - \mu_w = 1, \quad \Delta_a^+ > 0 \Rightarrow \mu_v - \mu_w = -1, \qquad \forall\, a \in A',\ a = (v,w). \tag{4.3.14b}$$

The equations (4.3.13a) and (4.3.13b) yield the constraints of the Lemma, while the tables follow from (4.3.14a) and (4.3.14b). From the complementarity constraints (4.3.1h) and (4.3.1i) we obtain

$$\begin{aligned} \pi_v < \overline{\pi}_v &\Rightarrow \lambda_v^+ = 0,\ \mu_v^+ \Delta_v^- = 0, \\ \pi_v > \underline{\pi}_v &\Rightarrow \lambda_v^- = 0,\ \mu_v^- \Delta_v^+ = 0, \end{aligned} \qquad \forall\, v \in V, \tag{4.3.15a}$$

$$\begin{aligned} q_a < \overline{q}_a &\Rightarrow \lambda_a^+ = 0,\ \mu_a^+ \Delta_a^- = 0, \\ q_a > \underline{q}_a &\Rightarrow \lambda_a^- = 0,\ \mu_a^- \Delta_a^+ = 0, \end{aligned} \qquad \forall\, a \in A'. \tag{4.3.15b}$$

We compare (4.3.13) with (4.2.7), the corresponding results for the domain relaxation (4.2.1). Most of the interpretation of (4.2.7) remains also valid for (4.3.13). In the following we will focus on the differences. Conditions (4.3.13a) correspond to the potential flow coupling (4.2.7a), and conditions (4.3.13b) correspond to the flow conservation (4.2.7b). The derived conditions (4.2.8a) state that in certain cases for the node potential values we have to fix the dual node flows. Complementary, the derived conditions (4.3.14a) state that in certain cases for the node flow slack values we have to fix the dual node potentials. Similarly, conditions (4.2.8b) state that in certain cases for the arc flow values we have to fix the dual variables λ_a^+ and λ_a^-. Complementary, the derived conditions (4.3.14b) state that in certain cases for the arc flow slack values we have to fix the dual node potential difference at the end nodes of the arc.

□

4.4 Relaxation of Potential-Flow-Coupling Constraints

A relaxation of the passive transmission problem (4.1.1) which is obtained by relaxing the potential-flow-coupling constraint (4.1.1b) writes as

$$\min \sum_{a \in A'} (\Delta_a^+ + \Delta_a^-) \tag{4.4.1a}$$

$$\text{s.t.}\quad \alpha_a\, q_a |q_a|^{k_a} - (\pi_v - \gamma_a \pi_w) - (\Delta_a^+ - \Delta_a^-) = \tilde{\beta}_a \qquad \forall\, a \in A',\ a = (v,w), \tag{4.4.1b}$$

$$\sum_{a \in \delta_{A'}^+(v)} q_a - \sum_{a \in \delta_{A'}^-(v)} q_a = d_v \qquad \forall\, v \in V, \tag{4.4.1c}$$

$$\pi_v - \overline{\pi}_v \le 0 \qquad \forall\, v \in V, \tag{4.4.1d}$$

$$\underline{\pi}_v - \pi_v \le 0 \qquad \forall\, v \in V, \tag{4.4.1e}$$

$$q_a - \overline{q}_a \le 0 \qquad \forall\, a \in A', \tag{4.4.1f}$$

$$\underline{q}_a - q_a \le 0 \qquad \forall\, a \in A', \tag{4.4.1g}$$

$$\pi_v \in \mathbb{R} \qquad \forall\, v \in V, \tag{4.4.1h}$$

$$q_a \in \mathbb{R} \qquad \forall\, a \in A', \tag{4.4.1i}$$

$$\Delta_a^+, \Delta_a^- \in \mathbb{R}_{\ge 0} \qquad \forall\, a \in A'. \tag{4.4.1j}$$

Lemma 4.4.1:
The optimization problem (4.4.1) *is a relaxation of the passive transmission problem* (4.1.1).

Proof. A solution (q^*, π^*, p^*) is feasible for the passive transmission problem (4.1.1) only if $(q^*, \pi^*, 0)$ is feasible for the nonlinear optimization problem (4.4.1). Hence (4.4.1) is a relaxation of (4.1.1). □

In the following we will show that this relaxation (4.4.1) is a non-convex optimization problem having different KKT points with different objective values.

4.4.1 Conditions of the KKT System

In order to show that there exist multiple KKT points of (4.4.1) with different objective values we proceed as follows: The objective and all constraints of (4.4.1) are continuously differentiable. Now let us consider a KKT point $(q^*, \pi^*, \Delta^*, \mu^*, \lambda^*)$. Further let us write the conditions (2.4.2a) of the KKT system (2.4.2) for (4.4.1). We denote the Lagrange multipliers by $(\mu, \lambda) = (\mu_v, \mu_a, \lambda_v^+, \lambda_v^-, \lambda_a^+, \lambda_a^-, \tilde{\lambda}_a^+, \tilde{\lambda}_a^-)_{v \in V, a \in A'}$, such that $\mu_v, \mu_a \in \mathbb{R}$ and $\lambda_v^+, \lambda_v^-, \lambda_a^+, \lambda_a^-, \tilde{\lambda}_a^+, \tilde{\lambda}_a^- \in \mathbb{R}_{\geq 0}$. We set $\Phi_a(q_a) := \alpha_a \, q_a |q_a|^{k_a} - \tilde{\beta}_a$. Then the Lagrange function of problem (4.4.1) has the form

$$
\begin{aligned}
L(q, \pi, \Delta, \mu, \lambda) = & \sum_{a \in A'} \left(\Delta_a^+ + \Delta_a^- \right) \\
& + \sum_{a=(v,w) \in A'} \mu_a \left(\Phi_a(q_a) - (\pi_v - \gamma_a \pi_w) - (\Delta_a^+ - \Delta_a^-) \right) \\
& + \sum_{v \in V} \mu_v \left(d_v - \sum_{a \in \delta_{A'}^+(v)} q_a + \sum_{a \in \delta_{A'}^-(v)} q_a \right) \\
& + \sum_{v \in V} \left(\lambda_v^+ (\pi_v - \overline{\pi}_v) + \lambda_v^- (\underline{\pi}_v - \pi_v) \right) \\
& + \sum_{a \in A'} \left(\lambda_a^+ (q_a - \overline{q}_a) + \lambda_a^- (\underline{q}_a - q_a) \right) \\
& - \sum_{a \in A'} \tilde{\lambda}_a^+ \Delta_a^+ + \tilde{\lambda}_a^- \Delta_a^- .
\end{aligned}
$$

Now (2.4.2a) which are fulfilled by the KKT point $(q^*, \pi^*, \Delta^*, \mu^*, \lambda^*)$ write as

$$\frac{\partial L}{\partial q_a} = 0 \Rightarrow \quad \mu_a \frac{d\Phi_a}{dq_a}(q_a^*) + \lambda_a^+ - \lambda_a^- = \mu_v - \mu_w \quad \forall\, a \in A',\ a = (v, w), \tag{4.4.2a}$$

$$\frac{\partial L}{\partial \pi_v} = 0 \Rightarrow \quad \sum_{a \in \delta_{A'}^+(v)} \mu_a - \sum_{a \in \delta_{A'}^-(v)} \gamma_a \mu_a = \lambda_v^+ - \lambda_v^- \quad \forall\, v \in V, \tag{4.4.2b}$$

$$\frac{\partial L}{\partial \Delta_a^+} = 0 \Rightarrow \quad \mu_a + \tilde{\lambda}_a^+ = 1 \quad \forall\, a \in A', \tag{4.4.2c}$$

$$\frac{\partial L}{\partial \Delta_a^-} = 0 \Rightarrow \quad \mu_a - \tilde{\lambda}_a^- = -1 \quad \forall\, a \in A'. \tag{4.4.2d}$$

In combination with the complementarity conditions (2.4.2e) the subsequent conditions follow for $(q^*, \pi^*, \Delta^*, \mu^*, \lambda^*)$:

$$q_a < \overline{q}_a \Rightarrow \lambda_a^+ = 0, \quad q_a > \underline{q}_a \Rightarrow \lambda_a^- = 0 \quad \forall\, a \in A', \tag{4.4.3a}$$

$$\pi_v < \overline{\pi}_v \Rightarrow \lambda_v^+ = 0, \quad \pi_v = \overline{\pi}_v \Rightarrow \lambda_v^+ \geq 0 \quad \forall\, v \in V, \tag{4.4.3b}$$

$$\pi_v > \underline{\pi}_v \Rightarrow \lambda_v^- = 0, \quad \pi_v = \underline{\pi}_v \Rightarrow \lambda_v^- \leq 0 \quad \forall\, v \in V, \tag{4.4.3c}$$

$$\Delta_a^+ > 0 \Rightarrow \mu_a = 1, \quad \Delta_a^- > 0 \Rightarrow \mu_a = -1 \quad \forall\, a \in A'. \tag{4.4.3d}$$

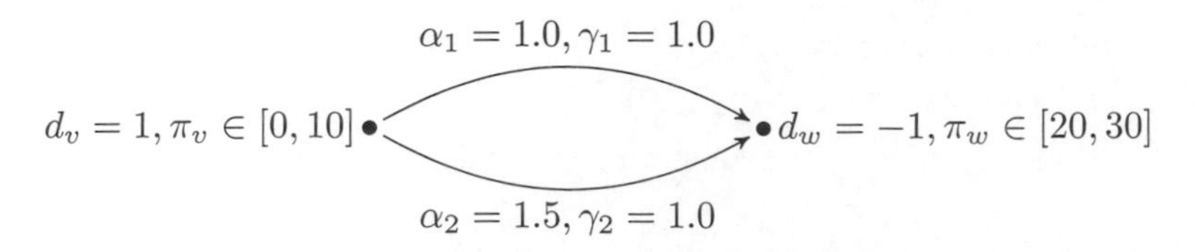

Figure 4.3: Example of an instance (network and nomination) demonstrating the nonconvexity of the relaxation (4.4.1) of the passive transmission problem (4.1.1). Two different KKT points of (4.4.1) having different objective values are presented in Example 4.4.2.

We compare (4.4.3) with (4.3.14) and (4.2.8). Recall that in the domain relaxation, (4.2.7), we had an enforcement of the variables $\lambda_v^{\pm}, \lambda_a^{\pm}$. In the flow conservation relaxation, (4.3.13), we derived an enforcement of the variables μ_v. Now in the potential-flow-coupling relaxation, (4.4.2), only an enforcement of the variables μ_a to nonzero values remains.

4.4.2 Different KKT Points

The following example shows that the feasible domain of (4.4.1) is non-convex in general, as there exist two different KKT points with different optimal objective values for a test instance being a planar graph. The convex combination of the primal parts is not feasible. Hence the feasible solution space is non-convex and a nonlinear solver, like IPOPT, which computes KKT points, cannot guarantee to compute the global optimal solution of (4.4.1).

Example 4.4.2:
Consider the network shown in Figure 4.3. It consists of two nodes v, w and two parallel arcs a_1, a_2 from v to w. We set $\Phi_a(q_a) := \alpha_a\, q_a|q_a|$ and assume arc constants $\alpha_1 = 1.0, \beta_1 = 0, \gamma_1 = 1.0$ and $\alpha_2 = 1.5, \beta_2 = 0, \gamma_2 = 1.0$, and node potential bounds $\underline{\pi}_v = 0, \overline{\pi}_v = 10, \underline{\pi}_w = 20, \overline{\pi}_w = 30$. Node v is an entry with flow $+1$, node w is an exit with flow -1. The following two solutions both fulfill the KKT system (4.4.2):

- *Let $q_1 = 3, q_2 = -2, \pi_v = 10, \pi_w = 20, \Delta_1^+ = 19, \Delta_2^+ = 4, \Delta_1^- = 0, \Delta_2^- = 0$. The objective function value is 23. The dual values are $\mu_1 = 1$, $\mu_2 = 1$, $\mu_v = 6$, $\mu_w = 0$.*

- *Let $q_1 = 0.6, q_2 = 0.4, \pi_v = 10, \pi_w = 20, \Delta_1^+ = 10.36, \Delta_2^+ = 10.24$, $\Delta_1^- = 0$, $\Delta_2^- = 0$. The objective function value is 20.6. The dual values are $\mu_1 = 1$, $\mu_2 = 1$, $\mu_v = 1.2$, $\mu_w = 0$.*

Thus we found two different KKT points. It is easy to see that a convex combination of both primal feasible solutions is not feasible for the relaxation (4.4.1) because of constraint (4.4.1b). *Hence the feasible solution space of* (4.4.1) *is not convex. As both solutions have different objective values this shows that even a nonlinear solver, like* IPOPT, *which computes KKT points, cannot guarantee to compute the global optimal solution of* (4.4.1).

4.5 Solving the Passive Transmission Problem

The convex relaxations (4.2.1) and (4.3.1) of the passive transmission problem (4.1.1) described in the previous Sections 4.2 and 4.3 neglect constraints (4.1.1c) and (4.1.1e). Hence they can be used to compute a solution for the passive transmission problem efficiently only if $A_a = 0, b_a = 0$ for each arc $a \in A'$. Furthermore it is required that there exists no arc $a = (v, w) \in A'$ that is modeled by the constraint $\pi_v = \pi_w$, which implies $\alpha_a > 0$ for all arcs $a \in A'$, see Section 3.2.

In this section we turn to the general case and describe how the relaxations can be used to compute a solution for the passive transmission problem without these restrictions. First we consider the case that the passive transmission problem contains an arc (v, w) which is modeled by $\pi_v = \pi_w$. Our strategy for solving this problem splits up into three steps as follows:

Step 1: First we apply a preprocessing technique: For each arc $a = (v, w) \in A'$ where constraint (4.1.1b) writes as $\pi_v = \pi_w$, we contract arc a and identify the end nodes v and w. This identification goes along with setting node potential bounds $[\underline{\pi}_v, \overline{\pi}_v] \cap [\underline{\pi}_w, \overline{\pi}_w]$ for the node which represents the contraction of v and w. If this intersection is empty, then the passive transmission problem (4.1.1) is infeasible because constraint (4.1.1b) is in contradiction with the node potential bounds at v and w. If we do not detect infeasibility here, then we continue with the next step.

Step 2: From the contraction we obtain the node set V' and the arc set A''. Now we solve the corresponding preprocessed passive transmission problem by making use of either the domain relaxation (4.2.1) presented in Section 4.2 or the flow conservation relaxation (4.3.1) described in Section 4.3. A solution of one of these relaxations with optimal objective zero yields a feasible solution for the preprocessed passive transmission problem. Both problems are convex by Theorem 4.2.9 and Theorem 4.3.6 and so can be solved efficiently. The relaxation also means that an optimal solution with positive objective value implies that the preprocessed passive transmission problem is infeasible.

If the preprocessed problem turns out to be infeasible, then again, the passive transmission problem (4.1.1) is infeasible. This is proven as follows: For every solution (q', π') of the passive transmission problem it holds that $(q'_a)_{a \in A''}$ is a feasible flow for the preprocessed passive transmission problem. Hence the preprocessed passive transmission problem cannot be infeasible if the non-preprocessed passive transmission problem is feasible. If we do not detect infeasibility here, then, similar to step 1, we continue with the next step.

Step 3: For an optimal solution (q^*, π^*) of the preprocessed passive transmission problem, we solve the non-preprocessed passive transmission problem and fix the flow on all arcs $a \in A''$ to q_a^* before. We define $\Phi_a(q_a) := \alpha_a \, q_a |q_a|^{k_a} - \tilde{\beta}_a$. The arising problem is a linear program and writes as follows:

$$
\begin{aligned}
& \exists \; q, \pi && && (4.5.1) \\
\text{s. t.} \quad & \pi_v - \gamma_a \pi_w = \Phi_a(q_a^*) && \forall \, a = (v,w) \in A'', \\
& \pi_v - \pi_w = 0 && \forall \, a = (v,w) \in A' \setminus A'', \\
& q_a = q_a^* && \forall \, a = (v,w) \in A'', \\
& \sum_{a \in \delta^+_{A'}(v)} q_a - \sum_{a \in \delta^-_{A'}(v)} q_a = d_v && \forall \, v \in V, \\
& \underline{\pi}_v \le \pi_v \le \overline{\pi}_v && \forall \, v \in V, \\
& \underline{q}_a \le q_a \le \overline{q}_a && \forall \, a \in A', \\
& \pi_v \in \mathbb{R} && \forall \, v \in V, \\
& q_a \in \mathbb{R} && \forall \, a \in A'.
\end{aligned}
$$

For every solution flow q' of the passive transmission problem it holds that $q'_a = q_a^*$ for every arc $a \in A''$. This can be seen as follows: A solution q' of the passive transmission problem yields a flow vector $(q'_a)_{a \in A''}$ which is feasible for the preprocessed passive transmission problem and hence feasible for its domain relaxation. By Lemma 4.2.7 this flow vector is unique. As q^* is also a feasible flow vector for the domain relaxation we obtain $q'_a = q_a^*$ for every arc $a \in A''$. We conclude that LP (4.5.1) is feasible if and only if the passive transmission problem is feasible. Hence (4.5.1) is either feasible and the optimal solution $(\tilde{q}, \tilde{\pi})$ yields a feasible solution $(\tilde{q}, \tilde{\pi}, \tilde{p})$ with $\tilde{p} := \operatorname{sgn}(\pi)\sqrt{|\pi|}$ for the passive transmission problem, or otherwise it is infeasible, which implies that the passive transmission problem is infeasible, too.

The discussion above describes how to solve the passive transmission problem, if it contains arcs $a = (v,w) \in A'$ that are modeled by $\pi_v = \pi_w$. Now we turn to the more general case $A_a, b_a \neq 0$ for an arc $a \in A'$. In order to solve the passive transmission problem in this case, we proceed as described in the following two steps:

Step 1: We ignore the constraints $A_a\left(q_a, p_v, p_w\right)^T \leq b_a$ for each arc $a \in A'$ and solve the arising passive transmission problem as described above. If we detect infeasibility, then the passive transmission problem (4.1.1) is infeasible because neglecting constraints (4.1.1c) means to consider a relaxation of (4.1.1). Otherwise, if we obtain a feasible solution (q^*, π^*, p^*), then it is not guaranteed that (q^*, π^*, p^*) is a feasible solution for the passive transmission problem (4.1.1). In this case we proceed with the next step.

Step 2: From the previous discussion in Step 3 we know that the arc flow q_a has a unique solution for every arc $a = (v, w) \in A'$ which is not modeled by $\pi_v = \pi_w$. Consequently, for a feasible solution (q^*, π^*, p^*) of the passive transmission problem without constraints (4.1.1c) and a feasible solution (q', π', p') of the passive transmission problem (including constraints (4.1.1c)) it holds $q'_a = q^*_a$ for every arc $a = (v, w) \in A'$ not modeled by $\pi_v = \pi_w$.

Let us now analyze the restrictions induced by constraints (4.1.1c) on the solution space of the passive transmission problem without constraints (4.1.1c). Recall that (4.1.1c) writes as $0 \leq 0$ for all arcs $a = (v, w) \in A'$ which are modeled by $\pi_v = \pi_w$ as $A_a = 0$ and $b_a = 0$, see Section 3.2. For all other arcs $a \in A'$ the previous argumentation implies that the feasible flow values are given by q^*_a. Let us now consider an arc $a = (v, w) \in A'$ and analyze the constraints

$$\begin{aligned} \Phi(q^*_a) = \pi_v - \gamma_a \pi_w, \quad A_a\left(q^*_a, p_v, p_w\right)^T \leq 0, \\ p_v|p_v| = \pi_v, \quad p_w|p_w| = \pi_w. \end{aligned} \tag{4.5.2}$$

Below we are going to reformulate this as

$$\Phi(q^*_a) = \pi_v - \gamma_a \pi_w, \quad (\pi_v, \pi_w) \in \bigcup_i I^{(i)}_{a,v} \times I^{(i)}_{a,w}. \tag{4.5.3}$$

where $I^{(i)}_{a,v}, i = 1, \ldots$ and $I^{(i)}_{a,w}, i = 1, \ldots$ are disjoint intervals, respectively. Then, after fixing the flows of all arcs $a \in A''$ to q^*_a and neglecting the pressure variables $p_v, v \in V$, we solve the equivalent formulation of the passive transmission problem (4.1.1) which writes as

$$\exists\ q, \pi \tag{4.5.4}$$

$$\begin{aligned} \text{s.t.} \quad & \pi_v - \gamma_a \pi_w = \Phi_a(q^*_a) && \forall\, a = (v, w) \in A'', \\ & \pi_v - \pi_w = 0 && \forall\, a = (v, w) \in A' \setminus A'', \end{aligned}$$

$$\begin{aligned}
q_a &= q_a^* && \forall\, a = (v,w) \in A'', \\
\sum_{a \in \delta^+_{A'}(v)} q_a - \sum_{a \in \delta^-_{A'}(v)} q_a &= d_v && \forall\, v \in V, \\
\bigvee_i (\pi_v, \pi_w) &\in I^{(i)}_{a,v} \times I^{(i)}_{a,w} && \forall\, a = (v,w) \in A', \\
\underline{\pi}_v \le \pi_v &\le \overline{\pi}_v && \forall\, v \in V, \\
\underline{q}_a \le q_a &\le \overline{q}_a && \forall\, a \in A', \\
\pi_v &\in \mathbb{R} && \forall\, v \in V, \\
q_a &\in \mathbb{R} && \forall\, a \in A'.
\end{aligned}$$

This problem is a disjunctive programming problem and its optimal solution $(\tilde{q}, \tilde{\pi})$ yields a feasible solution $(\tilde{q}, \tilde{\pi}, \tilde{p})$ with $\tilde{p} := \operatorname{sgn}(\pi)\sqrt{|\pi|}$ for the passive transmission problem (4.1.1) if and only if the passive transmission problem is feasible.

To complete this step, we show how (4.5.2) can be reformulated as (4.5.3). First we rewrite (4.5.2) and reformulate the equalities to obtain equivalent formulas

$$\begin{aligned}
A_a\left(q_a^*, p_v, p_w\right)^T &\le 0, \\
p_w &= \operatorname{sgn}(p_v|p_v| - \Phi_a(q_a^*))\sqrt{\left|p_v|p_v| - \Phi_a(q_a^*)\right|},
\end{aligned} \tag{4.5.5}$$

describing the feasible pressures p_v, p_w of (4.5.2). We observe that the constraints $A_a\left(q_a^*, p_v, p_w\right)^T \le 0$ form a polyhedron $\mathcal{P}_a^*$. According to (4.5.5) the feasible solution space of p_v and p_w is then given as an intersection of this polyhedron $\mathcal{P}_a^*$ with a curve defined by

$$g : p_v \mapsto \operatorname{sgn}(p_v|p_v| - \Phi_a(q_a^*))\sqrt{\left|p_v|p_v| - \Phi_a(q_a^*)\right|}. \tag{4.5.6}$$

This intersection, as illustrated in Figure 4.4, leads to intervals $\tilde{I}^{(i)}_{a,v}$ and $\tilde{I}^{(i)}_{a,w}$ such that

$$(p_v, p_w) \in \mathcal{P}_a^* : g(p_v) = p_w \quad \Leftrightarrow \quad (p_v, p_w) \in \bigcup_i \tilde{I}^{(i)}_{a,v} \times \tilde{I}^{(i)}_{a,w} : g(p_v) = p_w$$

The computation of these intervals can be performed as follows: First we compute all vertices $u_1, \ldots, u_n$ of the polyhedron $\mathcal{P}_a^*$. We assume an order such that u_i and u_{i+1} are neighbored vertices for all $i = 1, \ldots, n-1$. Considering two neighbored vertices we check whether their connecting line segment intersects the function (4.5.6). If this is the case, then we store the intersection point. As a result we get a set of

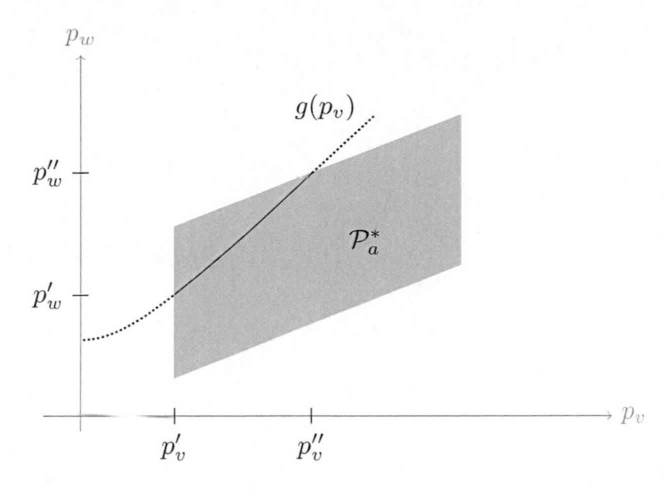

Figure 4.4: The figure is used to analyze the impact of the constraints (4.1.1c) on a feasible solution of the active transmission problem without (4.1.1c). For this solution especially the flow value is available. Assume that we fix this value and concentrate on the constraints (4.1.1b), (4.1.1c) and (4.1.1e). For a single arc they write as (4.5.2). The figure shows the intersection of the polyhedron $\mathcal{P}^*_a$ describing the feasible operation range for $q_a = q^*_a$ with the curve $g(p_v)$ determined by the equation $\Phi(q^*_a) = \pi_v - \pi_w = p_v|p_v| - p_w|p_w|$. In this case the intersection intervals yield $A_a(q^*_a, p_v, p_w)^T \leq 0 : g(p_v) = p_w \Leftrightarrow (p_v, p_w) \in [p'_v, p''_v] \times [p'_w, p''_w] : g(p_v) = p_w$.

intersection points that we sort in non-decreasing order. Two points consecutively limit feasible and infeasible regions alternately. From a pair of these nodes limiting a feasible region we derive the intervals $\tilde{I}^{(i)}_{a,v} \times \tilde{I}^{(i)}_{a,w}$. From these intervals which limit the pressure variables p we derive intervals $I^{(i)}_{a,v} \times I^{(i)}_{a,w}$ limiting the node potentials π by taking the relation $p|p| = \pi$ into account.

4.6 Integration and Computational Results

In this chapter we focused on the topology optimization problem (3.2.1) arising from the first type of network that we consider in this thesis. Recall that these networks consist of pipes and valves only. In this case it holds $\underline{y} = \overline{y}$ for our model (3.2.1). The outline of the solution framework that we apply is shown in Figure 4.5. We solve (3.2.1) by SCIP as described in Section 2.2. For each node of the branching tree we check whether the solution of the LP relaxation yields integral values for the integral variables. If this is the case, then we consider the corresponding passive transmission problem (4.1.1). We solve this nonlinear optimization problem to global optimality and further classify the current MILP feasible node. If the passive transmission problem is feasible, then we obtain a global optimal solution for the current node of the branching tree and SCIP itself prunes the node. Otherwise, if the passive transmission problem is infeasible, and all integral variables are fixed by branching,

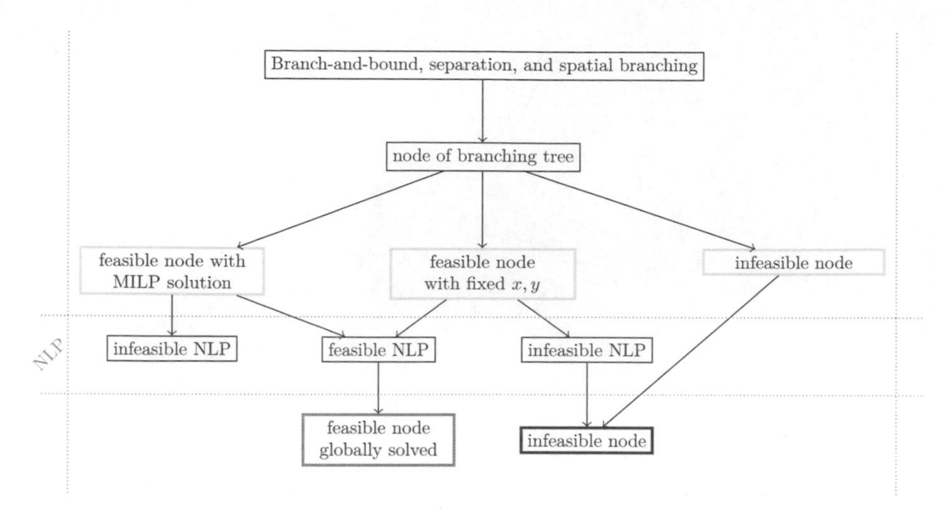

Figure 4.5: Solution framework presented in Section 4.6. The topology optimization problem (3.2.1) is solved with SCIP essentially by branch-and-bound, separation, and spatial branching, see Section 2.2. We adapt this framework and solve globally the passive transmission problem (4.1.1) as discussed in Section 4.5, classify the current node of the branching tree and prune it if possible.

then we prune the current node of the branch-and-bound tree manually. If it is not possible to solve the passive transmission problem globally due to numerical troubles then we continue with branching.

We implemented the algorithms described in the previous section in C, i.e., domain relaxation (4.2.1), flow conservation relaxation (4.3.1), and the disjunctive problem (4.5.4). The computational setup is described in Section 3.5. We compare four strategies for solving the passive transmission problem (4.1.1).

1. The first strategy is to solve the topology optimization problem by SCIP without any adaptations on the solver settings. All branching decisions are up to the solver, and the topology optimization problem (3.2.1) is basically solved by branch-and-bound, separation and spatial branching.

2. The second strategy is to solve the topology optimization problem by SCIP and enforce a certain branching priority rule, so that SCIP first branches on binary variables x. Only after all these variables are fixed, it is allowed to perform spatial branching on continuous variables.

3. The third strategy implements the domain relaxation from Section 4.2 for solving the passive transmission problem (4.1.1). We consider the passive transmission problem and apply presolve and then solve the LP relaxation. If these solution methods detect infeasibility, then the passive transmission problem is infeasible. Otherwise we proceed as described in Section 4.5 and use

the nonlinear solver IPOPT for solving the convex domain relaxation (4.2.1). Additionally we set branching priorities according to the second strategy.

4. The fourth strategy uses the relaxation of the flow conservation constraints from Section 4.3 for solving the passive transmission problem (4.1.1). Again, we first presolve the passive transmission problem and then solve the LP relaxation. If infeasibility is detected, then the passive transmission problem is infeasible. Otherwise we proceed as described in Section 4.5 and use the nonlinear solver IPOPT for solving the convex flow conservation relaxation (4.3.1). Additionally we set branching priorities according to the second strategy.

We did not implement the flow-coupling-constraint relaxation (4.4.1) described in Section 4.4, because the relaxation is non-convex, and thus a local solver cannot guarantee to find a global optimal solution. This is necessary for pruning nodes of the branch-and-bound tree. Further we note that presolving and solving the LP relaxation is done efficiently by SCIP and hence included in strategy 3 and 4.

Computational Results

We consider those test instances described in Section 3.5 which belong to the first type of network. These networks are `net4` and `net5` while we contract all arcs which represent compressors and control valves. This means that we identify the end nodes of these arcs. This way the networks consist of pipes and valves only. Note that the instances become harder at first sight because the deleted active elements do not enforce a fixed relation between the flow and the node potentials at their end nodes compared to pipes. But on the other hand a possible flow bound enforcing a positive arc flow through a compressor is removed by this contraction. In this sence the instances are relaxed and easier to solve.

We have 52 nominations in total. For every pipe a of these networks it holds $\alpha_a > 0$ and $A_a = 0, b_a = 0$. Hence solving the passive transmission problem in the third and fourth strategy means solving the convex domain relaxation (4.2.1) and flow conservation relaxation (4.3.1), respectively. It is not necessary to solve (4.5.4) in advance as described in Section 4.5. Furthermore these networks do not contain any compressors and control valves (as they are contracted). So fixing all binary variables of the topology optimization problem (3.2.1) yields the passive transmission problem (4.1.1) as assumed in this chapter.

For the computations we imposed a time limit of 39 600 s and used the computational setup described in Section 3.5. The results are available in Table A.4 and Table A.5. A summary is shown in Table 4.1 and Table 4.2. Here we use the

strategy	1	2	3	4	all
solved instances	24	30	45	46	46

Table 4.1: Summary of the Tables A.4 and A.5 showing the globally solved instances out of 52 nominations in total. The third strategy globally solves all instances which are solved to global optimality by the second and the first strategy.

	(A,B) = (2,3)			(A,B) = (4,3)		
	solved(30)		incomp.(1)	solved(45)		incomp.(4)
	time [s]	nodes	gap [%]	time [s]	nodes	gap [%]
strategy A	25.9	1,038	15	66.2	827	141
strategy B	7.2	147	15	33.1	746	122
shifted geom. mean	−72 %	−86 %	0 %	−50 %	−10 %	−13 %

Table 4.2: Run time, number of branch-and-bound nodes and gap comparison for the strategies 2 and 3 and additionally 4 and 3 (aggregated results). The columns "solved" contain mean values for those instances globally solved by both strategies A and B. The columns "incomplete" show mean values for those instances having a primal feasible solution available but were not globally solved by both strategies A and B. The underlying data are available in Tables A.4 and A.5.

geometric mean of run time, number of branch-and-bound nodes and gap as described in Section 3.5.

Table 4.1 shows a clear order of the four strategies: Strategy 1 solves less instances than strategy 2. We conclude that branching priorities as set by the second strategy are a first step to improve the solving performance of SCIP. Approximately 57 % of the instances (30 out of 52) are solved to global optimality by strategy 2. These instances are also globally solved by strategy 3. Additionally around 29 % more instances of the test set (15 out of 52) are solved to global optimality compared to the second strategy. The fourth strategy solves one more instance than the third strategy.

Table 4.2 shows that the third strategy is two times faster than the fourth one while the the number of nodes is only reduced by 10 %. Figure 4.6 yields an explanation. One can see that the run time for the domain relaxation is one order of magnitude lower than for the flow conservation relaxation. We depict only those instances that were solved by IPOPT, but not those that were detected as infeasible during the presolve. In average the gap is reduced by 13 % following strategy 3 in comparison to strategy 4 on those instances which remain with a finite positive gap value following both strategies. As both strategies solve globally nearly the same number of instances we conclude that the domain relaxation (4.2.1) described in Section 4.2 for solving the passive transmission problem (4.1.1) yields the most

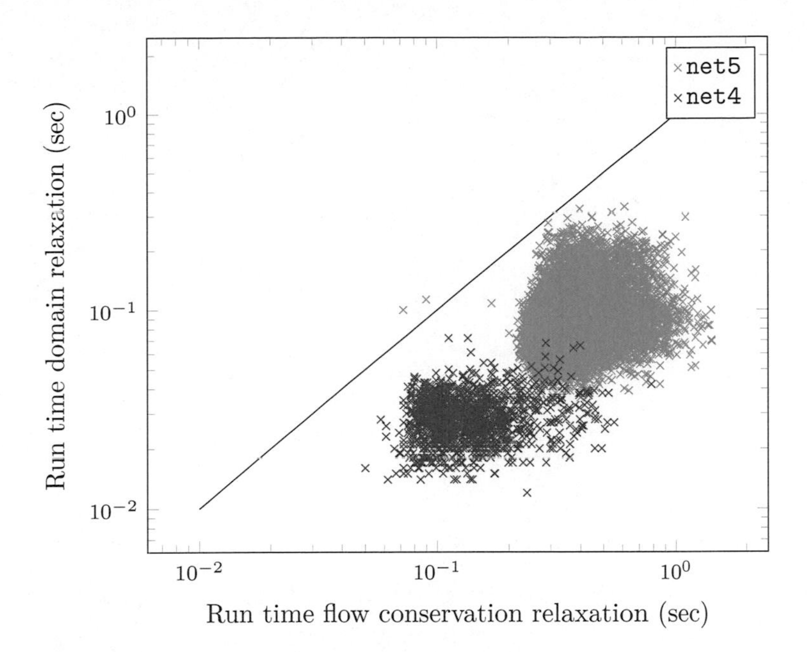

Figure 4.6: Run time comparison for the domain relaxation (4.2.1) and the flow conservation relaxation (4.3.1) on instances of the networks `net4` and `net5`. The two crosses above the straight line belong to passive transmission problems where the domain relaxation had numerical troubles.

efficient results comparing strategy 3 and 4. Hence we decided to use strategy 3 in our practical application.

In order to demonstrate the benefit of strategy 3 we finally compare it with the second strategy. Recall from the previous analysis that the third strategy solves approximately 29 % more instances to global optimality. Table 4.2 shows that strategy 3 saves 72 % of run time and 86 % of nodes in comparison to strategy 2. The savings in terms of number of nodes are higher than the run time reduction due to the following reason: the relaxations are set up and the nonlinear solver IPOPT has to be called. In summary we conclude that more instances are globally solved and less run time is needed.

Figure 4.7 summarizes the run time results of the four strategies in a performance plot. The four graphs show the share of instances (in per cent) that could be solved within a time limit of 39 600 s. Evaluating Figure 4.7 shows a result coherent to our previous observations: the graphs for SCIP (strategy 1 and 2) are below the graph for the flow conservation relaxation, which is below the graph for the domain relaxation. Figure 4.8 shows a consistent result. It shows a scatter plot comparing the run times of the second and the third strategy (the best strategy following spatial branching vs. the best strategy using the convex relaxations of this chapter). The run time of many instances is reduced when using the third strategy.

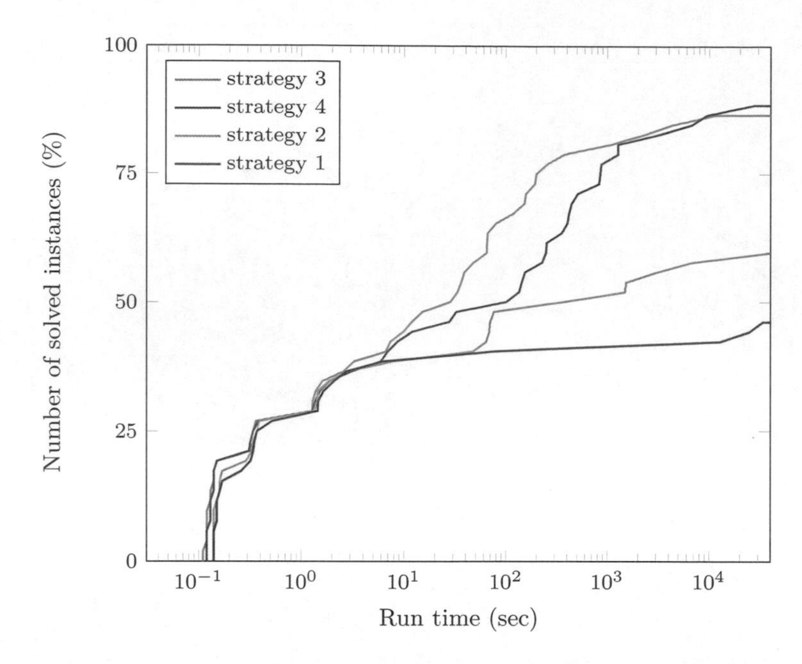

Figure 4.7: Performance plot for different nominations on the networks `net4` and `net5` (aggregated) and a time limit of 39 600 s. The different strategies are described in Section 4.6. Strategy 1 and 2 mainly consist of SCIP. Strategies 3 and 4 also correspond to SCIP together with our elaborated solution methods presented in this chapter. The underlying data are available in Tables A.4 and A.5.

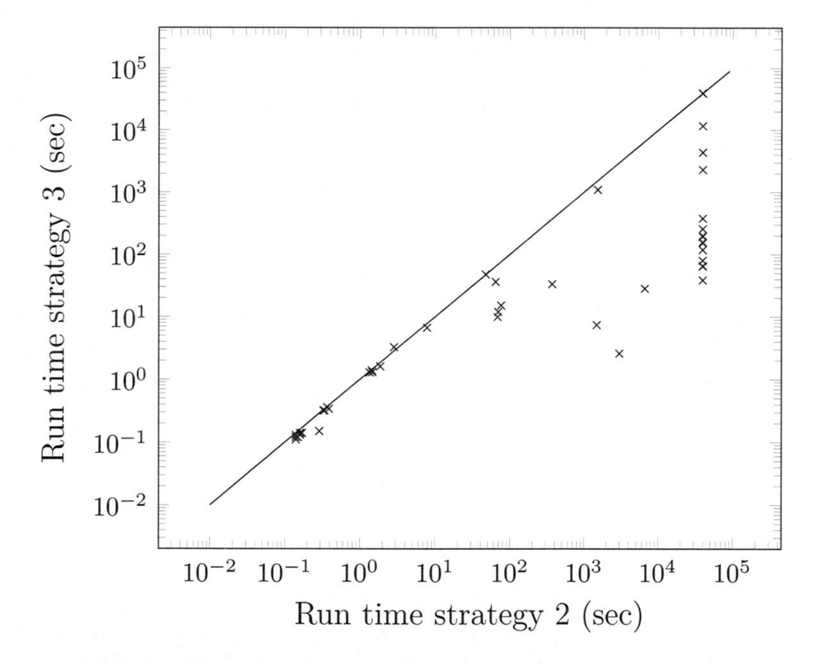

Figure 4.8: Run time comparison for different nominations on the networks `net4` and `net5`. Each cross ($\times$) corresponds to a single instance of the test set. Note that multiple crosses are drawn in the upper right corner of the plots that cannot be differed. They represent those instances that ran into the time limit of 39 600 s for both strategies. The underlying data are available in Tables A.4 and A.5.

We refer to Humpola et al. (2015b) where we carried out a computational study for different nominations on the real-world network `net6` by applying the results of this chapter. This network contains compressors and control valves. Therefor we discretized the feasible solution space of these active elements so that $y_{a,i} \in \mathbb{Z}$ holds for every arc $(a, i) \in A_X$. Recall that y are continuous variables in our model. So fixing binary and discrete variables x and y in the topology optimization problem (3.2.1) yields the passive transmission problem. Additionally a method is presented which computes the coefficients α_a for each compressor and control valve $a \in A$ according to this discretization. The computational results show large running times that increase further with the number of discretizations. A large number of discretizations is in turn necessary to reduce the approximation error for compressors and control valves. We conclude that the algorithms of this chapter are mainly useful for networks containing only pipelines and valves.

Summary

We presented different solution methods for the passive transmission problem. They allow to speed up the solution process of the topology optimization problem in the case $\underline{y} = \overline{y}$. This case includes the first type of network of our test instances, namely those which contain only pipes and valves. From the computational results we conclude that a strategy which is based on solving the domain relaxation yields the most convincing results. This strategy shows faster computation times in comparison to SCIP. On average the run time is reduced by 72 % in comparison to SCIP with branching priorities set. Approximately 29 % more instances of the test set are also solved to global optimality compared to the solver SCIP. Therefore in our practical applications we use the domain relaxation method for networks that contain only pipes and valves.

Let us now try to understand the reason for the performance of SCIP. Therefor we considered a specific part of the solution process for a test instance. We analyzed the computational effort for handling the nonlinear equations of our model that are associated with pipes. As a result we collected all cutting planes that are generated during the branch-and-bound process for each pipe individually. The result for one single pipe is visualized in Figure 4.9.

Many inequalities are required for the approximation of the convex part of the function $q_a \mapsto \alpha_a\, q_a |q_a|^{k_a}$. We conclude that the use of nonlinear solvers on specific nodes of branch-and-bound reduces the computational effort in handling nonlinear equations compared to spatial branching.

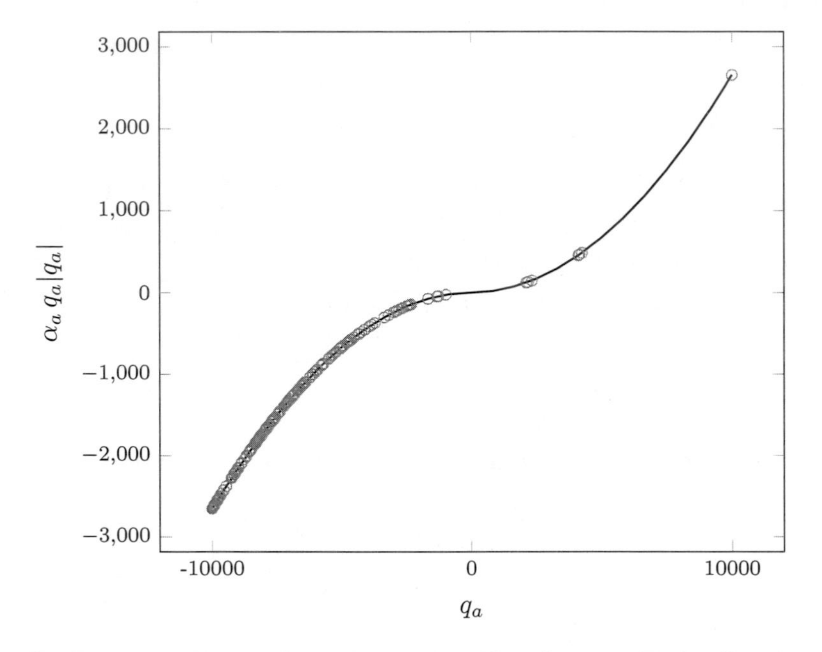

Figure 4.9: Handling the nonlinearity during the branch-and-bound process. For handling the nonlinear function $q_a \mapsto \alpha_a\, q_a|q_a|$ of a single pipe a with $\alpha_a = 1/37\,700$, cutting planes were generated for all marked points.

Chapter 5

An Improved Benders Cut for the Topology Optimization Problem

In Chapter 4 we presented two convex relaxations (4.2.1) and (4.3.1) of the passive transmission problem (4.1.1). Both allowed us to solve the passive transmission problem efficiently and thereby reduce the use of time consuming spatial branching when solving the topology optimization problem (3.2.1). We focused on the case $\underline{y} = \overline{y}$ which includes networks containing only pipes and valves. These networks are the first type of network which we consider in this thesis. We now focus on the second type of gas network which additionally contains loops or rather parallel pipes. Again we restrict to the case $\underline{y} = \overline{y}$.

In this chapter we show that a KKT point of the previously mentioned relaxations can be used to generate a new linear inequality for (3.2.1). By fixing all integral variables in (3.2.1) we obtain the passive transmission problem. Whenever one of the relaxations (4.2.1) and (4.3.1) of the passive transmission problem is solved during the branch-and-bound process and when we conclude that the passive transmission problem is infeasible, then we dynamically add such an inequality to the problem formulation. This act allows to solve approximately 13 % more instances of the second type of network to global optimality within our given time limit. For those instances which are already solvable by SCIP the run time is reduced by approximately 33 %.

The linear inequality that we are going to present contains only binary variables. When fixing the discrete variables to those values leading to the passive transmission problem from which the inequality is derived, then it reflects the infeasibility of the corresponding passive transmission problem, i.e., it is violated if and only if the passive transmission problem is infeasible. As the inequality does not contain any continuous variable we are reminded of generalized Benders decomposition (see Geoffrion 1972). Here the master problem would consist of the integral variables and the constraints containing only integral variables and we would write it as

$\min\{cx \mid x \in \mathcal{X}, x \in \{0,1\}^{A_X}\}$. Each subproblem would be a feasibility problem equal to the passive transmission problem (4.1.1). Nevertheless, we do not follow this decomposition approach, but restart the overall branch-and-bound process after a predefined number of inequalities is generated.

The inequality presented in this chapter is obtained by a Benders argument from the Lagrange function of the domain relaxation (4.2.1) augmented by a specially tailored *pc-regularization*. This regularization is necessary to derive a globally valid cut. A certain choice of the Lagrange multipliers finally yields the described properties. Hence we consider our inequality as an improved Benders cut.

The outline of this chapter is as follows: In Section 5.1 we present a linear inequality for the passive transmission problem (4.1.1) which represents its feasibility. In Section 5.2 we derive an extended formulation of this inequality which is then valid for the topology optimization problem (3.2.1). Computational results are given in Section 5.3. They show the benefit of our inequalities when added to the topology optimization problem (3.2.1).

5.1 Valid Inequalities for the Passive Transmission Problem

Let us consider the passive transmission problem which is obtained from the topology optimization problem (3.2.1) by fixing all integral and continuous variables x and y. Recall that the variables y are fixed for the networks considered in this chapter. In Section 5.1.1 we first formulate a valid nonlinear inequality for this problem. This inequality bases on the definition of a *dual transmission flow* (see Definition 5.1.1). In Section 5.1.2 we explain how to obtain a linear inequality, which is constant on both sides and valid for the passive transmission problem. The choice of the parameters of the linear inequality is discussed in Section 5.1.3. In Section 5.1.4 we explain the relation between the linear inequality that we derive in this section and the Lagrange function of the domain relaxation (4.2.1).

Throughout this section let A' contain all arcs such that the flow is not fixed to zero, i.e., $A' := \{(a,i) \in A_X \mid x_{a,i} = 1, i > 0\}$. We review the domain relaxation (4.2.1) to improve readability:

$$\min \sum_{v \in V} \Delta_v + \sum_{a \in A'} \Delta_a \tag{5.1.1a}$$

$$\text{s.t.} \quad \alpha_a\, q_a |q_a|^{k_a} - \tilde{\beta}_a - (\pi_v - \gamma_a \pi_w) = 0 \qquad \forall\, a = (v,w) \in A', \tag{5.1.1b}$$

$$\sum_{a\in\delta^+_{A'}(v)} q_a - \sum_{a\in\delta^-_{A'}(v)} q_a = d_v \qquad \forall\, v \in V, \tag{5.1.1c}$$

$$\pi_v - \Delta_v \leq \overline{\pi}_v \qquad \forall\, v \in V, \tag{5.1.1d}$$

$$\pi_v + \Delta_v \geq \underline{\pi}_v \qquad \forall\, v \in V, \tag{5.1.1e}$$

$$q_a - \Delta_a \leq \overline{q}_a \qquad \forall\, a \in A', \tag{5.1.1f}$$

$$q_a + \Delta_a \geq \underline{q}_a \qquad \forall\, a \in A', \tag{5.1.1g}$$

$$\pi_v \in \mathbb{R} \qquad \forall\, v \in V, \tag{5.1.1h}$$

$$q_a \in \mathbb{R} \qquad \forall\, a \in A', \tag{5.1.1i}$$

$$\Delta_v \in \mathbb{R}_{\geq 0} \quad \forall\, v \in V, \tag{5.1.1j}$$

$$\Delta_a \in \mathbb{R}_{\geq 0} \quad \forall\, a \in A'. \tag{5.1.1k}$$

Let us turn to the definition of a dual transmission flow. Therefor we write $\Phi_a(q_a) := \alpha_a\, q_a|q_a|^{k_a} - \tilde{\beta}_a$ and recall the Lagrange function of this problem as

$$\begin{aligned} L(q,\pi,\Delta,\mu,\lambda) = {} & \sum_{v\in V} \Delta_v + \sum_{a\in A'} \Delta_a \\ & + \sum_{a=(v,w)\in A'} \mu_a \left(\Phi_a(q_a) - (\pi_v - \gamma_a \pi_w)\right) \\ & + \sum_{v\in V} \mu_v \left(d_v - \sum_{a\in\delta^+_{A'}(v)} q_a + \sum_{a\in\delta^-_{A'}(v)} q_a \right) \\ & + \sum_{v\in V} \left(\lambda^+_v \left(\pi_v - \Delta_v - \overline{\pi}_v\right) + \lambda^-_v \left(\underline{\pi}_v - \pi_v - \Delta_v\right)\right) \\ & + \sum_{a\in A'} \left(\lambda^+_a \left(q_a - \Delta_a - \overline{q}_a\right) + \lambda^-_a \left(\underline{q}_a - q_a - \Delta_a\right)\right) \\ & - \sum_{v\in V} \lambda_v \Delta_v - \sum_{a\in A'} \lambda_a \Delta_a . \end{aligned} \tag{5.1.2}$$

In Section 4.2.3 we showed that the dual values of a KKT point $(q^*, \pi^*, \Delta^*, \mu^*, \lambda^*)$ of domain relaxation (4.2.1), which is indexed in this chapter by (5.1.1), form a network flow $(\mu^*_a)_{a\in A'}$, which is induced by dual node potentials $(\mu^*_v)_{v\in V}$. We recall the constraints (4.2.7) that are fulfilled by the dual variables (μ^*, λ^*):

$$\mu_a \frac{d\Phi_a}{dq_a}(q^*_a) + \lambda^+_a - \lambda^-_a = \mu_v - \mu_w \qquad \forall\, a = (v,w) \in A', \tag{5.1.3a}$$

$$\sum_{a \in \delta^+_{A'}(v)} \mu_a - \sum_{a \in \delta^-_{A'}(v)} \mu_a \gamma_a = \lambda^+_v - \lambda^-_v \qquad \forall v \in V. \tag{5.1.3b}$$

Relation (5.1.3b) is the basis for the inequalities that we consider in this chapter. Therefor we give the following definition:

Definition 5.1.1:
Every vector $(\mu, \lambda) = (\mu_v, \mu_a, \lambda^+_v, \lambda^-_v, \lambda^+_a, \lambda^-_a)_{v \in V, a \in A}$, *such that* $\mu_v, \mu_a \in \mathbb{R}$ *and* $\lambda^+_v, \lambda^-_v, \lambda^+_a, \lambda^-_a \in \mathbb{R}_{\geq 0}$, *which fulfills the constraints*

$$\sum_{a \in \delta^+_A(v)} \mu_a - \sum_{a \in \delta^-_A(v)} \gamma_a \mu_a = \lambda^+_v - \lambda^-_v \qquad \forall v \in V, \tag{5.1.4}$$

is called ***dual transmission flow****. We regard the vector* $(\mu_a)_{a \in A}$ *of this dual transmission flow as a generalized flow in the original network* (V, A).

In the remainder of this section we do not differ between an arc $a \in A$ and the corresponding arc $a' = (a, i) \in A'$. For a dual transmission flow (μ, λ) this implies that we speak of $\mu_{a'}$ which is defined as $\mu_{a'} := \mu_a$ for $a \in A$ and $a' = (a, i) \in A'$.

5.1.1 A Nonlinear Inequality

In the next two lemmata we derive two different inequalities for the passive transmission problem (4.1.1). They both only contain the flow variables q. The right-hand side of both inequalities is constant. The left-hand side of the first inequality is a non-convex function in q while the left-hand side of the second inequality is quasi-convex in q. We will consider a linear combination of both inequalities which then allows to project out the flow variables q as described in Section 5.1.2.

The first inequality is derived from any dual transmission flow.

Lemma 5.1.2:
For any dual transmission flow (μ, λ) *with* $\mu_a = 0$ *for all arcs* $a \notin A'$, *the inequality in* q

$$\sum_{a \in A'} \mu_a \, \Phi_a(q_a) \leq \sum_{v \in V} \left(\lambda^+_v \overline{\pi}_v - \lambda^-_v \underline{\pi}_v \right)$$

is valid for the passive transmission problem (4.1.1).

Proof. We prove the following estimation which is valid for any solution (q^*, π^*) of the passive transmission problem (4.1.1):

$$\sum_{a\in A'} \mu_a \Phi_a(q_a) = \sum_{u=(v,w)\in A'} \mu_a(\pi_v - \gamma_a \pi_w) \tag{5.1.5a}$$

$$= \sum_{v\in V} \pi_v \left(\sum_{a\in\delta^+_{A'}(v)} \mu_a - \sum_{a\in\delta^-_{A'}(v)} \gamma_a\mu_a \right) \tag{5.1.5b}$$

$$= \sum_{v\in V} \pi_v \left(\sum_{a\in\delta^+_{A}(v)} \mu_a - \sum_{a\in\delta^-_{A}(v)} \gamma_a\mu_a \right) \tag{5.1.5c}$$

$$= \sum_{v\in V} \pi_v \left(\lambda^+_v - \lambda^-_v\right) \tag{5.1.5d}$$

$$\leq \sum_{v\in V} (\lambda^+_v \overline{\pi}_v - \lambda^-_v \underline{\pi}_v). \tag{5.1.5e}$$

To obtain (5.1.5a) we multiply equation (4.1.1b) by μ_a and sum over all arcs $a \in A'$. Then we rewrite the right-hand side by changing the order of summation and obtain (5.1.5b). Note that all arcs $a \in A \backslash A'$ have $\mu_a = 0$, hence they can be added and we obtain (5.1.5c). We use equation (5.1.4) in Definition 5.1.1 of a dual transmission flow and obtain (5.1.5d). Finally we estimate the right-hand side and obtain (5.1.5e). □

Remark 5.1.3:

We briefly explain the practical meaning of the inequality of Lemma 5.1.2. Therefor let (μ, λ) be a dual transmission flow with $\mu_a = 0$ for all arcs $a \notin A'$. Furthermore we assume $\mu \geq 0$ and $\gamma_a = 1, a \in A$ for this explanation. We split the network flow μ into sets of flow along paths $P_1, \dots, P_m$ and flow along circuits $C_1, \dots, C_n$, see Theorem 4.2.5. This way we obtain flow values $\mu_{P_i} > 0, i = 1, \dots, m$ and $\mu_{C_i} > 0, i = 1, \dots, n$ such that

$$\mu_a = \sum_{\substack{i=1,\dots,m:\\ a\in A'(P_i)}} \mu_{P_i} + \sum_{\substack{i=1,\dots,n:\\ a\in A'(C_i)}} \mu_{C_i} \qquad \forall\, a \in A'$$

holds. We use this equation and denote the start and end node of P_i by $s(P_i)$ and $t(P_i)$ to write the inequality of Lemma 5.1.2 in a different way. For any primal solution (q^*, π^*) of the passive transmission problem it holds

$$\sum_{i=1}^{m} \mu_{P_i} \left(\pi_{s(P_i)} - \pi_{t(P_i)}\right)$$

$$
\begin{aligned}
&= \sum_{i=1}^{m} \mu_{P_i} \overbrace{\sum_{a \in A'(P_i)} \Phi_a(q_a)}^{=\pi_{s(P_i)} - \pi_{t(P_i)}} + \sum_{i=1}^{n} \mu_{C_i} \overbrace{\sum_{a \in A'(C_i)} \Phi_a(q_a)}^{=0} \\
&= \sum_{a \in A'} \Phi_a(q_a) \sum_{i: a \in A'(P_i)} \mu_{P_i} + \sum_{a \in A'} \Phi_a(q_a) \sum_{i: a \in A'(C_i)} \mu_{C_i} \\
&= \sum_{a \in A'} \Phi_a(q_a) \, \mu_a \\
&\leq \sum_{v \in V} \left(\lambda_v^+ \overline{\pi}_v - \lambda_v^- \underline{\pi}_v \right) = \sum_{i=1}^{n} \mu_{P_i} \left(\overline{\pi}_{s(P_i)} - \underline{\pi}_{t(P_i)} \right).
\end{aligned}
$$

In this argumentation we used the pressure conservation along circuits, i.e., the sum of potential differences along a circuit equals zero. We conclude that Lemma 5.1.2 presents a valid inequality for the passive transmission problem that requires the weighted sum of the potential losses $\pi^*_{s(P_i)} - \pi^*_{t(P_i)}$ along the paths P_i to be bounded by the weighted sum of the available potential losses $\overline{\pi}_{s(P_i)} - \underline{\pi}_{t(P_i)}$ along these paths.

Now we turn to the second inequality. For the next lemma we make use of $\gamma_{r,v}$ and the function π' from Definition 4.2.3, and recall the reformulation (4.2.5):

$$\pi'_v(\pi) - \pi'_w(\pi) = \gamma_{r,v} \left(\pi_v - \gamma_a \pi_w \right).$$

For abbreviations we defined $\overline{\pi}'_v := \pi'_v(\overline{\pi})$ and $\underline{\pi}'_v := \pi'_v(\underline{\pi})$ for each node $v \in V$. Furthermore we set

$$\Phi'_a(q_a) := \gamma_{r,v} \Phi_a(q_a) \tag{5.1.6}$$

for every arc $a = (v, w) \in A'$.

Lemma 5.1.4:
Let $q^* \in \mathbb{R}^{A'}$, $\Delta_v^\pm \in \mathbb{R}_{\geq 0}, v \in V$, and $\Delta_a^\pm \in \mathbb{R}_{\geq 0}, a \in A'$ be vectors such that the flow conservation

$$\sum_{a \in \delta^+_{A'}(v)} (q^*_a - (\Delta^+_a - \Delta^-_a)) - \sum_{a \in \delta^-_{A'}(v)} (q^*_a - (\Delta^+_a - \Delta^-_a)) - (\Delta^+_v - \Delta^-_v) = d_v$$

is fulfilled. Then the inequality in q

$$
\begin{aligned}
&\sum_{a=(v,w) \in A'} \gamma_{r,v} (q_a - q^*_a) \, \Phi_a(q_a) \\
&\leq \sum_{v \in V} (\Delta^-_v \overline{\pi}'_v - \Delta^+_v \underline{\pi}'_v) + \sum_{a \in A'} (\Delta^-_a \Phi'_a(\overline{q}_a) - \Delta^+_a \Phi'_a(\underline{q}_a))
\end{aligned}
$$

is valid for the passive transmission problem (4.1.1)

Proof. Let (q', π') be a feasible solution of the passive transmission problem (4.1.1). By (4.1.1d) the flow vector q' fulfills the flow conservation

$$\sum_{a \in \delta^+_{A'}(v)} q_a - \sum_{a \in \delta^-_{A'}(v)} q_a = d_v$$

for all nodes $v \in V$. From this we derive that q' is feasible for

$$\sum_{a \in \delta^+_{A'}(v)} (q_a - q^*_a) - \sum_{a \in \delta^-_{A'}(v)} (q_a - q^*_a) =$$
$$(\Delta^-_v - \Delta^+_v) + \sum_{a \in \delta^+_{A'}(v)} (\Delta^-_a - \Delta^+_a) - \sum_{a \in \delta^-_{A'}(v)} (\Delta^-_a - \Delta^+_a)$$

for all nodes $v \in V$. We multiply each side by $\pi'_v(\pi)$, take the sum over all nodes $v \in V$ and obtain:

$$\begin{aligned} &\sum_{a=(v,w) \in A'} (q_a - q^*_a)(\pi'_v(\pi) - \pi'_w(\pi)) \\ &= \sum_{v \in V} \pi'_v(\pi)(\Delta^-_v - \Delta^+_v) + \sum_{a=(v,w) \in A'} (\Delta^-_a - \Delta^+_a)(\pi'_v(\pi) - \pi'_w(\pi)) \\ &\leq \sum_{v \in V} (\pi'_v(\overline{\pi})\Delta^-_v - \pi'_v(\underline{\pi})\Delta^+_v) + \sum_{a \in A'} (\Delta^-_a - \Delta^+_a)\, \Phi'_a(q_a) \\ &\leq \sum_{v \in V} (\overline{\pi}'_v \Delta^-_v - \underline{\pi}'_v \Delta^+_v) + \sum_{a \in A'} (\Delta^-_a \Phi'_a(\overline{q}_a) - \Delta^+_a \Phi'_a(\underline{q}_a)). \end{aligned}$$

The estimations are obtained by taking the lower and upper bounds on $\pi'_v(\pi)$ and q_a into account. We use this estimation to obtain from (4.1.1b) and (4.2.5)

$$\begin{aligned} &\sum_{a=(v,w) \in A'} \gamma_{r,v}(q_a - q^*_a)\, \Phi_a(q_a) \\ &= \sum_{a=(v,w) \in A'} (q_a - q^*_a)\gamma_{r,v}(\pi_v - \gamma_a \pi_w) \\ &= \sum_{a=(v,w) \in A'} (q_a - q^*_a)(\pi'_v(\pi) - \pi'_w(\pi)) \\ &\leq \sum_{v \in V} (\overline{\pi}'_v \Delta^-_v - \underline{\pi}'_v \Delta^+_v) + \sum_{a \in A'} (\Delta^-_a \Phi'_a(\overline{q}_a) - \Delta^+_a \Phi'_a(\underline{q}_a)). \end{aligned}$$

Hence every primal solution of the passive transmission problem fulfills the inequality of the Lemma. □

Remark 5.1.5:

We briefly explain the practical meaning of the inequality of Lemma 5.1.4. Therefor let q^* fulfill the flow conservation constraint. Furthermore we assume $\gamma_a = 1, a \in A$ for this explanation. For any feasible solution $(\tilde{q}, \tilde{\pi})$ for the passive transmission problem we obtain that $\tilde{q} - q^*$ forms a circulation. In this case the inequality of Lemma 5.1.4 writes as

$$\sum_{a=(v,w)\in A'} (\tilde{q}_a - q^*_a)\,\Phi_a(q_a) \leq 0$$

Let us assume that $\tilde{q} \geq q^*$. We split this network flow into sets of flow along circuits $C_1, \dots, C_n$, see Theorem 4.2.5. We obtain flow values $\mu_{C_i} > 0, i = 1, \dots, n$ such that

$$\tilde{q}_a - q^*_a = \sum_{\substack{i=1,\dots,n:\\ a\in A'(C_i)}} \mu_{C_i} \qquad \forall\, a \in A'$$

holds. We use this equation to obtain

$$\begin{aligned}\sum_{a=(v,w)\in A'} (\tilde{q}_a - q^*_a)\,\Phi_a(\tilde{q}_a) &= \sum_{a\in A'} \Phi_a(\tilde{q}_a) \sum_{i:a\in A'(C_i)} \mu_{C_i}\\ &= \sum_{i=1}^{n} \mu_{C_i} \sum_{a\in A'(C_i)} \Phi_a(\tilde{q}_a).\end{aligned}$$

We conclude that Lemma 5.1.4 presents a valid inequality for the passive transmission problem which requires the weighted sum of the potential losses along the circuits C_i to be bounded by zero. Obviously a true statement because the loss of potential along a circuit equals zero.

Now we consider a linear combination of both inequalities from the previous two Lemmata 5.1.2 and 5.1.4:

Corollary 5.1.6:

Let (μ, λ) be a dual transmission flow with $\mu_a = 0$ for all arcs $a \notin A'$. Furthermore let $q^* \in \mathbb{R}^{A'}$, $\Delta^{\pm}_v \in \mathbb{R}_{\geq 0}, v \in V$, and $\Delta^{\pm}_a \in \mathbb{R}_{\geq 0}, a \in A'$ be vectors such that the flow conservation

$$\sum_{a\in\delta^{+}_{A'}(v)} (q^*_a - (\Delta^+_a - \Delta^-_a)) - \sum_{a\in\delta^{-}_{A'}(v)} (q^*_a - (\Delta^+_a - \Delta^-_a)) - (\Delta^+_v - \Delta^-_v) = d_v$$

is fulfilled. Then for any $\zeta \in [0,1]$ *the inequality in* q

$$\begin{aligned} &\sum_{a\in A'} (\zeta\, \gamma_{r,v}\,(q_a - q_a^*) + (1-\zeta)\mu_a)\, \Phi_a(q_a) \\ \leq \zeta &\sum_{v\in V} (\Delta_v^- \overline{\pi}_v' - \Delta_v^+ \underline{\pi}_v') + \zeta \sum_{a\in A'} (\Delta_a^- \Phi_a'(\overline{q}_a) - \Delta_a^+ \Phi_a'(\underline{q}_a)) \\ &+(1-\zeta) \sum_{v\in V} (\lambda_v^+\, \overline{\pi}_v - \lambda_v^-\, \underline{\pi}_v) \end{aligned} \tag{5.1.7}$$

is valid for the passive transmission problem (4.1.1)*.*

Proof. The inequality is a linear combination of the inequalities from Lemma 5.1.2 and Lemma 5.1.4. □

It is easy to see that the inequality from Lemma 5.1.4 is used to circumvent unboundedness of the left-hand-side of (5.1.7). Hence we regard this second inequality as a regularization. It allows us to use a linear underestimator for the left-hand-side of inequality (5.1.7) in the ongoing part of this chapter. As this regularization expresses the conservation of potential along circuits by Remark 5.1.5 we call it the *pc-regularization.*

5.1.2 A Linear Inequality

So far we derived a nonlinear inequality stated in Corollary 5.1.6 which is valid for the passive transmission problem (4.1.1). In principle, it would be possible to add this inequality to the topology optimization problem (3.2.1) at different nodes of the branch-and-bound tree. But this act would increase the number of nonlinearities of the model. Hence, in the following, we describe how to derive a linear inequality. Therefor we will use a certain linear underestimator for the left-hand side of inequality (5.1.7). This left-hand side is a sum of functions in q_a over the arcs $a \in A'$. We consider each of these functions

$$f_{\zeta,q^*,\mu}(q_a) := (\zeta\, \gamma_{r,v}(q_a - q_a^*) + (1-\zeta)\mu_a)\, \Phi_a(q_a)$$

separately and give a linear underestimator. For the underestimation we use the function

$$\ell_{\zeta,\pi,\mu,\lambda}(q_a) := \zeta\, (\pi_v'(\pi) - \pi_w'(\pi))\, q_a + (1-\zeta)(\mu_v - \mu_w - \lambda_a^+ + \lambda_a^-)\, q_a$$

for an arc $a = (v, w) \in A'$. This function is used to linearize inequality (5.1.7) as follows:

Lemma 5.1.7:

Let (μ, λ) be dual transmission flow with $\mu_a = 0$ for all arcs $a \notin A'$. Let $q^* \in \mathbb{R}^{A'}$, $\pi^* \in \mathbb{R}^V$, $\Delta_v^\pm \in \mathbb{R}_{\geq 0}, v \in V$, and $\Delta_a^\pm \in \mathbb{R}_{\geq 0}, a \in A'$ be vectors such that the flow conservation

$$\sum_{a \in \delta^+_{A'}(v)} (q_a^* - (\Delta_a^+ - \Delta_a^-)) - \sum_{a \in \delta^-_{A'}(v)} (q_a^* - (\Delta_a^+ - \Delta_a^-)) - (\Delta_v^+ - \Delta_v^-) = d_v$$

is fulfilled for each node $v \in V$. Furthermore let $\zeta \in [0,1]$. Then for constants $\tau_a := \inf\{f_{\zeta,q^*,\mu}(q_a) - \ell_{\zeta,\pi^*,\mu,\lambda}(q_a) \mid \underline{q}_a \leq q_a \leq \overline{q}_a\}$ for each arc $a \in A'$ the inequality

$$\begin{aligned} &\sum_{a \in A'} \tau_a \leq \\ &\zeta \left(\sum_{v \in V} (\Delta_v^- \overline{\pi}_v' - \Delta_v^+ \underline{\pi}_v') + \sum_{a \in A'} (\Delta_a^- \Phi_a'(\overline{q}_a) - \Delta_a^+ \Phi_a'(\underline{q}_a)) - \sum_{v \in V} d_v \pi_v'(\pi^*) \right) \\ &+ (1-\zeta) \left(\sum_{v \in V} \left(\lambda_v^+ \overline{\pi}_v - \lambda_v^- \underline{\pi}_v \right) + \sum_{a \in A'} \left(\lambda_a^+ \overline{q}_a - \lambda_a^- \underline{q}_a \right) - \sum_{v \in V} d_v \mu_v \right) \end{aligned} \tag{5.1.8}$$

is valid for the passive transmission problem (4.1.1).

Remark 5.1.8:

We note that the unfixed variables in the passive transmission problem are flow variables q and node potentials π. These variables are not contained in inequality (5.1.8). Thus inequality (5.1.8) is constant on both sides, left- and right-hand side, and hence linear. If the inequality is violated, then the passive transmission problem is infeasible.

Using the underestimator $\ell(q_a)$ to reformulate (5.1.7) we obtain inequality (5.1.8) which depends on the dual transmission flow (μ, λ) and other parameters q^*, π^*, Δ, and ζ. The choice of these parameters such that (5.1.8) represents the infeasibility of the passive transmission problem (4.1.1) will be discussed in Section 5.1.3.

Proof. [Lemma 5.1.7] Let $\tau_a := \inf\{f_{\zeta,q^*,\mu}(q_a) - \ell_{\zeta,\pi^*,\mu,\lambda}(q_a) \mid \underline{q}_a \leq q_a \leq \overline{q}_a\}$ for each arc $a \in A'$. We consider inequality (5.1.7) of Corollary 5.1.6. By the definition of τ_a we obtain the underestimator for each summand on the left-hand side as

$$\begin{aligned} f_{\zeta,q^*,\mu}(q_a) &\geq \tau_a + \ell_{\zeta,\pi^*,\mu,\lambda}(q_a) \\ &\geq \tau_a + \zeta\left(\pi_v'(\pi^*) - \pi_w'(\pi^*)\right) q_a + (1-\zeta)(\mu_v - \mu_w - \lambda_a^+ + \lambda_a^-)\, q_a. \end{aligned}$$

We rewrite the underestimator. Each flow vector q' which is feasible for (4.1.1) fulfills the flow conservation constraint

$$\sum_{a \in \delta^+_{A'}(v)} q_a - \sum_{a \in \delta^-_{A'}(v)} q_a = d_v.$$

Multiplying this equation with $\pi'_v(\pi^*)$ and summing over the nodes $v \in V$ we obtain that q' is feasible for

$$\zeta \sum_{a=(v,w) \in A'} (\pi'_v(\pi^*) - \pi'_w(\pi^*))\, q_a = \zeta \sum_{v \in V} d_v \pi'_v(\pi^*) \tag{5.1.9}$$

and multiplying the flow conservation constraint with μ_v and summing over the nodes $v \in V$ we derive

$$(1-\zeta) \sum_{a=(v,w) \in A'} (\mu_v - \mu_w)\, q_a = (1-\zeta) \sum_{v \in V} d_v \mu_v. \tag{5.1.10}$$

Using the reformulations (5.1.9), (5.1.10) we obtain that q' is feasible for

$$\begin{aligned} &\sum_{a \in A'} f_{\zeta,q^*,\mu}(q_a) \\ \geq &\sum_{a \in A'} \tau_a + \zeta \sum_{v \in V} d_v \pi'_v(\pi^*) + (1-\zeta) \sum_{v \in V} d_v \mu_v - (1-\zeta) \sum_{a \in A'} (\lambda^+_a - \lambda^-_a) q_a. \end{aligned} \tag{5.1.11}$$

We use the lower and upper bounds on q_a to obtain

$$(1-\zeta) \sum_{a \in A'} (\lambda^+_a - \lambda^-_a) q_a \leq (1-\zeta) \sum_{a \in A'} (\lambda^+_a \overline{q}_a - \lambda^-_a \underline{q}_a).$$

Now (5.1.11) writes as

$$\begin{aligned} &\sum_{a \in A'} \tau_a \leq \\ &\sum_{a \in A'} f_{\zeta,q^*,\mu}(q_a) + (1-\zeta) \sum_{a \in A'} (\lambda^+_a \overline{q}_a - \lambda^-_a \underline{q}_a) \\ &\quad -\zeta \sum_{v \in V} d_v \pi'_v(\pi^*) - (1-\zeta) \sum_{v \in V} d_v \mu_v. \end{aligned}$$

Inequality (5.1.7) of Corollary 5.1.6 yields an over estimator for $\sum_{a \in A'} f_{\zeta,q^*,\mu}(q_a)$. Applying this over estimator proves the lemma. □

5.1.3 Feasibility Characterization by a Linear Inequality

So far we proved that inequality (5.1.8) is valid for the passive transmission problem (4.1.1). The upcoming question is how to choose the dual transmission flow (μ, λ) and the parameters $q^*, \pi^*, \Delta, \zeta$ in order to obtain an inequality that represents the infeasibility of the passive transmission problem. We want to obtain an inequality that is violated if the passive transmission problem is infeasible. Therefor, as stated in Definition 5.1.11, we *derive* a dual transmission flow from a KKT point of the domain relaxation (4.2.1) or the flow conservation relaxation (4.3.1) and discuss a suitable choice of the parameters $q^*, \pi^*, \Delta, \zeta$.

The first lemma describes the choice of the parameter ζ for inequality (5.1.8) for the case that the dual transmission flow (μ, λ) and the parameters q^* and π^* are given. The results are important for the subsequent lemmata of this section.

Lemma 5.1.9:
Let $(q^*, \pi^*, \Delta^*, \mu^*, \lambda^*)$ be a KKT point of the domain relaxation (5.1.1) or the flow conservation relaxation (4.3.1). Let $\zeta \in]0,1]$ such that the following conditions hold for every arc $a = (v,w) \in A'$:

1. *if $\mu_a\, q_a^* > 0$, then $(1-\zeta)|\mu_a^*| < \zeta\, \gamma_{r,v} |q_a^*|$,*
2. *if $\mu_a\, q_a^* < 0$, then $(1-\zeta)\,|\mu_v^* - \mu_w^* - \lambda_a^{*+} + \lambda_a^{*-}| < \zeta\, \gamma_{r,v} |\pi_v^* - \gamma_a \pi_w^* - \tilde{\beta}_a|$,*
3. *if $\mu_a\, q_a^* = 0$, then $(1-\zeta)\, \mu_a^* = 0$.*

Then the minimum of the function

$$f_{\zeta, q^*, \mu^*}(q_a) - \ell_{\zeta, \pi^*, \mu^*, \lambda^*}(q_a) \tag{5.1.12}$$

is attained at q_a^* for every arc $a \in A'$.

Remark 5.1.10:
It is not guaranteed that a value $\zeta \in]0,1]$ exists which fulfills the conditions of Lemma 5.1.9.

Proof. [Lemma 5.1.9] Let $(q^*, \pi^*, \Delta^*, \mu^*, \lambda^*)$ be a KKT point of the domain relaxation (5.1.1) or the flow conservation relaxation (4.3.1). Let $\zeta \in]0,1]$ such that conditions 1, 2 and 3 are fulfilled for every arc $a \in A'$. We consider an arc $a = (v, w) \in A'$ and write the derivative of function (5.1.12) as

$$\begin{aligned}
& f'_{\zeta,q^*,\mu^*}(q_a) - \ell'_{\zeta,\pi^*,\mu^*,\lambda^*}(q_a) \\
& = \zeta\, \gamma_{r,v} \left((q_a - q_a^*) \frac{d\Phi_a}{dq_a}(q_a) + \Phi_a(q_a) \right) + (1-\zeta)\mu_a^* \frac{d\Phi_a}{dq_a}(q_a) \\
& \quad - \zeta\, (\pi'_v(\pi^*) - \pi'_w(\pi^*)) - (1-\zeta)(\mu_v^* - \mu_w^* - \lambda_a^{*+} + \lambda_a^{*-}).
\end{aligned}$$

We set $q_a = q_a^*$ and obtain from (5.1.1b), (4.3.1b), Lemma 4.2.11 and Lemma 4.3.7 and the relation (4.2.5)

$$\begin{aligned}
& f'_{\zeta,q^*,\mu^*}(q_a^*) - \ell'_{\zeta,\pi^*,\mu^*,\lambda^*}(q_a^*) \\
& = \zeta\, (\gamma_{r,v} \Phi_a(q_a^*) - (\pi'_v(\pi^*) - \pi'_w(\pi^*))) \\
& \quad + (1-\zeta) \left(\mu_a^* \frac{d\Phi_a}{dq_a}(q_a^*) - (\mu_v^* - \mu_w^* - \lambda_a^{*+} + \lambda_a^{*-}) \right) \\
& = \zeta\, \gamma_{r,v} (\Phi_a(q_a^*) - \pi_v^* + \gamma_a \pi_w^*) \\
& \quad + (1-\zeta) \left(\mu_a^* \frac{d\Phi_a}{dq_a}(q_a^*) - (\mu_v^* - \mu_w^* - \lambda_a^{*+} + \lambda_a^{*-}) \right) \\
& = 0 + 0.
\end{aligned}$$

This implies that function (5.1.12) has an extreme point for $q_a = q_a^*$.

We still have to prove that the point $q_a = q_a^*$ is a global minimum of (5.1.12) for the choice of ζ. We write function (5.1.12) as $g(q_a) - h(q_a)$ where g and h are defined as

$$\begin{aligned}
g(q_a) : &= (\overbrace{\zeta\, \gamma_{r,v}}^{=:b>0} q_a + \overbrace{(1-\zeta)\mu_a^* - \zeta\, \gamma_{r,v} q_a^*}^{=:c}) \alpha_a\, q_a |q_a|^{k_a} - \overbrace{((1-\zeta)\mu_a^* - \zeta \gamma_{r,v} q_a^*)\tilde{\beta}_a}^{=:d} \\
&= (b q_a + c)\, \alpha_a\, q_a |q_a|^{k_a} - d
\end{aligned}$$

and

$$h(q_a) := \zeta \gamma_{r,v} (\pi_v^* - \gamma_a \pi_w^* + \tilde{\beta}_a)\, q_a + (1-\zeta)(\mu_v^* - \mu_w^* - \lambda_a^{+*} + \lambda_a^{-*})\, q_a.$$

In the case $\alpha_a = 0$ it follows that $g(q_a)$ is constant and (5.1.1b) and (5.1.3a) imply $h(q_a) \equiv 0$. This implies that q_a^* is a global minimum of (5.1.12). To prove this for the case $\alpha_a > 0$ we briefly show that the choice of ζ means that $g(q_a)$ and $h(q_a)$ have the form as indicated in Figure 5.1 depending on the value of q_a^*. This implies that $g(q_a) - h(q_a)$ is a convex function. In combination with our previous analysis we obtain that $g(q_a) - h(q_a)$ has a global optimum at $q_a = q_a^*$.

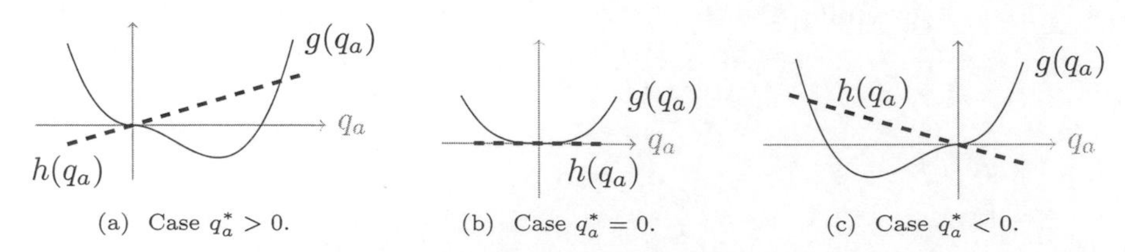

Figure 5.1: Visualization of the functions $g(q_a)$ and $h(q_a)$ defined in the proof of Lemma 5.1.9 for different values of q_a^* and $\alpha_a > 0$.

In the following we characterize $g(q_a)$ and show $c < 0$ if $q_a^* > 0$, $c > 0$ if $q_a^* < 0$ and $c = 0$ if $q_a^* = 0$. This then proves that $g(q_a)$ has the form as shown in Figure 5.1. We distinguish three cases:

Case $q_a^* > 0$: It holds $c = (1 - \zeta)\mu_a^* - \zeta\,\gamma_{r,v} q_a^* < 0$:

- If $\mu_a^* \leq 0$ then it holds $(1 - \zeta)\mu_a^* \leq 0, -\zeta\,\gamma_{r,v} q_a^* < 0$. Note that $\zeta \neq 0$.
- If $\mu_a^* > 0$ then it follows from assumption 1 that $(1 - \zeta)\mu_a^* = (1 - \zeta)|\mu_a^*| < |\zeta\,\gamma_{r,v} q_a^*| = \zeta\,\gamma_{r,v} q_a^*$.

Case $q_a^* = 0$: It holds $c = (1 - \zeta)\mu_a^* - \zeta\,\gamma_{r,v} q_a^* = 0$ because $(1 - \zeta)\mu_a^* = 0$ holds by assumption 3.

Case $q_a^* < 0$: It holds $c = (1 - \zeta)\mu_a^* - \zeta\,\gamma_{r,v} q_a^* > 0$:

- If $\mu_a^* \geq 0$ then it holds $(1 - \zeta)\mu_a^* \geq 0, -\zeta\,\gamma_{r,v} q_a^* > 0$. Note that $\zeta \neq 0$.
- If $\mu_a^* < 0$ then it follows from assumption 1 that $-(1 - \zeta)\mu_a^* = (1 - \zeta)|\mu_a^*| < |\zeta\,\gamma_{r,v} q_a^*| = -\zeta\,\gamma_{r,v} q_a^*$ and hence $(1 - \zeta)\mu_a^* > \zeta\,\gamma_{r,v} q_a^*$.

We now turn to the analysis of $h(q_a)$.

Case $q_a^* > 0$: It holds $\zeta\gamma_{r,v}(\pi_v^* - \gamma_a\pi_w^* - \tilde{\beta}_a) + (1 - \zeta)(\mu_v^* - \mu_w^* - \lambda_a^{+*} + \lambda_a^{-*}) > 0$: From (5.1.1b) it follows $\pi_v^* - \gamma_a\pi_w^* - \tilde{\beta}_a > 0$.

- If $\mu_a^* \geq 0$ then it holds $\mu_v^* - \mu_w^* - \lambda_a^{+*} + \lambda_a^{-*} \geq 0$ by (5.1.3a).
- If $\mu_a^* < 0$ then it follows from (5.1.3a) and assumption 2 that $(1 - \zeta)|\mu_v^* - \mu_w^* - \lambda_a^{+*} + \lambda_a^{-*}| < \zeta\gamma_{r,v}|\pi_v^* - \gamma_a\pi_w^* - \tilde{\beta}_a| = \zeta\gamma_{r,v}(\pi_v^* - \gamma_a\pi_w^* - \tilde{\beta}_a)$.

Case $q_a^* = 0$: It holds $\zeta\gamma_{r,v}(\pi_v^* - \gamma_a\pi_w^* - \tilde{\beta}_a) + (1 - \zeta)(\mu_v^* - \mu_w^* - \lambda_a^{+*} + \lambda_a^{-*}) = 0$: From (5.1.1b) it follows $\pi_v^* - \gamma_a\pi_w^* - \tilde{\beta}_a = 0$. From (5.1.3a) we obtain $\mu_v^* - \mu_w^* - \lambda_a^{+*} + \lambda_a^{-*} = 0$. Hence $g(q_a) \equiv 0$.

Case $q_a^* < 0$: It holds $\zeta\gamma_{r,v}(\pi_v^* - \gamma_a\pi_w^* - \tilde{\beta}_a) + (1 - \zeta)(\mu_v^* - \mu_w^* - \lambda_a^{+*} + \lambda_a^{-*}) < 0$: From (5.1.1b) it follows $\pi_v^* - \gamma_a\pi_w^* - \tilde{\beta}_a < 0$.

- If $\mu_a^* \leq 0$ then it holds $\mu_v^* - \mu_w^* - \lambda_a^{+*} + \lambda_a^{-*} \leq 0$ by (5.1.3a).
- If $\mu_a^* > 0$ then it follows from (5.1.3a) and assumption 2 that $-(1-\zeta)(\mu_v^* - \mu_w^* - \lambda_a^{+*} + \lambda_a^{-*}) = -(1-\zeta)|\mu_v^* - \mu_w^* - \lambda_a^{+*} + \lambda_a^{-*}| > \zeta\gamma_{r,v}(\pi_v^* - \gamma_a\pi_w^* - \tilde{\beta}_a)$.

These cases show that $h(q_a)$ is a linear function with positive slope if $q_a^* > 0$, negative slope if $q_a^* < 0$ and slope zero if $q_a^* = 0$. Hence $h(q_a)$ has the form as shown in Figure 5.1. □

The dual transmission flow which is necessary to write inequality (5.1.8) is obtained from a KKT point of the domain relaxation (5.1.1) or the flow conservation relaxation (4.3.1) as follows:

Definition 5.1.11:
A dual transmission flow (μ, λ) is ***derived*** *from a KKT point $(q^*, \pi^*, \Delta^*, \mu^*, \lambda^*)$ of domain relaxation (5.1.1) as follows: The dual values fulfill the dual flow conservation*

$$\sum_{a \in \delta^+_{A'}(v)} \mu_a^* - \sum_{a \in \delta^-_{A'}(v)} \gamma_a \mu_a^* = \lambda_v^{+*} - \lambda_v^{-*}$$

for all nodes $v \in V$ by Lemma 4.2.11. We extend this network flow to all arcs $a \in A$ by setting zero flow values to those arcs that are contained in A but not in A'. Hence we obtain a dual transmission flow (μ, λ) in (V, A) by setting $\mu_v := \mu_v^*$ and $\lambda_v^\pm := \lambda_v^{\pm *}$ for each node $v \in V$ and $\mu_a := \mu_a^*, \lambda_a^\pm := \lambda_a^{\pm *}$ if $a \in A'$ and $\mu_a := 0, \lambda_a^\pm := 0$ otherwise for each arc $a \in A \setminus A'$.
A dual transmission flow (μ, λ) is ***derived*** *from a KKT point $(q^*, \pi^*, \Delta^*, \mu^*, \lambda^*)$ of the flow conservation relaxation (4.3.1) as follows: The dual values fulfill the dual flow conservation*

$$\sum_{a \in \delta^+_{A'}(v)} \mu_a^* - \sum_{a \in \delta^-_{A'}(v)} \gamma_a \mu_a^* = (\lambda_v^{+*} - \lambda_v^{-*}) + (\mu_v^{+*}\Delta_v^{-*} - \mu_v^{-*}\Delta_v^{+*})$$

for each node $v \in V$. Setting $\mu_v := \mu_v^*$ and $\lambda_v^+ := \lambda_v^{+*} + \mu_v^{+*}\Delta_v^{-*}$ and further $\lambda_v^- := \lambda_v^{-*} + \mu_v^{-*}\Delta_v^{+*}$ for each node $v \in V$ and $\mu_a := \mu_a^*, \lambda_a^\pm := \lambda_a^{\pm *}$ if $a \in A'$ and $\mu_a := 0, \lambda_a^\pm := 0$ otherwise for each arc $a \in A \setminus A'$ yields a dual transmission flow (μ, λ) in (V, A).

In the next lemma we derive a dual transmission flow from a KKT point of the domain relaxation (5.1.1). We use the primal solution of this KKT point and describe a suitable choice of ζ such that inequality (5.1.8) is violated if and only if the passive transmission problem (4.1.1) is infeasible. For the choice of ζ we refer to the previous Lemma 5.1.9.

Lemma 5.1.12:
Let $(q^*, \pi^*, \Delta^*, \mu^*, \lambda^*)$ be a KKT point of the domain relaxation (5.1.1) and let (μ, λ) be a dual transmission flow derived from this KKT point. Let $\zeta \in]0,1[$ such that the conditions of Lemma 5.1.9 are fulfilled. Then we obtain inequality (5.1.8) from Lemma 5.1.7 with dual transmission flow (μ, λ), and parameters q^*, π^*, ζ and $\Delta = 0$ which is violated if and only if the passive transmission problem (4.1.1) is infeasible. The violation (i.e., the absolute difference of the left-hand and the right-hand side of inequality (5.1.8)) is greater than or equal to $(1-\zeta)$ times the objective value of the KKT point.

Proof. First we write inequality (5.1.8) of Lemma 5.1.7. It is valid for the passive transmission problem and we only need to concentrate on the violation of this inequality. Therefor we simply rewrite the inequality in the remainder of this proof. Note that we do not need to refer to Theorem 4.2.9 stating the convexity of the domain relaxation under some assumptions.

Since q_a^* realizes the nomination, we derive from

$$\sum_{a \in \delta^+_{A'}(v)} (q_a^* - (\Delta_a^+ - \Delta_a^-)) - \sum_{a \in \delta^-_{A'}(v)} (q_a^* - (\Delta_a^+ - \Delta_a^-)) - (\Delta_v^+ - \Delta_v^-) = d_v$$

for each node $v \in V$, that the slack variables are zero, i.e., $\Delta_v^+ = \Delta_v^- = 0$ and $\Delta_a^+ = \Delta_a^- = 0$. Then inequality (5.1.8) reduces to

$$\begin{aligned} &\sum_{a \in A'} \tau_a \leq -\zeta \sum_{v \in V} d_v \pi_v'(\pi^*) \\ &+(1-\zeta) \left(\sum_{v \in V} \left(\lambda_v^+ \overline{\pi}_v - \lambda_v^- \underline{\pi}_v \right) + \sum_{a \in A'} \left(\lambda_a^+ \overline{q}_a - \lambda_a^- \underline{q}_a \right) - \sum_{v \in V} d_v \mu_v \right). \end{aligned} \tag{5.1.13}$$

We recall the definition of $\tau_a = \inf\{f_{\zeta,q^*,\mu}(q_a) - \ell_{\zeta,\pi^*,\mu,\lambda}(q_a) \mid \underline{q}_a \leq q_a \leq \overline{q}_a\}$ from Lemma 5.1.7. By Lemma 5.1.9 the minimum of the function

$$f_{\zeta,q^*,\mu}(q_a) - \ell_{\zeta,\pi^*,\mu,\lambda}(q_a)$$

is attained at q_a^*, which is not necessarily contained in $[\underline{q}_a, \overline{q}_a]$. Hence the left-hand-side of (5.1.13) is estimated as follows:

$$\begin{aligned}\sum_{a\in A'} \tau_a \geq & \sum_{a=(v,w)\in A'} \Big((1-\zeta)\mu_a \Phi_a(q_a^*) - \zeta\, \gamma_{r,v}(\pi_v^* - \gamma_a \pi_w^*)\, q_a^* \Big) \\ & - \sum_{a=(v,w)\in A'} (1-\zeta)(\mu_v - \mu_w - \lambda_a^+ + \lambda_a^-)\, q_a^* \\ = & (1-\zeta) \sum_{a\in A'} \mu_a \Phi_a(q_a^*) - \zeta \sum_{v\in V} d_v \pi_v'(\pi^*) \\ & - (1-\zeta) \sum_{v\in V} d_v \mu_v + (1-\zeta) \sum_{a\in A'} (\lambda_a^+ - \lambda_a^-) q_a^* .\end{aligned} \tag{5.1.14}$$

Combining (5.1.13) with (5.1.14) results in

$$\begin{aligned} & (1-\zeta) \sum_{a\in A'} \mu_a \Phi_a(q_a^*) + (1-\zeta) \sum_{a\in A'} (\lambda_a^+ - \lambda_a^-) q_a^* \\ \leq & (1-\zeta) \sum_{v\in V} \Big(\lambda_v^+ \overline{\pi}_v - \lambda_v^- \underline{\pi}_v \Big) + (1-\zeta) \sum_{a\in A'} \Big(\lambda_a^+ \overline{q}_a - \lambda_a^- \underline{q}_a \Big).\end{aligned} \tag{5.1.15}$$

In the remainder of this proof we analyze this inequality (5.1.15) and its violation. First we prove

$$\sum_{a\in A'} \mu_a \Phi_a(q_a^*) - \sum_{v\in V} \Big(\lambda_v^+ \overline{\pi}_v - \lambda_v^- \underline{\pi}_v \Big) = \sum_{v\in V} (\lambda_v^+ + \lambda_v^-) \Delta_v^*, \tag{5.1.16}$$

$$\sum_{a\in A'} (\lambda_a^+ - \lambda_a^-) q_a^* - \sum_{a\in A'} \Big(\lambda_a^+ \overline{q}_a - \lambda_a^- \underline{q}_a \Big) = \sum_{a\in A'} (\lambda_a^+ + \lambda_a^-) \Delta_a^* . \tag{5.1.17}$$

Equality (5.1.17) follows from Lemma 4.2.11 because the vector $(q^*, \pi^*, \Delta^*, \mu^*, \lambda^*)$ is a KKT point of the domain relaxation (5.1.1). By this lemma it holds $\lambda_a^+ = \lambda_a^{+*} > 0$ only if $q_a^* \geq \overline{q}_a$ and $\lambda_a^- = \lambda_a^{-*} > 0$ only if $q_a^* \leq \underline{q}_a$. Furthermore $\Delta_a^* = \max\{0, q_a^* - \overline{q}_a, \underline{q}_a - q_a^*\}, a \in A'$ because other values imply that Δ^* is not optimal and hence that (q^*, π^*, Δ^*) is not the primal part of a KKT point.

To show equality (5.1.16) we proceed as follows. From the proof of Lemma 5.1.2 we obtain the equality

$$\sum_{a\in A'} \mu_a \Phi_a(q_a^*) = \sum_{v\in V} (\lambda_v^+ - \lambda_v^-) \pi_v^* .$$

By Lemma 4.2.11 it holds that $\lambda_v^+ = \lambda_v^{+*} > 0$ only if $\pi_v^* \geq \overline{\pi}_v$ and $\lambda_v^- = \lambda_v^{-*} > 0$ only if $\pi_v^* \leq \underline{\pi}_v$. Furthermore $\Delta_v^* = \max\{0, \pi_v^* - \overline{\pi}_v, \underline{\pi}_v - \pi_v^*\}, v \in V$ because other values imply that Δ^* is not optimal and hence that (q^*, π^*, Δ^*) is not a primal part of a KKT point. This implies

$$\sum_{v \in V} (\lambda_v^+ - \lambda_v^-) \pi_v^* - \sum_{v \in V} \left(\lambda_v^+ \overline{\pi}_v - \lambda_v^- \underline{\pi}_v \right) = \sum_{v \in V} (\lambda_v^+ + \lambda_v^-) \Delta_v^*.$$

It follows from the reformulations (5.1.16) and (5.1.17) that (5.1.15) equivalently rewrites to

$$(1 - \zeta) \left(\sum_{v \in V} (\lambda_v^+ + \lambda_v^-) \Delta_v^* + \sum_{a \in A'} (\lambda_a^+ + \lambda_a^-) \Delta_a^* \right) \leq 0. \tag{5.1.18}$$

Now it follows from (4.2.8) that $\Delta_v^* > 0$ (which means $\pi_v^* < \underline{\pi}_v$ or $\pi_v^* > \overline{\pi}_v$) implies $\lambda_v^+ + \lambda_v^- = 1$ for each node $v \in V$ and $\Delta_a^* > 0$ implies $\lambda_a^+ + \lambda_a^- = 1$ for each arc $a \in A'$. Thus we rewrite (5.1.18) equivalently as

$$(1 - \zeta) \left(\sum_{v \in V} \Delta_v^* + \sum_{a \in A'} \Delta_a^* \right) \leq 0.$$

This inequality is violated if and only if the primal solution (q^*, π^*, Δ^*) has positive slack. □

Next we turn to the flow conservation relaxation (4.3.1). Similar to the previous lemma we derive a dual transmission flow from a KKT point of the relaxation (4.3.1). We describe a suitable choice of the parameters $q^*, \pi^*, \Delta, \zeta$ such that we obtain an inequality by Lemma 5.1.7 which is violated if and only if the passive transmission problem (4.1.1) is infeasible. Again we refer to Lemma 5.1.9 for the choice of ζ.

Lemma 5.1.13:
Let $(q^*, \pi^*, \Delta^*, \mu^*, \lambda^*)$ *be a KKT point of the flow conservation relaxation* (4.3.1) *and let* (μ, λ) *be a dual transmission flow derived from this KKT point. Let* $\zeta \in]0, 1[$ *such that the conditions of Lemma 5.1.9 are fulfilled. Then we obtain inequality* (5.1.8) *from Lemma 5.1.7 with dual transmission flow* (μ, λ), *and parameters* $q^*, \pi^*, \Delta^*, \zeta$ *which is violated if and only if the passive transmission problem* (4.1.1) *is infeasible. The violation is greater than or equal to* $(1 - \zeta)$ *times the optimal objective value of the relaxation* (4.3.1).

Proof. We rewrite inequality (5.1.8) of Lemma 5.1.7. It is valid for the passive transmission problem and we only need to concentrate on the violation of this inequality. Therefor we simply rewrite the inequality in the remainder of this proof.

Note that we do not need to refer to Theorem 4.3.6 stating the convexity of the flow conservation relaxation under some assumptions.

We recall the definition of

$$\tau_a = \inf\{f_{\zeta,q^*,\mu}(q_a) - \ell_{\zeta,\pi^*,\mu,\lambda}(q_a) \mid \underline{q}_a \leq q_a \leq \overline{q}_a\}$$

from Lemma 5.1.7. By Lemma 5.1.9 the minimum of the function

$$f_{\zeta,q^*,\mu}(q_a) - \ell_{\zeta,\pi^*,\mu,\lambda}(q_a)$$

is attained at q_a^*, which is not necessarily contained in $[\underline{q}_a, \overline{q}_a]$. Hence, similar as in the previous proof of Lemma 5.1.12, $\sum_{a\in A'} \tau_a$ rewrites as

$$\begin{aligned}\sum_{a\in A'} \tau_a \geq \sum_{a=(v,w)\in A'} \Big(&(1-\zeta)\mu_a \Phi_a(q_a^*) \\ &- \zeta\,(\pi_v'(\pi^*) - \pi_w'(\pi^*))q_a^* \\ &- (1-\zeta)(\mu_v - \mu_w - \lambda_a^+ + \lambda_a^-)q_a^*\Big).\end{aligned}$$

After rearranging we obtain

$$\begin{aligned}\sum_{a\in A'} \tau_a \geq &(1-\zeta)\sum_{a\in A'} \mu_a \Phi_a(q_a^*) + (1-\zeta)\sum_{a\in A'} (\lambda_a^+ - \lambda_a^-)q_a^* \\ &+ \zeta \sum_{v\in V} \pi_v'(\pi^*)\Big(d_v - \sum_{a\in\delta_{A'}^+(v)} q_a^* + \sum_{a\in\delta_{A'}^-(v)} q_a^*\Big) \\ &+ (1-\zeta)\sum_{v\in V} \mu_v \Big(d_v - \sum_{a\in\delta_{A'}^+(v)} q_a^* + \sum_{a\in\delta_{A'}^-(v)} q_a^*\Big) \\ &- \zeta\sum_{v\in V} d_v \pi_v'(\pi^*) - (1-\zeta)\sum_{v\in V} d_v\mu_v.\end{aligned} \tag{5.1.19}$$

We rewrite the first three lines of this equality separately as follows:

1. From (4.3.15a) and (4.3.15b) we obtain

$$\begin{aligned}\sum_{a\in A'} \mu_a \Phi_a(q_a^*) &= \sum_{v\in V} (\lambda_v^+ \overline{\pi}_v - \lambda_v^- \underline{\pi}_v), \\ \sum_{a\in A'} (\lambda_a^+ - \lambda_a^-)q_a^* &= \sum_{a\in A'} (\lambda_a^+ \overline{q}_a - \lambda_a^- \underline{q}_a).\end{aligned}$$

2. Setting $\Delta_v^\pm := \Delta_v^{\pm *}$ for all nodes $v \in V$ and $\Delta_a^\pm := \Delta_a^{\pm *}$ for all arcs $a \in A'$ the vector q^* fulfills the flow conservation

$$\sum_{a\in\delta^+_{A'}(v)} (q_a^* - (\Delta_a^+ - \Delta_a^-)) - \sum_{a\in\delta^-_{A'}(v)} (q_a^* - (\Delta_a^+ - \Delta_a^-)) - (\Delta_v^+ - \Delta_v^-) = d_v$$

for all nodes $v \in V$. We multiply each side by $\pi'_v(\pi^*)$, take the sum over all nodes $v \in V$ and obtain

$$\sum_{v\in V} \pi'_v(\pi^*) \left(d_v - \sum_{a\in\delta^+_{A'}(v)} q_a^* + \sum_{a\in\delta^-_{A'}(v)} q_a^* \right) =$$
$$\sum_{v\in V} (\Delta_v^- - \Delta_v^+)\pi'_v(\pi^*) + \sum_{a=(v,w)\in A'} (\Delta_a^- - \Delta_a^+)(\pi'_v(\pi^*) - \pi'_w(\pi^*))$$

From the value of Δ from (4.3.1h), (4.3.1i), (4.3.1j) and (4.3.1k) we derive the equalities

$$\sum_{v\in V} (\Delta_v^- - \Delta_v^+)\pi'_v(\pi^*) = \sum_{v\in V} (\Delta_v^- \pi'_v(\overline{\pi}) - \Delta_v^+ \pi'_v(\underline{\pi}))$$
$$\sum_{a=(v,w)\in A'} (\Delta_a^- - \Delta_a^+)(\pi'_v(\pi^*) - \pi'_w(\pi^*)) = \sum_{a\in A'} (\Delta_a^- \Phi'_a(\overline{q}_a) - \Delta_a^+ \Phi'_a(\underline{q}_a))$$

3. With the same definition of Δ from the previous item and similar reasoning we obtain (by multiplication by μ_v instead of $\pi'_v(\pi^*)$)

$$\sum_{v\in V} \mu_v \left(d_v - \sum_{a\in\delta^+_{A'}(v)} q_a^* + \sum_{a\in\delta^-_{A'}(v)} q_a^* \right)$$
$$= \sum_{v\in V} (\Delta_v^- - \Delta_v^+)\mu_v + \sum_{a=(v,w)\in A'} (\Delta_a^- - \Delta_a^+)(\mu_v - \mu_w).$$

Using all these reformulations together we obtain from (5.1.19)

$$\sum_{a\in A'} \tau_a \geq$$
$$(1-\zeta)\sum_{v\in V} \left(\lambda_v^+ \overline{\pi}_v - \lambda_v^- \underline{\pi}_v\right) + (1-\zeta)\sum_{a\in A'} \left(\lambda_a^+ \overline{q}_a - \lambda_a^- \underline{q}_a\right)$$
$$+\zeta\left(\sum_{v\in V}(\Delta_v^- \pi'_v(\overline{\pi}) - \Delta_v^+ \pi'_v(\underline{\pi})) + \sum_{a\in A'}(\Delta_a^- \Phi'_a(\overline{q}_a) - \Delta_a^+ \Phi'_a(\underline{q}_a)) - \sum_{v\in V} d_v \pi'_v(\pi^*)\right)$$
$$+(1-\zeta)\left(\sum_{v\in V}(\Delta_v^- - \Delta_v^+)\mu_v + \sum_{a=(v,w)\in A'} (\Delta_a^- - \Delta_a^+)(\mu_v - \mu_w) - \sum_{v\in V} d_v \mu_v\right).$$

We take this inequality and substitute the left-hand side of inequality (5.1.8) which then reduces to

$$(1-\zeta)\Big(\sum_{v\in V}(\Delta_v^- - \Delta_v^+)\mu_v + \sum_{a=(v,w)\in A'}(\Delta_a^- - \Delta_a^+)(\mu_v - \mu_w)\Big) \le 0 \qquad (5.1.20)$$

Now it follows from Lemma 4.3.7 that $\Delta_v^- > 0$ implies $\mu_v = \mu_v^* = 1$ and $\Delta_v^+ > 0$ implies $\mu_v = \mu_v^* = -1$ for each node $v \in V$. Furthermore by Lemma 4.3.7 it holds that $\Delta_a^- > 0$ implies $\mu_v - \mu_w = \mu_v^* - \mu_w^* = 1$ and $\Delta_a^+ > 0$ implies $\mu_v - \mu_w = \mu_v^* - \mu_w^* = -1$ for each arc $a \in A'$. Thus we rewrite (5.1.20) equivalently as

$$(1-\zeta)\Big(\sum_{v\in V}(\Delta_v^- + \Delta_v^+) + \sum_{a\in A'}(\Delta_a^- + \Delta_a^+)\Big) \le 0$$

This inequality is violated if and only if the primal solution (q^*, π^*, Δ^*) has positive slack. □

The next corollary summarizes Lemma 5.1.12 and Lemma 5.1.13. It presents a choice of the parameters for inequality (5.1.8) such that it is violated if and only if the passive transmission problem (4.1.1) is infeasible. Thus the corollary enhances Lemma 5.1.7.

Corollary 5.1.14:
Let $(q^*, \pi^*, \Delta^*, \mu^*, \lambda^*)$ be a KKT point of relaxation (5.1.1) or (4.3.1), and let (μ, λ) be a dual transmission flow derived from this KKT point. Furthermore let $\Delta = 0$ if the KKT point belongs to the domain relaxation and $\Delta = \Delta^*$ otherwise. Let $\zeta \in]0,1[$ such that the conditions of Lemma 5.1.9 are fulfilled. Then for constants $\tau_a := \inf\{f_{\zeta,q^*,\mu}(q_a) - \ell_{\zeta,\pi^*,\mu,\lambda}(q_a) \mid \underline{q}_a \le q_a \le \overline{q}_a\}$ for each arc $a \in A'$ the inequality

$$\sum_{a\in A'}\tau_a \le \qquad (5.1.21)$$

$$\zeta\left(\sum_{v\in V}(\Delta_v^-\overline{\pi}_v' - \Delta_v^+\underline{\pi}_v') + \sum_{a\in A'}(\Delta_a^-\Phi_a'(\overline{q}_a) - \Delta_a^+\Phi_a'(\underline{q}_a)) - \sum_{v\in V} d_v\pi_v'(\pi^*)\right)$$
$$+(1-\zeta)\left(\sum_{v\in V}\left(\lambda_v^+\overline{\pi}_v - \lambda_v^-\underline{\pi}_v\right) + \sum_{a\in A'}\left(\lambda_a^+\overline{q}_a - \lambda_a^-\underline{q}_a\right) - \sum_{v\in V} d_v\mu_v\right)$$

is violated if and only if the passive transmission problem (4.1.1) is infeasible. The violation is greater than or equal to $(1-\zeta)$ times the optimal objective value of the relaxations (4.2.1) or (4.3.1) respectively.

Proof. This follows from Lemma 5.1.7, Lemma 5.1.12 and Lemma 5.1.13. □

We note that it is not guaranteed that ζ as assumed in Corollary 5.1.14 exists. If it does not exist, then inequality (5.1.21) cannot be set up.

5.1.4 A Linear Inequality derived from the Lagrange Function of the Domain Relaxation

We show that inequality (5.1.21) of the previous Corollary 5.1.14 with $\zeta = 0$ is motivated by the Lagrange function (5.1.2) of the domain relaxation (5.1.1). However, a restriction of the flow directions of the passive transmission problem is necessary for the validity of the inequality.

Lemma 5.1.15:
Let $L(q, \pi, \Delta, \mu, \lambda)$ be the Lagrange function (5.1.2) *of the nonlinear domain relaxation* (5.1.1). *Let $(q^*, \pi^*, \Delta^*, \mu^*, \lambda^*)$ be a KKT point of this relaxation. Assume that the flow directions of the passive transmission problem* (4.1.1) *are restricted by $q_a\mu_a^* \geq 0$ for each arc $a \in A'$, i.e., $\underline{q}_a \geq 0$ if $\mu_a^* \geq 0$ and $\overline{q}_a \leq 0$ if $\mu_a^* \leq 0$.*

If primal and dual arc flows have the same direction, i.e., $q_a^*\mu_a^* \geq 0$ holds for each arc $a \in A'$, and $\underline{q}_a \leq q_a \leq \overline{q}_a, a \in A'$ then the inequality

$$\inf_{q,\pi,\Delta} L(q, \pi, \Delta, \mu^*, \lambda^*) \leq 0$$

is equal to inequality (5.1.21) *of Corollary 5.1.14 with $\zeta = 0$ and the KKT point $(q^*, \pi^*, \Delta^*, \mu^*, \lambda^*)$.*

Proof. Let $(q^*, \pi^*, \Delta^*, \mu^*, \lambda^*)$ be a KKT point of the domain relaxation (5.1.1). We write the Lagrange function (5.1.2):

$$\begin{aligned}
L(q, \pi, \Delta, \mu^*, \lambda^*) = &\sum_{v \in V} \Delta_v + \sum_{a \in A'} \Delta_a \\
&+ \sum_{\substack{a \in A' \\ a=(v,w)}} \mu_a^* \left(\Phi_a(q_a) - (\pi_v - \gamma_a \pi_w)\right) \\
&+ \sum_{v \in V} \mu_v^* \left(d_v - \sum_{a \in \delta^+_{A'}(v)} q_a + \sum_{a \in \delta^-_{A'}(v)} q_a \right) \\
&+ \sum_{v \in V} \left(\lambda_v^{+*} \left(\pi_v - \Delta_v - \overline{\pi}_v\right) + \lambda_v^{-*} \left(\underline{\pi}_v - \pi_v - \Delta_v\right)\right) \\
&+ \sum_{a \in A'} \left(\lambda_a^{+*} \left(q_a - \Delta_a - \overline{q}_a\right) + \lambda_a^{-*} \left(\underline{q}_a - q_a - \Delta_a\right)\right)
\end{aligned}$$

$$-\sum_{v\in V}\lambda_v^*\Delta_v-\sum_{a\in A'}\lambda_a^*\Delta_a.$$

By Lemma 4.2.11 the dual solution of the KKT point forms a general network flow in (V,A'), i.e.,

$$\sum_{a\in\delta^+_{A'}(v)}\mu_a^*-\sum_{a\in\delta^-_{A'}(v)}\mu_a^*\gamma_a=\lambda_v^{+*}-\lambda_v^{-*}\qquad\forall v\in V.$$

This implies

$$\sum_{\substack{a\in A'\\a=(v,w)}}\mu_a^*\,(\pi_v-\gamma_a\pi_w)=\sum_{v\in V}(\lambda_v^{+*}\,\pi_v-\lambda_v^{-*}\,\pi_v).$$

This means that the Lagrange function $L(q,\pi,\Delta,\mu^*,\lambda^*)$ is independent of the values of π. Recall the conditions (4.2.7c) and (4.2.7d) from the proof of Lemma 4.2.11 which state that the dual solution (μ^*,λ^*) also fulfills

$$\lambda_v^{+*}+\lambda_v^{-*}+\lambda_v{}^*=1\quad\forall v\in V,$$
$$\lambda_a^{+*}+\lambda_a^{-*}+\lambda_a{}^*=1\quad\forall a\in A'.$$

This implies

$$\lambda_v^{+*}\Delta_v+\lambda_v^{-*}\Delta_v+\lambda_v{}^*\Delta_v=\Delta_v\quad\forall v\in V,$$
$$\lambda_a^{+*}\Delta_a+\lambda_a^{-*}\Delta_a+\lambda_a{}^*\Delta_a=\Delta_a\quad\forall a\in A'.$$

Again we conclude that the Lagrange function $L(q,\pi,\Delta,\mu^*,\lambda^*)$ is independent of the values of Δ. From these observations we conclude that $L(q,\pi,\Delta,\mu^*,\lambda^*)$ writes as follows:

$$\begin{aligned}L(q,\pi,\Delta,\mu^*,\lambda^*)&=\sum_{a\in A'}\mu_a^*\,\Phi_a(q_a)\\&\quad+\sum_{v\in V}\mu_v^*\left(d_v-\sum_{a\in\delta^+_{A'}(v)}q_a+\sum_{a\in\delta^-_{A'}(v)}q_a\right)\\&\quad-\sum_{v\in V}\left(\lambda_v^{+*}\,\overline{\pi}_v-\lambda_v^{-*}\,\underline{\pi}_v\right)\\&\quad+\sum_{a\in A'}\left(\lambda_a^{+*}\,(q_a-\overline{q}_a)+\lambda_a^{-*}\,(\underline{q}_a-q_a)\right).\end{aligned}$$

We rearrange this as

$$\begin{aligned} L(q,\pi,\Delta,\mu^*,\lambda^*) = &\sum_{a\in A'} \left(\mu_a^*\,\Phi_a(q_a) - (\mu_v^* - \mu_w^* - \lambda_a^{+*} + \lambda_a^{-*})\,q_a\right) \\ &+ \sum_{v\in V} \mu_v^* d_v \\ &- \sum_{v\in V} \left(\lambda_v^{+*}\,\overline{\pi}_v + \lambda_v^{-*}\,\underline{\pi}_v\right) \\ &- \sum_{a\in A'} \left(\lambda_a^{+*}\,\overline{q}_a - \lambda_a^{-*}\,\underline{q}_a\right). \end{aligned}$$

We derive a dual transmission flow (μ,λ) from the KKT point $(q^*,\pi^*,\Delta^*,\mu^*,\lambda^*)$ and define τ_a for each arc $a\in A'$ by

$$\tau_a := \inf\left\{\mu_a\,\Phi_a(q_a) - (\mu_v - \mu_w - \lambda_a^+ + \lambda_a^-)\,q_a \mid \underline{q}_a \le q_a \le \overline{q}_a, \mu_a q_a \ge 0\right\}.$$

Then it follows from (5.1.3a) of the KKT conditions and $\mu_a^* q_a^* \ge 0$ and $\underline{q}_a \le q_a \le \overline{q}_a$ that

$$\tau_a = \mu_a^*\,\Phi_a(q_a^*) - (\mu_v^* - \mu_w^* - \lambda_a^{+*} + \lambda_a^{-*})\,q_a^*$$

holds. Then we have:

$$\begin{gathered} \inf_{\underline{q}\le q\le\overline{q},\underline{\pi}\le\pi\le\overline{\pi},\Delta\ge 0} L(q,\pi,\Delta,\mu^*,\lambda^*) \le 0 \\ \Leftrightarrow \\ \sum_{a\in A'} \tau_a \le \sum_{v\in V}\left(\lambda_v^+\,\overline{\pi}_v + \lambda_v^-\,\underline{\pi}_v\right) + \sum_{a\in A'}\left(\lambda_a^+\,\overline{q}_a - \lambda_a^-\,\underline{q}_a\right) - \sum_{v\in V}\mu_v d_v. \end{gathered} \tag{5.1.22}$$

This inequality equals inequality (5.1.21) of Corollary 5.1.14 for $\zeta = 0$. □

5.2 A Valid Inequality for the Topology Optimization Problem

Corollary 5.1.14 states the main result of the previous Section 5.1. The corollary states that inequality (5.1.21) is violated if and only if the passive transmission problem is infeasible. This inequality contains no variables and is valid for the passive transmission problem (4.1.1). In this section we extend this inequality such that it is valid for the topology optimization problem (3.2.1), and not only for the passive transmission problem. Recall that the passive transmission problem is obtained by fixing all binary variables x as we focus on $\underline{y} = \overline{y}$ in this chapter. Hence x and y are the variables of our extended inequality.

We recall the definition of the functions $f_{\zeta,q^*,\mu}$ and $\ell_{\zeta,\pi^*,\mu,\lambda}$ and extend them to the more general case of a flow value $q_{a,i}$ for an arc $(a,i) \in A_X, i > 0$ of the extended graph:

$$\tilde{f}_{\zeta,q^*,\mu}(q_{a,i}, y_{a,i}) := (\zeta\, \gamma_{r,v}(q_{a,i} - q_a^*) + (1-\zeta)\mu_a)\,(\alpha_{a,i}\, q_{a,i}|q_{a,i}|^{k_a} - \beta_{a,i} y_{a,i})$$
$$\tilde{\ell}_{\zeta,\pi^*,\mu,\lambda}(q_{a,i}) := \zeta\,(\pi_v'(\pi^*) - \pi_w'(\pi^*))\, q_{a,i} + (1-\zeta)(\mu_v - \mu_w - \lambda_a^+ + \lambda_a^-)\, q_{a,i}$$

For abbreviations we set, similarly as in (5.1.6),

$$\Phi_{a,i}(q,y) := \alpha_{a,i}\, q|q|^{k_a} - \beta_{a,i}\, y \quad \text{ and } \quad \Phi_{a,i}'(q,y) := \gamma_{r,v}\Phi_{a,i}(q,y)$$

In the following we extend Lemma 5.1.7.

Theorem 5.2.1:
Let (μ,λ) be a dual transmission flow. Let $\pi^* \in \mathbb{R}^V$, $q^* \in \mathbb{R}^A$, $\Delta_v^\pm \geq 0, v \in V$ and $\Delta_a^\pm \geq 0, a \in A$ be vectors such that the flow conservation

$$\sum_{a \in \delta_A^+(v)} (q_a^* - (\Delta_a^+ - \Delta_a^-)) - \sum_{a \in \delta_A^-(v)} (q_a^* - (\Delta_a^+ - \Delta_a^-)) - (\Delta_v^+ - \Delta_v^-) = d_v$$

is fulfilled for each node $v \in V$. Furthermore let $\zeta \in [0,1]$. Then for constants $\tau_{a,i}(y_{a,i}) := \inf\{\tilde{f}_{\zeta,q^*,\mu}(q_{a,i}, y_{a,i}) - \tilde{\ell}_{\zeta,\pi^*,\mu,\lambda}(q_{a,i}) \mid \underline{q}_{a,i} \leq q_{a,i} \leq \overline{q}_{a,i}\}$ for each arc $(a,i) \in A_X, i \neq 0$ and each value $y_{a,i}$ the inequality in binary and continuous but fixed variables x and y

$$\sum_{\substack{(a,i)\in A_X\\ i\neq 0}} x_{a,i}\, \tau_{a,i}(y_{a,i}) \leq -\zeta \sum_{v\in V} d_v \pi_v'(\pi^*) \tag{5.2.1}$$
$$+\zeta\left(\sum_{v\in V}(\Delta_v^- \overline{\pi}_v' - \Delta_v^+ \underline{\pi}_v') + \sum_{\substack{(a,i)\in A_X\\ i\neq 0}} x_{a,i}(\Delta_a^- \Phi_{a,i}'(\overline{q}_{a,i}, y_{a,i}) - \Delta_a^+ \Phi_{a,i}'(\underline{q}_{a,i}, y_{a,i}))\right)$$
$$+(1-\zeta)\left(\sum_{v\in V}\left(\lambda_v^+ \overline{\pi}_v - \lambda_v^- \underline{\pi}_v\right) + \sum_{\substack{(a,i)\in A_X\\ i\neq 0}} x_{a,i}\left(\lambda_a^+ \overline{q}_{a,i} - \lambda_a^- \underline{q}_{a,i}\right) - \sum_{v\in V} d_v \mu_v\right)$$
$$+\zeta \sum_{a=(v,w)\in A} x_{a,0} \max\left\{(q_a^* - (\Delta_a^+ - \Delta_a^-))(\overline{\pi}_v' - \underline{\pi}_w'), (q_a^* - (\Delta_a^+ - \Delta_a^-))(\underline{\pi}_v' - \overline{\pi}_w')\right\}$$

$$+(1-\zeta) \sum_{a=(v,w)\in A} x_{a,0} \max\left\{\mu_a(\overline{\pi}_v - \underline{\pi}_w), \mu_a(\underline{\pi}_v - \overline{\pi}_w)\right\}$$

is valid for the topology optimization problem (3.2.1).

Remark 5.2.2:
Note that the basic difference of this inequality (5.2.1)*, which is valid for the topology optimization problem* (3.2.1)*, and inequality* (5.1.8)*, which is valid for the passive transmission problem* (4.1.1)*, are the last two lines of* (5.2.1)*. The other parts of these inequalities almost coincide except the addition of the binary variables x and the continuous but fixed variables y.*

Proof. We are going to apply Lemma 5.1.7 that yields a valid inequality for the passive transmission problem corresponding to the arc set A'. We write $A' = A'(x)$ depending on the binary vector x by $A'(x) := \{(a,i) \in A_X \mid x_{a,i} = 1, i > 0\}$. In order to apply Lemma 5.1.7 we define Δ variables in dependence on $A'(x)$ by considering the following equality which is valid for each active configuration x:

$$d_v = \sum_{a\in\delta^+_A(v)} x_{a,0}(q^*_a - (\Delta^+_a - \Delta^-_a)) - \sum_{a\in\delta^-_A(v)} x_{a,0}(q^*_a - (\Delta^+_a - \Delta^-_a))$$
$$+ \sum_{(a,i)\in\delta^+_{A'(x)}(v)} (q^*_a - (\Delta^+_a - \Delta^-_a)) - \sum_{(a,i)\in\delta^-_{A'(x)}(v)} (q^*_a - (\Delta^+_a - \Delta^-_a)) - (\Delta^+_v - \Delta^-_v).$$

We define values $\tilde{\Delta}^+_v(x) \geq 0$ and $\tilde{\Delta}^-_v(x) \geq 0$ for each node $v \in V$ where at least one of both values equals zero by

$$\tilde{\Delta}^-_v(x) := \max\left\{0, \sum_{a\in\delta^+_A(v)} x_{a,0}\left(q^*_a - (\Delta^+_a - \Delta^-_a)\right) - \sum_{a\in\delta^-_A(v)} x_{a,0}\left(q^*_a - (\Delta^+_a - \Delta^-_a)\right)\right\},$$
$$\tilde{\Delta}^+_v(x) := \max\left\{0, \sum_{a\in\delta^-_A(v)} x_{a,0}\left(q^*_a - (\Delta^+_a - \Delta^-_a)\right) - \sum_{a\in\delta^+_A(v)} x_{a,0}\left(q^*_a - (\Delta^+_a - \Delta^-_a)\right)\right\}.$$

From this we obtain

$$\sum_{(a,i)\in\delta^+_{A'(x)}(v)} (q^*_a - (\Delta^+_a - \Delta^-_a)) - \sum_{(a,i)\in\delta^-_{A'(x)}(v)} (q^*_a - (\Delta^+_a - \Delta^-_a))$$
$$-(\tilde{\Delta}^+_v(x) + \Delta^+_v) + (\tilde{\Delta}^-_v(x) + \Delta^-_v) = d_v \tag{5.2.2}$$

for every node $v \in V$. We proceed analogously for μ in order to obtain a dual transmission flow for $A'(x)$. For each active configuration x it holds

$$\sum_{a\in\delta_A^+(v)} x_{a,0}\,\mu_a - \sum_{a\in\delta_A^-(v)} x_{a,0}\,\mu_a + \sum_{(a,i)\in\delta^+_{A'(x)}(v)} \mu_a - \sum_{(a,i)\in\delta^-_{A'(x)}(v)} \mu_a = \lambda_v^+ - \lambda_v^-$$

for every node $v \in V$. We define values $\tilde{\lambda}_v^+(x) \geq 0$ and $\tilde{\lambda}_v^-(x) \geq 0$ for each node $v \in V$ with at least one of both values being equal to zero by

$$\tilde{\lambda}_v^-(x) := \max\left\{0, \sum_{a\in\delta_A^+(v)} x_{a,0}\,\mu_a - \sum_{a\in\delta_A^-(v)} x_{a,0}\,\mu_a\right\}$$

$$\tilde{\lambda}_v^+(x) := \max\left\{0, \sum_{a\in\delta_A^-(v)} x_{a,0}\,\mu_a - \sum_{a\in\delta_A^+(v)} x_{a,0}\,\mu_a\right\}$$

and obtain

$$\sum_{(a,i)\in\delta^+_{A'(x)}(v)} \mu_a - \sum_{(a,i)\in\delta^-_{A'(x)}(v)} \mu_a = \left(\tilde{\lambda}_v^+(x) + \lambda_v^+\right) - \left(\tilde{\lambda}_v^-(x) - \lambda_v^-\right). \tag{5.2.3}$$

We take the definition of the constant $\tau_{a,i}$ for each arc $(a,i) = (v,w,i) \in A_X, i \neq 0$ based on Lemma 5.1.7 as follows:

$$\tau_{a,i}(y_{a,i}) := \inf\{\tilde{f}_{\zeta,q^*,\mu}(q_{a,i}, y_{a,i}) - \tilde{\ell}_{\zeta,\pi^*,\mu,\lambda}(q_{a,i}) \mid \underline{q}_{a,i} \leq q_{a,i} \leq \overline{q}_{a,i}\}$$

Now assume a fixation of the binary vector x and recall that y is fixed. Inequality (5.1.8) is valid for the corresponding passive transmission problem by Lemma 5.1.8. Using equation (5.2.2) and (5.2.3) the following inequality is equal to (5.1.8) and hence valid for the topology optimization problem (3.2.1):

$$\sum_{(a,i)\in A'(x)} \tau_{a,i}(y_{a,i}) \leq$$

$$\zeta\left(\sum_{v\in V}((\Delta_v^- + \tilde{\Delta}_v^-(x))\overline{\pi}'_v - (\Delta_v^+ + \tilde{\Delta}_v^+(x))\underline{\pi}'_v)\right)$$

$$+\zeta\left(\sum_{(a,i)\in A'(x)} (\Delta_a^-\Phi'_{a,i}(\overline{q}_{a,i}, y_{a,i}) - \Delta_a^+\Phi'_{a,i}(\underline{q}_{a,i}, y_{a,i})) - \sum_{v\in V} d_v \pi'_v(\pi^*)\right)$$

$$+(1-\zeta)\left(\sum_{v\in V}\left(\left(\lambda_v^+ + \tilde{\lambda}_v^+(x)\right)\overline{\pi}_v - \left(\lambda_v^- + \tilde{\lambda}_v^-(x)\right)\underline{\pi}_v\right)\right)$$
$$+(1-\zeta)\left(\sum_{(a,i)\in A'(x)}\left(\lambda_a^+\overline{q}_{a,i} - \lambda_a^-\underline{q}_{a,i}\right) - \sum_{v\in V} d_v\mu_v\right).$$

We use

$$\sum_{(a,i)\in A'(x)} \tau_{a,i}(y_{a,i}) = \sum_{\substack{(a,i)\in A_X\\ i\neq 0}} x_{a,i}\tau_{a,i}(y_{a,i}).$$

We complete the proof by rewriting and estimating the right-hand side. We write

$$\sum_{(a,i)\in A'(x)}\left(\lambda_a^+\overline{q}_{a,i} - \lambda_a^-\underline{q}_{a,i}\right) = \sum_{\substack{(a,i)\in A_X\\ i\neq 0}} x_{a,i}\left(\lambda_a^+\overline{q}_{a,i} - \lambda_a^-\underline{q}_{a,i}\right)$$

and

$$\sum_{(a,i)\in A'}(\Delta_a^-\Phi'_{a,i}(\overline{q}_{a,i}) - \Delta_a^+\Phi'_{a,i}(\underline{q}_{a,i})) = \sum_{\substack{(a,i)\in A_X\\ i\neq 0}} x_{a,i}(\Delta_a^-\Phi'_{a,i}(\overline{q}_{a,i}) - \Delta_a^+\Phi'_{a,i}(\underline{q}_{a,i})).$$

Using the previous definitions of $\tilde{\Delta}_v^+(x)$ and $\tilde{\Delta}_v^-(x)$ and setting

$$\tilde{\pi}_v := \begin{cases} \overline{\pi}'_v & \text{if } \tilde{\Delta}_v^-(x) - \tilde{\Delta}_v^+(x) \geq 0,\\ \underline{\pi}'_v & \text{else,}\end{cases}$$

we obtain

$$\sum_{v\in V}\left(\tilde{\Delta}_v^-(x)\overline{\pi}'_v - \tilde{\Delta}_v^+(x)\underline{\pi}'_v\right) = \sum_{v\in V}\left(\tilde{\Delta}_v^-(x) - \tilde{\Delta}_v^+(x)\right)\tilde{\pi}_v$$
$$= \sum_{a=(v,w)\in A} x_{a,0}(q_a^* - (\Delta_a^+ - \Delta_a^-))(\tilde{\pi}_v - \tilde{\pi}_w) \leq$$
$$\sum_{a=(v,w)\in A} x_{a,0}\max\left\{(q_a^* - (\Delta_a^+ - \Delta_a^-))(\overline{\pi}'_v - \underline{\pi}'_w), (q_a^* - (\Delta_a^+ - \Delta_a^-))(\underline{\pi}'_v - \overline{\pi}'_w)\right\}.$$

Similarly, using the definition of $\lambda_v^+(x)$ and $\lambda_v^-(x)$ we derive the estimation

$$\sum_{v\in V}\left(\tilde{\lambda}_v^+(x)\overline{\pi}_v - \tilde{\lambda}_v^-(x)\underline{\pi}_v\right) \leq$$
$$\sum_{a=(v,w)\in A} x_{a,0}\max\left\{\mu_a(\overline{\pi}_v - \underline{\pi}_w), \mu_a(\underline{\pi}_v - \overline{\pi}_w)\right\}.$$

□

Now we turn to the domain relaxation (5.1.1) and explain how the parameters of inequality (5.2.1) have to be set such that (5.2.1) represents the feasibility of a passive transmission problem. We derive the dual transmission flow (μ, λ) from a KKT point of relaxation (5.1.1) and apply Lemma 5.1.12. The binary and continuous values x^* and y^* which correspond to this relaxation are then infeasible for the linear inequality (5.2.1) if and only if the corresponding passive transmission problem is infeasible.

Theorem 5.2.3:
Let $(q^*, \pi^*, \Delta^*, \mu^*, \lambda^*)$ be a KKT point of the domain relaxation (5.1.1) for arc set A' and let (μ, λ) be a dual transmission flow derived from this KKT point. Denote by x^* and y^* the binary and continuous values which yield the domain relaxation (5.1.1). Let $\zeta \in]0,1[$ as mentioned in Lemma 5.1.9. Then for constants $\tau_{a,i} = \tau_{a,i}(y_{a,i})$ defined by $\tau_{a,i}(y_{a,i}) := \inf\{f_{\zeta,q^*,\mu}(q_{a,i}, y_{a,i}) - \ell_{\zeta,\pi^*,\mu,\lambda}(q_{a,i}) \mid \underline{q}_{a,i} \leq q_{a,i} \leq \overline{q}_{a,i}\}$ for each arc $(a,i) \in A_X, i \neq 0$ and the value $y_{a,i} = y^*_{a,i}$ the inequality in binary variables x

$$\begin{aligned}
\sum_{\substack{(a,i)\in A_X \\ i\neq 0}} x_{a,i}\, \tau_{a,i}(y_{a,i}) \leq &-\zeta \sum_{v\in V} d_v \pi'_v(\pi^*) \qquad (5.2.4)\\
&+(1-\zeta)\left(\sum_{v\in V}\left(\lambda^+_v \overline{\pi}_v - \lambda^-_v \underline{\pi}_v\right) + \sum_{\substack{(a,i)\in A_X \\ i\neq 0}} x_{a,i}\left(\lambda^+_a \overline{q}_{a,i} - \lambda^-_a \underline{q}_{a,i}\right) - \sum_{v\in V} d_v \mu_v\right)\\
&+\zeta \sum_{a=(v,w)\in A'} x_{a,0} \max\left\{q^*_a(\overline{\pi}'_v - \underline{\pi}'_w), q^*_a(\underline{\pi}'_v - \overline{\pi}'_w)\right\}\\
&+(1-\zeta) \sum_{a=(v,w)\in A} x_{a,0} \max\left\{\mu_a(\overline{\pi}_v - \underline{\pi}_w), \mu_a(\underline{\pi}_v - \overline{\pi}_w)\right\}
\end{aligned}$$

is valid for the topology optimization problem (3.2.1). This inequality cuts off the passive transmission problem corresponding to the arc set A' if and only if it is infeasible. For the corresponding decision vector $x = x^*$ and $y = y^* = \underline{y} = \overline{y}$ the violation of inequality (5.2.4) is greater than or equal to $(1-\zeta)$ times the optimal objective value of the domain relaxation.

Proof. We define $q^*_a := 0$ for all arcs $a \in A \setminus A'$ and obtain

$$\sum_{a\in\delta^+_A(v)} q^*_a - \sum_{a\in\delta^-_A(v)} q^*_a = d_v.$$

Furthermore we set $\Delta^+_v := \Delta^-_v := 0$ for all $v \in V$ and $\Delta^+_a := \Delta^-_a := 0$ for all $a \in A$. Now the validity of (5.2.4) as a globally valid inequality for the topology optimization problem (3.2.1) follows from Theorem 5.2.1.

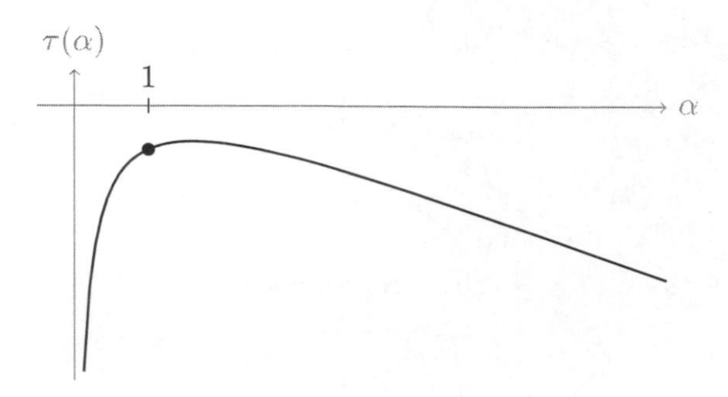

Figure 5.2: Visualization of $\tau(\alpha)$ defined by (5.2.5) for $q_a^* = 5, \mu_a = 1, \gamma_a = 1, \pi_v^* - \pi_w^* = 25, \mu_v - \mu_w = 10, \lambda_a^+ - \lambda_a^- = 0, k_a = 1, \zeta = 0.7, \gamma_{r,v} = 1, \tilde{\beta} = 0$. The depicted point corresponds to the original value $\alpha = 1$ of the passive transmission problem (4.1.1). The function $\tau(\alpha)$ corresponds to the term $\sum_{(a,i)\in A_X, i\neq 0} x_{a,i}\, \tau_{a,i}(y_{a,i})$ for a fixed arc $a \in A$ in (5.2.4), a valid inequality for the topology optimization problem (3.2.1) by Theorem 5.2.3.

If x and y are the binary and continuous values, respectively, that correspond to the passive transmission problem of the relaxation (5.1.1), for arc set A', i.e., $x = x^*$ and $y = y^*$, then (5.2.4) can be rewritten as (5.1.21). So the theorem follows from the special choice of ζ and Corollary 5.1.14. □

Inequality (5.2.4) forms our improved Benders cut because of the properties stated in Theorem 5.2.3. A similar result as in Theorem 5.2.3 can be obtained for the flow conservation relaxation (4.3.1) but we do not give any further details here. As discussed in Section 4.6 the flow conservation relaxation turned out to be less efficient than the domain relaxation (4.2.1). That is why we decided to use the domain relaxation in our computations.

Theorem 5.2.3 states an inequality which represents the infeasibility of a certain passive transmission problem. More precisely, if it is infeasible, then the inequality is violated for the specific values of the binary and continuous variables x and y which lead to the passive transmission problem. Let us now concentrate on the violation. Therefor we visualize $\tau_{a,i}$ for an arc $a = (v, w) \in A$. Note that different binary values for x correspond to different values $\tau_{a,i}$ in inequality (5.2.4). The value of $\tau_{a,i}$ depends on $\alpha_{a,i}$ and $y_{a,i}$. Thus we consider $\tau_{a,i}(y_{a,i})$ as a function $\tau(\alpha)$ for fixed $y_{a,i}$ defined by

$$\begin{aligned}\tau(\alpha) := \min_{q\in\mathbb{R}} \big\{ &(\zeta\, \gamma_{r,v}(q - q_a^*) + (1-\zeta)\mu_a)\,(\alpha\, q|q|^{k_a} - \tilde{\beta}) \\ &- \zeta\, \gamma_{r,v}(\pi_v^* - \gamma_a \pi_w^*)\, q - (1-\zeta)(\mu_v - \mu_w - \lambda_a^+ + \lambda_a^-)\, q \big\}.\end{aligned} \tag{5.2.5}$$

A visualization of this function is shown in Figure 5.2. From this image we conclude, that either increasing or decreasing the diameter of a pipe reduces the violation of inequality (5.2.1) of the previous Theorem 5.2.1. However, this is not coherent with the technical properties of a pipeline: decreasing the diameter of a pipe means to reduce the capacity of the network which, in general, means to increase the infeasibility of the passive transmission problem.

5.3 Integration and Computational Results

In this chapter we focused on the topology optimization problem (3.2.1) arising from the second type of network that we consider in this thesis. Recall that these networks consist only of pipes, loops and valves. In this case it holds $\underline{y} = \overline{y}$ for our model (3.2.1). Let us present our solution framework for the topology optimization problem. We solve the model (3.2.1) by SCIP as described in Section 2.2. Additionally we generate the cut (5.2.4) which is valid for the topology optimization problem by Theorem 5.2.3 as follows: Consider a node of the branching tree where the optimal solution of the LP relaxation yields integral values for the binary variables. If this is the case, then we consider the corresponding passive transmission problem (4.1.1). Since this problem might be infeasible, we use the domain relaxation (5.1.1) to solve it. Recall that this relaxation has better solving performance than the flow conservation relaxation (4.3.1) as discussed in the previous Chapter 4. Moreover it follows from Lemma 4.2.10 that the domain relaxation is feasible and convex. After solving this relaxation to global optimality by IPOPT, one of the following two cases occurs:

1. In the first case, the optimal solution has a zero objective function value. We derive a feasible solution for the passive transmission problem and add it to the solution pool of the solver and continue with the branching process.

2. In the second case, the optimal solution has a positive objective function value. This means that some slack variables are nonzero. In this case we use the KKT point computed by IPOPT, i.e., primal and dual solution values, and try to derive inequality (5.2.4) where the parameters are chosen as described in Theorem 5.2.3. Note that a KKT point exists by Lemma 4.2.11. It might be impossible to generate inequality (5.2.4) if there does not exist $\zeta \in]0,1[$ fulfilling the conditions of Lemma 5.1.9. Otherwise, if there exists such a value ζ, then we take a minimal ζ and generate inequality (5.2.4), which is valid for the topology optimization problem (3.2.1) by Theorem 5.2.3. Furthermore it is violated for the current choice of binary values. We integrate the inequality in a cut pool of the solver SCIP. If all binary variables are fixed, then we prune the current node of the branch-and-bound tree and continue with the branching process. This is feasible due to the convexity of the domain relaxation by Lemma 4.2.10. Otherwise we continue without pruning.

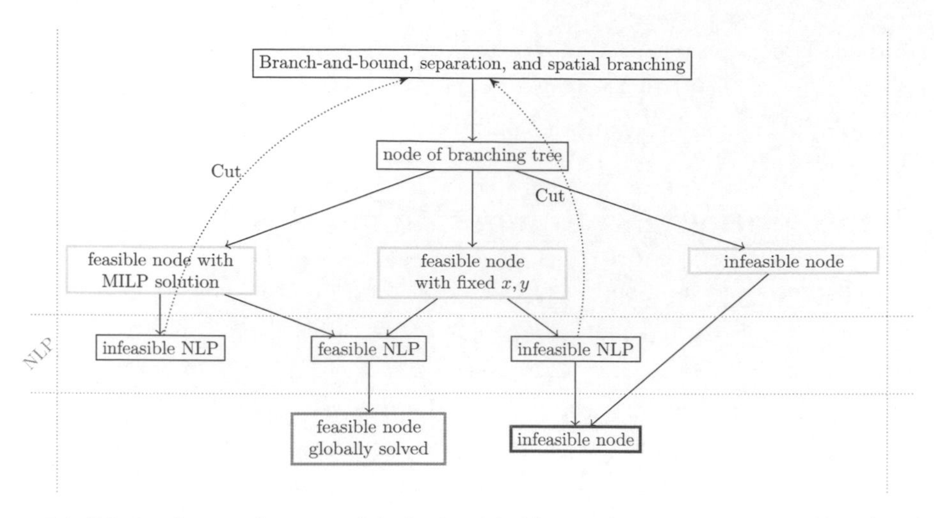

Figure 5.3: Solution framework presented in Section 5.3. The topology optimization problem (3.2.1) is solved with SCIP essentially by branch-and-bound, separation, and spatial branching, see Section 2.2. We adapt this framework and solve globally the passive transmission problem (4.1.1) as discussed in Section 4.5, classify the current node of the branching tree and prune it if possible. If the current passive transmission problem is infeasible, then we generate (as described in Section 5.3) a linear inequality in the binary and continuous but fixed variables x and y which is dynamically added to the topology optimization problem. This inequality is violated for the binary values which lead to the considered passive transmission problem.

We conclude that the described method is an extension of the solution method presented in Chapter 4 based on the domain relaxation.

The algorithmic scheme of this solution approach is shown in Figure 5.3. This figure includes the computation of inequalities compared to Figure 4.5. Recall that Figure 4.5 shows the use of the relaxations (4.2.1) and (4.3.1) within the branch-and-bound tree for solving the topology optimization problem (3.2.1).

We implemented the cut generation algorithm in C. In our initial implementation we added all obtained cuts directly to the branch-and-bound process. It turned out that this strategy was not very efficient. In the final implementation we do not add them immediately. They are instead stored in a cut pool until a predefined number of inequalities is reached (experimentally, a pool size of 40 inequalities turned out to be a good value). Then we restart the branch-and-bound solution process and multiply the cut pool size by 1.5. Additionally, we also restart if a new primal feasible solution is found with a better objective function value compared to the current best solution. For the restart, only the best feasible solution and the valid inequalities are kept. When restarting the solver SCIP the cuts are provided from the very beginning of the solution process. Hence they are available during presolve and SCIP might further strengthen our cuts (in combination with all other model inequalities).

We compare five different strategies for solving the topology optimization problem:

1. The first strategy is to use SCIP for solving our model (3.2.1) and to enforce a certain branching priority rule, so that SCIP first branches on binary decision variables x. Only after all discrete variables are fixed it is allowed to perform spatial branching on continuous variables.

2. The second strategy is similar to the first strategy. Additionally, whenever a node of the branching tree has fixed binary variables, and a global solution for the corresponding domain relaxation (5.1.1) is computed by IPOPT, then we prune the current node. Note that this strategy equals strategy 3 from Section 4.6.

3. The third strategy implements the cut generation as described above, but restarts are not enforced. Additionally we set branching priorities according to the first strategy and prune nodes of the branching tree as in 2.

4. The fourth strategy implements the cut generation together with restarts as described above. Additionally we set branching priorities according to the first strategy. Further, we prune nodes of the branching tree as in 2.

5. The fifth strategy is equal to the fourth one, but the added cut simply forbids the current binary vector. For the current integral solution values x^* this inequality is

$$\sum_{\substack{(a,i)\in A_X \\ x^*_{a,i}=0}} x_{a,i} + \sum_{\substack{(a,i)\in A_X \\ x^*_{a,i}=1}} (1 - x_{a,i}) \geq 1.$$

   Hence this fifth strategy basically represents the technique of the previous Chapter 4 using the domain relaxation (5.1.1) implemented by strategy 3 from Section 4.6 and storing the information about infeasible passive transmission problems when restarting.

Computational Results

For the computational study we consider the transport networks net1 and net2. For net2 we contract all compressors and control valves. These networks have different topologies and consist of pipelines only. All pipes (a, i) have constants $\alpha_{a,i} > 0$, $\beta_{a,i} = 0$ and $k_{a,i} = 1$. A visualization of these networks is shown in Figure 3.5 and Figure 3.6. All extensions are parallel arcs (or loops) as described in the next paragraph. Therefore they are not visible in the pictures. The dimensions of the networks are given in Table 3.1. The considered networks belong to the second type of network which we consider in this thesis. They only contain pipelines, loops and valves.

strategy	1	2	3	4	5	all
solved instances	53	53	54	64	51	64

Table 5.1: Summary of the Tables A.6 – A.9 showing the globally solved instances out of 82 nominations in total. All instances solved by strategy 2 are also solved by strategy 3.

	(A,B) = (1,2)		
	solved(52)		incomp.(28)
	time [s]	nodes	gap [%]
strategy A	161.6	44,146	113
strategy B	174.3	38,194	114
shifted geom. mean	+8 %	−13 %	+1 %

Table 5.2: Run time, number of branch-and-bound nodes and gap comparison for the strategies 1 and 2 (aggregated results). The columns "solved" contain mean values for those instances globally solved by both strategies A and B. The columns "incomplete" show mean values for those instances having a primal feasible solution available but were not globally solved by both strategies A and B. The underlying data are available in Tables A.6 – A.9.

For each network we are given a balanced flow demand at the entry and the exit nodes. There exists a feasible flow in the network for this given demand. Now we scale up this demand, that is, we multiply each entry and exit value by the same scalar > 1. For a certain value 2.0 the instance is no longer feasible, i.e., there is no valid flow which fulfills all model constraints. In order to obtain a feasible flow again, the network capacity needs to be extended, for which we introduced a number of parallel arcs (loops). For the network `net1` we consider different instances where we allow to build up to 7 loops, up to 8 loops, continuing up to 11 loops for each arc of the network. For the instance `net2` we similarly allow between 2 and 4 loops, respectively. That is, each original pipeline can be extended at most by this number of pipelines having the same characteristics as the original one.

We set a time limit of 39 600 s and used the computational setup described in Section 3.5. The computational results are shown in Tables A.6 – A.9. Especially the number of inequalities (5.2.4) that are generated and added to the topology optimization problem is available in these tables. A summary is given in Tables 5.1 – 5.3. We use the geometric mean of run time, number of branch-and-bound nodes and gap as described in Section 3.5.

The first and second strategy almost show the same solving performance in terms of number of solved instances. We remark that the number of nodes is reduced by 13 %, while the solving time increases by 8 %, see Table 5.2. We conclude that the domain relaxation presented in Chapter 4 allows to reduce the necessary number of branch-and-bound nodes, but the computation time increases. The first and the third

	(A,B) = (1,3)			(A,B) = (1,4)		
	solved(52)		incomp.(27)	solved(53)		incomp.(18)
	time [s]	nodes	gap [%]	time [s]	nodes	gap [%]
strategy A	161.6	44,146	116	180.0	49,219	159
strategy B	175.5	31,661	102	120.0	8,681	31
shifted geom. mean	+9 %	−28 %	−12 %	−33 %	−82 %	−81 %

Table 5.3: Run time, number of branch-and-bound nodes and gap comparison for the strategies 1 and 3 and additionally 1 and 4 (aggregated results). The columns "solved" contain mean values for those instances globally solved by both strategies A and B. The columns "incomplete" show mean values for those instances having a primal feasible solution available but were not globally solved by both strategies A and B. The underlying data are available in Tables A.6 – A.9.

strategy also show almost the same solving performance in terms of number of solved instances. The number of nodes is reduced by 28 %, while the solving time increases by 9 %, see Table 5.3. Hence the inequalities presented in this chapter allow to reduce the number of branch-and-bound nodes. But we conclude that the computation time being saved due to the reduction of branch-and-bound nodes is used to generate the inequalities. This is different when restarts are allowed as implemented by strategy 4. A comparison of the fourth and the first strategy shows that additionally 13 % more instances of the test set (11 out of 82) are solved to global optimality within the time limit of 39 600 s. Furthermore the run time is reduced by 33 %, the number of nodes by 82 % and the gap by 81 % for those instances which remain with a positive gap value following both strategies. Again the number of nodes decreases more than the run time. This is due to the time which is necessary for solving the domain relaxation and generating the inequalities.

Figure 5.4 shows an aggregated performance plot of the five strategies. The fifth strategy clearly shows the worst results. It does not even improve the first strategy. The graphs of the third and fourth strategy are coherent with the previous discussion. The graph of the fourth strategy dominates the other strategies for those instances that have a solution time of more than 200 s following strategy 4. This order is different for the other instances. Here the first strategy dominates. This solving behavior can be clearly attributed to the restarts that we apply whenever a primal solution is found. Figure 5.5 shows a coherent result. It is a scatterplot of the run time comparing the second and the fourth strategy. Finally Figure 5.6 shows a scatterplot of the gap comparing both strategies. The gap is reduced for every instance.

We do not generate the cut for the first type of network which we consider in this thesis. Recall that they consist of pipes and valves only. In this case the binary variables are associated with valves only. If an inequality (5.2.4) was generated for a

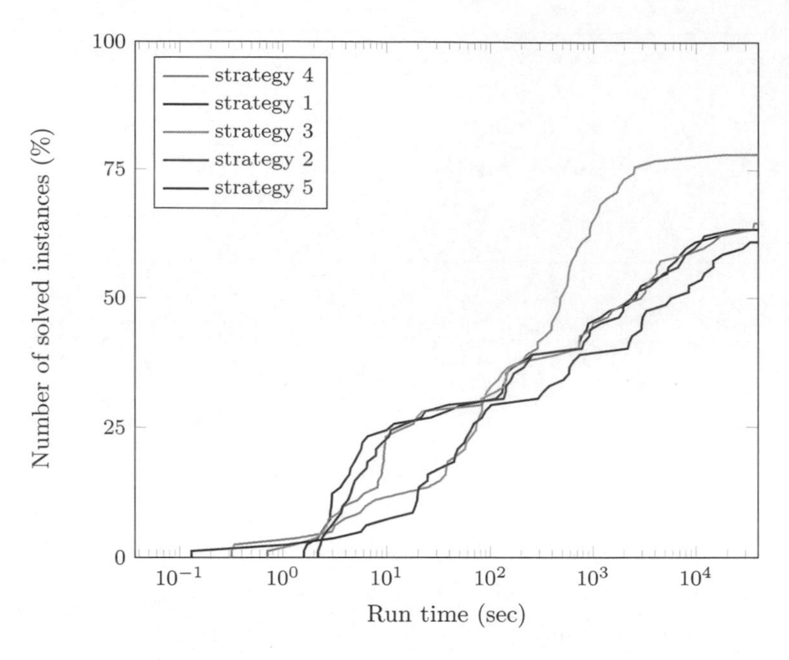

Figure 5.4: Performance plot for different nominations on the networks net1 and net2 (aggregated) and a time limit of 39 600 s. The different strategies are described in Section 5.3. Strategy 1 mainly consist of SCIP. Strategies 2 and 3 also correspond to SCIP in combination with our solution methods presented in this chapter, i.e., generating the cuts (5.2.4) and adding them to the problem formulation. Both strategies only differ in the restarting policy. The underlying data are available in Tables A.6 – A.9.

closed valve a, then it would be easy to see that the inequality might get inactive when a is opened. This is due to a possible large coefficient of $x_{a,1}$. The same observation holds for an opened valve getting closed. In this case it is due to a possible large coefficient of $x_{a,0}$. Both cases imply that the inequality typically has a similar characteristic as the cut used in strategy 5.

Summary

In this chapter we presented a valid inequality for the topology optimization problem (3.2.1) restricted to the case $\underline{y} = \overline{y}$. This case includes the second type of network of our test instances which we consider in this thesis, namely those which contain pipes, loops and valves only. A first step for the definition of the cut was the inequality of Corollary 5.1.6 being a linear combination of two different nonlinear inequalities. In Section 5.1.4 we explained that the first one can be obtained from the Lagrange function of the domain relaxation (4.2.1). The second one is the so-called *pc-regularization* which was necessary to derive a globally valid cut. In a second step we gave a linear underestimator in Section 5.1.2 which then led to an inequality without any variables representing the infeasibility of a passive transmission problem,

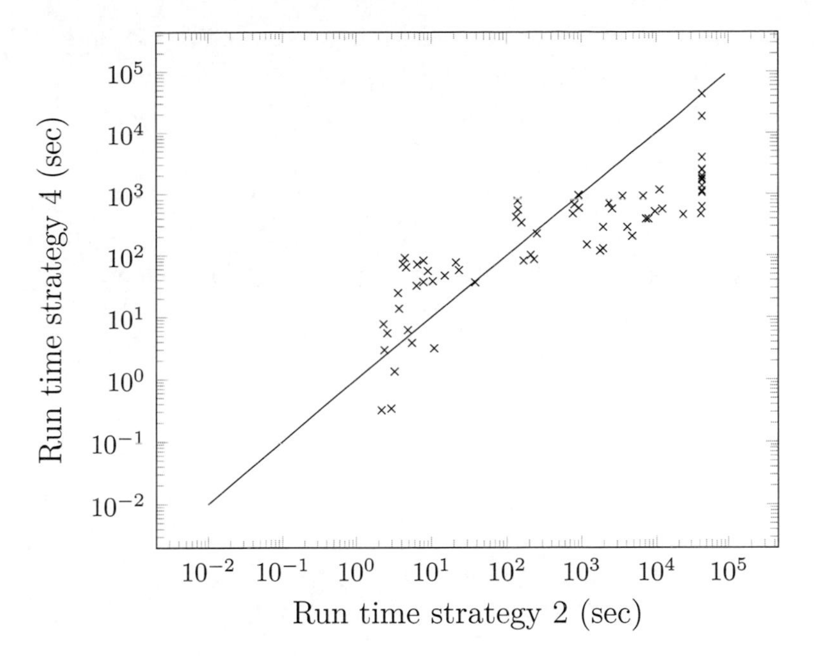

Figure 5.5: Run time comparison for different nominations on the networks `net1` and `net2`. Each cross (×) corresponds to a single instance of the test set. Note that multiple crosses are drawn in the upper right corner of the plots that cannot be differentiated. They represent those instances that ran into the time limit of 39 600 s for SCIP without and with using cuts. The underlying data are available in Tables A.6 – A.9.

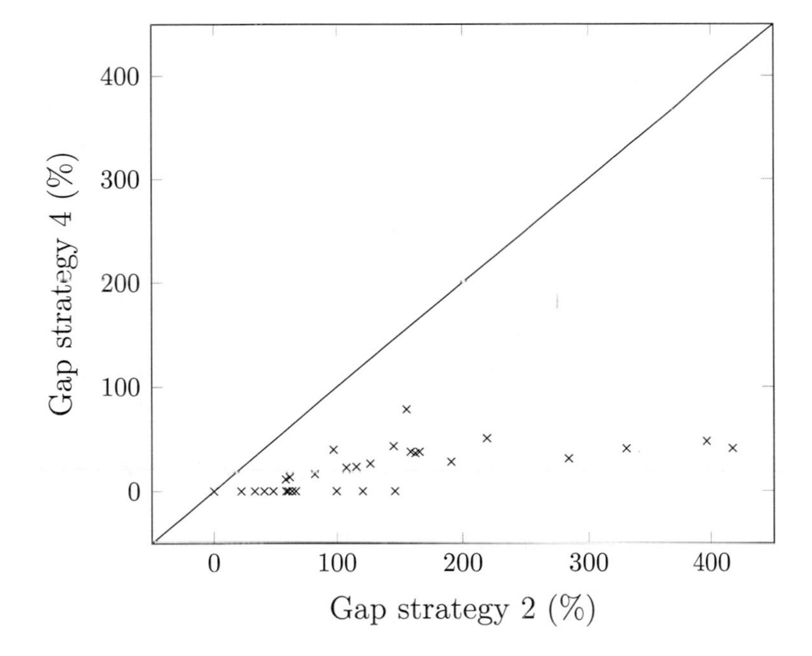

Figure 5.6: Gap comparison for different nominations on the networks `net1` and `net2`. Each cross (×) corresponds to a single instance of the test set. Those instances where at least one gap value is infinity are not depicted here. The underlying data are available in Tables A.6 – A.9.

see Section 5.1.3. In Section 5.2 we explained how to extend this inequality to a valid cut for the topology optimization problem. Adding the inequality to the problem is reminiscent of a Benders decomposition approach as the inequality represents the infeasibility of a certain passive transmission problem. That is why we call the inequality an improved Benders cut. From the computational results we conclude that the separation of the new inequalities (5.2.4) leads to significant smaller branch-and-bound trees and thus lower overall running times. On average the run time is reduced by 33 % in comparison to SCIP. In addition the test set is solved to global optimality in 13 % more instances.

Chapter 6

Sufficient Conditions for Infeasibility of the Active Transmission Problem

In the previous Chapters 4 and 5 we focused on networks without compressor and control valves. We considered the topology optimization problem (3.2.1) in the case $\underline{y} = \overline{y}$. Especially in Chapter 4 we presented different relaxations of the passive transmission problem (4.1.1). They allow to solve (4.1.1) efficiently. The passive transmission problem occurs at nodes of the branching tree of the topology optimization problem being solved by SCIP after branching on discrete variables. Time consuming spatial branching can be reduced when solving these nodes by applying the relaxations. Hence, the relaxations allow to solve the topology optimization problem efficiently. In this chapter we turn to the more general case of networks containing compressors and control valves, i.e., $\underline{y} \neq \overline{y}$. They belong to the third type of network which we consider in this thesis. We present an algorithm which allows to proceed in a similar fashion as in the restricted case of $\underline{y} = \overline{y}$, i.e., we solve subproblems of the topology optimization problem (3.2.1) by a specialized method.

The subproblem considered in this chapter is the active transmission problem (3.4.2). As previously mentioned this problem arises from the topology optimization problem after fixing all binary variables. Our strategy for solving the active transmission problem is as follows: We relax it in a similar way as done for the passive transmission problem (4.1.1) as described in Chapter 4. Then we compute a locally optimal solution for the upcoming relaxation by IPOPT. Now this may provide a solution that is also feasible for the active transmission problem. In this case we solved it to global optimality. However, if it does not yield a feasible solution, then we try to verify that there exists no solution for the active transmission problem. Therefor we present conditions for the infeasibility of the active transmission problem.

In the end, our approach consists of a primal heuristic together with infeasibility conditions of the active transmission problem.

We include this solution method into the branch-and-bound tree which is used for solving the topology optimization problem (3.2.1). Whenever an active transmission problem arises, we use the heuristic to solve it. If the primal heuristic yields a feasible solution for the topology optimization problem, then we add this solution to the solution pool of the solver. If we can prove that the active transmission problem is infeasible, then we prune the corresponding node of the branch-and-bound tree. This allows to cut off nodes from the branch-and-bound tree. This in turn leads to an increase of 20 % for globally solved topology expansion instances on the third type of network considered in this thesis.

The outline of this chapter is as follows: In Section 6.1 we present two relaxations of the active transmission problem which are obtained in a similar way as the domain relaxation (4.2.1) and the flow conservation relaxation (4.3.1). We prove that both relaxations are non-convex nonlinear optimization problems. In Section 6.2 we present an MILP which represents the conditions of infeasibility of the active transmission problem. Its definition depends on a local optimum of the relaxations. In Section 6.3 we present our initial motivation for the definition of the MILP. Computational results are given in Section 6.4.

6.1 Non-Convex Relaxations for the Active Transmission Problem

The active transmission problem considered in this chapter is obtained from the topology optimization problem (3.2.1) by fixing all binary variables x. Recall that y is a vector of real variables, hence fixing x means to concentrate on the subproblem of (3.2.1) where all discrete decisions are fixed. Let $A' := \{(a,i) \in A_X : x_{a,i} = 1, i > 0\}$ denote the set of arcs so that the flow is not fixed to zero. The arising problem is as follows:

$$\exists\, q, \pi, p, y \tag{6.1.1a}$$

$$\text{s.t.} \quad \alpha_a\, q_a |q_a|^{k_a} - \beta_a y_a - (\pi_v - \gamma_a \pi_w) = 0 \quad \forall\, a = (v,w) \in A', \tag{6.1.1b}$$

$$\sum_{a \in \delta^+_{A'}(v)} q_a - \sum_{a \in \delta^-_{A'}(v)} q_a = d_v \quad \forall\, v \in V, \tag{6.1.1c}$$

$$A_a\,(q_a,p_v,p_w)^T \leq b_a \quad \forall\, a=(v,w)\in A', \tag{6.1.1d}$$

$$p_v|p_v| - \pi_v = 0 \quad \forall\, v\in V, \tag{6.1.1e}$$

$$\pi_v \leq \overline{\pi}_v \quad \forall\, v\in V, \tag{6.1.1f}$$

$$\pi_v \geq \underline{\pi}_v \quad \forall\, v\in V, \tag{6.1.1g}$$

$$q_a \leq \overline{q}_a \quad \forall\, a\in A', \tag{6.1.1h}$$

$$q_a \geq \underline{q}_a \quad \forall\, a\in A', \tag{6.1.1i}$$

$$y_a \leq \overline{y}_a \quad \forall\, a\in A', \tag{6.1.1j}$$

$$y_a \geq \underline{y}_a \quad \forall\, a\in A', \tag{6.1.1k}$$

$$p_v,\pi_v \in \mathbb{R} \quad \forall\, v\in V, \tag{6.1.1l}$$

$$q_a,y_a \in \mathbb{R} \quad \forall\, a\in A'. \tag{6.1.1m}$$

In order to solve the active transmission problem (6.1.1) we proceed in a similar way as for the passive transmission problem (4.1.1) described in Chapter 4. For this we consider two different relaxations which are presented in Section 6.1.1 and Section 6.1.2. Unfortunately it turns out that both relaxations are non-convex optimization problems. So far we are not aware of a convex one. In the following we assume that the graph (V,A') is connected because otherwise we consider an active transmission problem for every connected component separately.

6.1.1 Relaxation of Domains

The first relaxation we consider is similar to the domain relaxation (4.2.1) of the passive transmission problem. It turns out to be a non-convex optimization problems and writes as follows:

$$\min \sum_{v\in V} \Delta_v + \sum_{a\in A'} (\Delta_a + \|\Delta'_a\|) \tag{6.1.2a}$$

$$\text{s.t.}\quad \alpha_a\, q_a|q_a|^{k_a} - \beta_a y_a - (\pi_v - \gamma_a \pi_w) = 0 \quad \forall\, a=(v,w)\in A', \tag{6.1.2b}$$

$$\sum_{a\in\delta^+_{A'}(v)} q_a - \sum_{a\in\delta^-_{A'}(v)} q_a = d_v \quad \forall\, v\in V, \tag{6.1.2c}$$

$$A_a\,(q_a,p_v,p_w)^T - \Delta'_a \leq b_a \quad \forall\, a=(v,w)\in A', \tag{6.1.2d}$$

$$p_v|p_v| - \pi_v = 0 \qquad \forall v \in V, \tag{6.1.2e}$$

$$\pi_v - \Delta_v \leq \overline{\pi}_v \qquad \forall v \in V, \tag{6.1.2f}$$

$$\pi_v + \Delta_v \geq \underline{\pi}_v \qquad \forall v \in V, \tag{6.1.2g}$$

$$q_a - \Delta_a \leq \overline{q}_a \qquad \forall a \in A', \tag{6.1.2h}$$

$$q_a + \Delta_a \geq \underline{q}_a \qquad \forall a \in A', \tag{6.1.2i}$$

$$y_a \leq \overline{y}_a \qquad \forall a \in A', \tag{6.1.2j}$$

$$y_a \geq \underline{y}_a \qquad \forall a \in A', \tag{6.1.2k}$$

$$p_v, \pi_v \in \mathbb{R} \qquad \forall v \in V, \tag{6.1.2l}$$

$$q_a, y_a \in \mathbb{R} \qquad \forall a \in A', \tag{6.1.2m}$$

$$\Delta_v \in \mathbb{R}_{\geq 0} \qquad \forall v \in V, \tag{6.1.2n}$$

$$\Delta_a \in \mathbb{R}_{\geq 0} \qquad \forall a \in A', \tag{6.1.2o}$$

$$\Delta'_a \in \mathbb{R}_{\geq 0}^{\nu_a} \qquad \forall a \in A'. \tag{6.1.2p}$$

Lemma 6.1.1:
The optimization problem (6.1.2) *is a feasible relaxation of the active transmission problem* (6.1.1).

Proof. Every feasible solution of the active transmission problem (6.1.1) can be extended to a feasible solution of (6.1.2) by considering $\Delta := 0$ as slack values. Hence (6.1.2) is a relaxation of the active transmission problem (6.1.1).

When fixing y and neglecting constraints (6.1.2d) and (6.1.2e) the optimization problem (6.1.2) is equal to the domain relaxation (4.2.1). Contracting those arcs $a = (v, w) \in A'$ that are modeled by $\pi_v = \pi_w$ yields a modified version of the domain relaxation with $\alpha > 0$. This problem is always feasible by Theorem 4.2.9. Transforming a solution to the original problem and removing the fixation of y keeps the feasibility. By adding the neglected constraints (6.1.2e) the feasibility is kept again because there are no bounds on the pressure variables p. The same holds for constraints (6.1.2d) because the values Δ'_a can be chosen such that (6.1.2d) is valid for any values q_a, p_v, p_w for each arc $a = (v, w) \in A'$. □

Let us consider a local optimal solution of this problem (6.1.2). In the case that the objective value of this solution equals zero, we derive an optimal solution for the active transmission problem by neglecting the slack variables Δ. Otherwise,

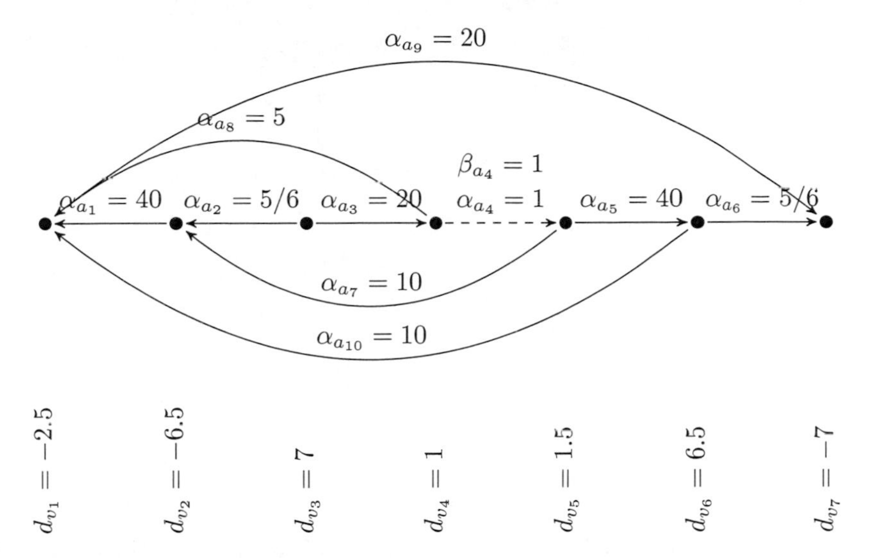

Figure 6.1: An example of a planar network. It is used in the discussion in Example 6.1.2 for showing that the domain relaxation (6.1.2) is a non-convex optimization problem. The dashed arc is a control valve with constraint $\alpha_{a_4} q_{a_4}|q_{a_4}| - y_{a_4} = \pi_{v_4} - \pi_{v_5}$ and $-50 \leq y_{a_4} \leq 0$. All the other arcs $a = (v, w) \in A'$ are passive pipes, i.e., $\alpha_a q_a |q_a| = \pi_v - \pi_w$. All primal node potential values are bounded by 0 and 100, except v_3 and v_7, where upper and lower bound are $\overline{\pi}_{v_7} = 29$ and $\underline{\pi}_{v_3} = 91$, respectively. The absolute value of the arc flow is bounded by 100.

i.e., if the objective value is greater than zero, then it is not guaranteed that the local optimal solution is a global one. Unfortunately it turns out, as discussed in Example 6.1.2, that the relaxation (6.1.2) is a non-convex optimization problem. So we cannot conclude that the active transmission problem (6.1.1) is infeasible in this case.

Example 6.1.2:
The domain relaxation (6.1.2) *for the network shown in Figure 6.1 has two different KKT points. The network is planar and consists of 7 nodes and 10 arcs (9 pipes and 1 active control valve). The parameters of the active transmission problem are available from Figure 6.1. Those that are not shown are defined as follows:*

$$\beta_a := \begin{cases} 1 & \textit{if } a = a_4, \\ 0 & \textit{else} \end{cases} \qquad \forall\, a \in A',$$

$$\gamma_a := 1 \qquad \forall\, a \in A',$$

$$k_a := 1 \qquad \forall\, a \in A',$$

$$\overline{\pi}_v := \begin{cases} 29 & \textit{if } v = v_7, \\ 100 & \textit{else} \end{cases} \qquad \forall\, v \in V,$$

$$\underline{\pi}_v := \begin{cases} 91 & \textit{if } v = v_3, \\ 0 & \textit{else} \end{cases} \qquad \forall\, v \in V,$$

$$\overline{q}_a := 100 \qquad \forall\, a \in A',$$

$$\underline{q}_a := -100 \qquad \forall\, a \in A',$$

$$\overline{y}_a := 0 \qquad \forall\, a \in A',$$

$$\underline{y}_a := \begin{cases} -50 & \textit{if } a = a_4, \\ 0 & \textit{else} \end{cases} \qquad \forall\, a \in A'.$$

There are no additional constraints for the arcs, i.e., $A_a = 0, b_a = 0$ *for every arc* $a \in A'$. *Hence we can neglect the coupling constraints* (6.1.1e) *and the pressure variables* p *and constraints* (6.1.2d).

In this context the domain relaxation (6.1.2) *is as follows:*

$$\min \sum_{v \in V} \Delta_v + \sum_{a \in A'} \Delta_a \tag{6.1.3}$$

$$\begin{aligned}
\text{s.t.} \quad & \sum_{a \in \delta^+_{A'}(v)} q_a - \sum_{a \in \delta^-_{A'}(v)} q_a = d_v && \forall\, v \in V, \\
& \alpha_a\, q_a |q_a| - (\pi_v - \pi_w) = 0 && \forall\, a = (v,w) \in A', a \neq a_4, \\
& \alpha_{a_4}\, q_{a_4} |q_{a_4}| - y_{a_4} - (\pi_{v_4} - \pi_{v_5}) = 0, && \\
& q_a - \Delta_a \leq 100 && \forall\, a \in A', \\
& q_a + \Delta_a \geq -100 && \forall\, a \in A', \\
& \pi_v - \Delta_v \leq 100, && \forall\, v \in V, v \neq v_7 \\
& \pi_v + \Delta_v \geq 0, && \forall\, v \in V, v \neq v_3 \\
& \pi_{v_7} - \Delta_{v_7} \leq 29, && \\
& \pi_{v_3} + \Delta_{v_3} \geq 91, && \\
& 0 \leq y_a \leq 0, && \forall\, a \in A', a \neq a_4
\end{aligned}$$

$$
\begin{aligned}
-50 \le y_{a_4} &\le 0, \\
\pi_v &\in \mathbb{R} && \forall\, v \in V, \\
q_a, y_a &\in \mathbb{R} && \forall\, a \in A', \\
\Delta_v &\in \mathbb{R}_{\ge 0} && \forall\, v \in V, \\
\Delta_a &\in \mathbb{R}_{\ge 0} && \forall\, a \in A'.
\end{aligned}
$$

In order to show that there exist different KKT points of (6.1.3) *we consider the KKT conditions for this problem. The objective and all constraints of* (6.1.3) *are continuously differentiable. We write the Lagrange function of problem* (6.1.3) *in order to derive the KKT conditions for* (6.1.3)*:*

$$
\begin{aligned}
L(q, \pi, y, \Delta, \mu, \lambda) = & \sum_{v \in V} \Delta_v + \sum_{a \in A'} \Delta_a \\
& + \sum_{a=(v,w) \in A'} \mu_a \left(\alpha_a\, q_a |q_a|^{k_a} - \beta_a y_a - (\pi_v - \pi_w) \right) \\
& + \sum_{v \in V} \mu_v \left(d_v - \sum_{a \in \delta^+_{A'}(v)} q_a + \sum_{a \in \delta^-_{A'}(v)} q_a \right) \\
& + \sum_{v \in V} \left(\lambda^+_v \left(\pi_v - \Delta_v - \overline{\pi}_v \right) + \lambda^-_v \left(\underline{\pi}_v - \pi_v - \Delta_v \right) \right) \\
& + \sum_{a \in A'} \left(\tilde{\lambda}^+_a \left(q_a - \Delta_a - \overline{q}_a \right) + \tilde{\lambda}^-_a \left(\underline{q}_a - q_a - \Delta_a \right) \right) \\
& + \sum_{a \in A'} \left(\lambda^+_a \left(y_a - \overline{y}_a \right) + \lambda^-_a \left(\underline{y}_a - y_a \right) \right) \\
& - \sum_{v \in V} \lambda_v \Delta_v - \sum_{a \in A'} \tilde{\lambda}_a \Delta_a.
\end{aligned}
$$

We obtain from the complementarity constraints (2.4.2a) *of the KKT conditions* (2.4.2) *of the domain relaxation* (6.1.2) *that a KKT point* $(q^*, \pi^*, y^*, \Delta^*, \mu^*, \lambda^*)$ *of* (6.1.3) *is feasible for*

$$
\begin{aligned}
\frac{\partial L}{\partial q_a} = 0 \Rightarrow\ & \mu_a\, 2\alpha_a |q_a| + \tilde{\lambda}^+_a - \tilde{\lambda}^-_a = \mu_v - \mu_w && \forall\, a = (v, w) \in A', \\
\frac{\partial L}{\partial \pi_v} = 0 \Rightarrow\ & \sum_{a \in \delta^+_{A'}(v)} \mu_a - \sum_{a \in \delta^-_{A'}(v)} \mu_a = \lambda^+_v - \lambda^-_v && \forall\, v \in V,
\end{aligned}
$$

	a_1	a_2	a_3	a_4	a_5	a_6	a_7	a_8	a_9	a_{10}
q_a	0.5	6	1	0	0.5	6	1	2	1	1
μ_a	$-1/3$	$-2/3$	$-1/3$	0	$-1/3$	$-2/3$	$1/3$	$-1/3$	$-1/3$	$1/3$
$\tilde{\lambda}_a$	1	1	1	1	1	1	1	1	1	1
	v_1	v_2	v_3	v_4	v_5	v_6	v_7			
π_v	50	60	90	70	70	60	30			
Δ_v	0	0	1	0	0	0	1			
μ_v	80/3	40/3	20/3	60/3	60/3	100/3	120/3			
λ_v^+	0	0	0	0	0	0	1			
λ_v^-	0	0	1	0	0	0	0			
λ_v	1	1	0	1	1	1	0			

Table 6.1: A KKT point of the domain relaxation (6.1.2) with positive objective value. Those variables not depicted have zero values, i.e., $y_a, \Delta_a, \lambda_a^{\pm}, \tilde{\lambda}_a^{\pm} = 0, a \in A$. The KKT point is used for a discussion in Example 6.1.2.

	a_1	a_2	a_3	a_4	a_5	a_6	a_7	a_8	a_9	a_{10}
q_a	0.34	6.38	0.61	-0.87	0.16	5.91	0.45	2.49	1.08	0.75
y_a	0	0	0	-25	0	0	0	0	0	0
$\tilde{\lambda}_a$	1	1	1	1	1	1	1	1	1	1
	v_1	v_2	v_3	v_4	v_5	v_6	v_7			
π_v	52.46	57.13	91.08	83.47	59.23	58.11	28.95			
λ_v	1	1	1	1	1	1	1			

Table 6.2: A KKT point of the domain relaxation (6.1.2) with zero objective value. All numbers are depicted up to a precision of 2 digits. Those variables not depicted have zero values, i.e., $\mu_a, \Delta_a, \lambda_a^{\pm}, \tilde{\lambda}_a^{\pm} = 0, a \in A$ and $\Delta_v, \mu_v, \lambda_v^{\pm} = 0, v \in V$. The KKT point is used for a discussion in Example 6.1.2.

$$\begin{aligned}
\frac{\partial L}{\partial \Delta_v} &= 0 \Rightarrow \quad \lambda_v^+ + \lambda_v^- + \lambda_v = 1 \quad \forall v \in V, \\
\frac{\partial L}{\partial \Delta_a} &= 0 \Rightarrow \quad \tilde{\lambda}_a^+ + \tilde{\lambda}_a^- + \tilde{\lambda}_a = 1 \quad \forall a \in A', \\
\frac{\partial L}{\partial y_a} &= 0 \Rightarrow \lambda_a^+ - \lambda_a^- - \beta_a \mu_a = 0 \quad \forall a \in A'.
\end{aligned} \tag{6.1.4}$$

Similar to Lemma 4.2.11 and the discussion in Section 4.2.3 we observe that the Lagrange multipliers $(\mu_a^*)_{a \in A'}$ *of a KKT point* $(q^*, \pi^*, y^*, \Delta^*, \mu^*, \lambda^*)$ *of* (6.1.3) *form a network flow in* (V, A').

Now it is easy to see that the solution values shown in Table 6.1 and Table 6.2 fulfill the KKT conditions (2.4.2) *of* (6.1.3), *i.e., they are feasible for* (6.1.3), (6.1.4) *and the complementarity constraints* (2.4.2e), *and hence are KKT points of the domain relaxation* (6.1.3)*. The primal part shown in Table 6.2 has zero objective value while this is not the case for the primal part shown in Table 6.1. A convex combination of both primal feasible solutions is not feasible for the relaxation* (6.1.3) *because of constraint* (6.1.2b)*. We conclude that* (6.1.2) *is a non-convex optimization problem. As both solutions have different objective values this shows that a nonlinear solver like* IPOPT*, which computes KKT points, cannot guarantee to compute the global optimal solution of* (6.1.2)*.*

6.1.2 Relaxation of Flow Conservation Constraints

Unfortunately the domain relaxation (6.1.3) presented in Section 6.1.1 is a non-convex optimization problem. Now we try to find another relaxation of the active transmission problem (6.1.1). Therefor we extend the flow conservation relaxation (4.3.1) presented in Chapter 4 to the more general case of this chapter. Again it turns out that the obtained relaxation is non-convex. It is as follows:

$$\min \sum_{v\in V}\left(\Delta_v^+ + \Delta_v^-\right) + \sum_{a\in A'}\left(\Delta_a^+ + \Delta_a^- + \|\Delta_a'\|\right) \tag{6.1.5}$$

$$\begin{aligned}
\text{s. t.} \quad & \alpha_a\, q_a|q_a|^{k_a} - \beta_a y_a - (\pi_v - \gamma_a \pi_w) = 0 && \forall\, a = (v,w) \in A', \\
& \sum_{a\in\delta^+_{A'}(v)} (q_a - (\Delta_a^+ - \Delta_a^-)) \\
& - \sum_{a\in\delta^-_{A'}(v)} (q_a - (\Delta_a^+ - \Delta_a^-)) - (\Delta_v^+ - \Delta_v^-) = d_v && \forall\, v \in V, \\
& A_a\,(q_{a,i}, p_v, p_w)^T - \Delta_a' \le b_a && \forall\, a = (v,w) \in A', \\
& p_v|p_v| - \pi_v = 0 && \forall\, v \in V, \\
& \pi_v \le \overline{\pi}_v && \forall\, v \in V, \\
& \pi_v \ge \underline{\pi}_v && \forall\, v \in V, \\
& q_a \le \overline{q}_a && \forall\, a \in A', \\
& q_a \ge \underline{q}_a && \forall\, a \in A', \\
& y_a \le \overline{y}_a && \forall\, a \in A', \\
& y_a \ge \underline{y}_a && \forall\, a \in A', \\
& \Delta_v^- \,(\overline{\pi}_v - \pi_v) \le 0 && \forall\, v \in V, \\
& \Delta_v^+ \,(\pi_v - \underline{\pi}_v) \le 0 && \forall\, v \in V, \\
& \Delta_a^- \,(\overline{q}_a - q_a) \le 0 && \forall\, a \in A', \\
& \Delta_a^+ \,(q_a - \underline{q}_a) \le 0 && \forall\, a \in A', \\
& p_v, \pi_v \in \mathbb{R} && \forall\, v \in V, \\
& q_a \in \mathbb{R} && \forall\, a \in A',
\end{aligned}$$

$$\Delta_v^{\pm} \in \mathbb{R}_{\geq 0} \quad \forall v \in V,$$

$$\Delta_a^{\pm} \in \mathbb{R}_{\geq 0} \quad \forall a \in A',$$

$$\Delta_a' \in \mathbb{R}_{\geq 0}^{\nu_a} \quad \forall a \in A'.$$

Lemma 6.1.3:
The optimization problem (6.1.5) *is a relaxation of the active transmission problem* (6.1.1).

Proof. Every feasible solution of the active transmission problem (6.1.1) yields a feasible solution of (6.1.5) in combination with $\Delta := 0$. □

In the next example we show that this relaxation is a non-convex optimization problem. Therefore, from a local optimal solution (6.1.5) with positive slack we cannot conclude that the active transmission problem (6.1.1) is infeasible.

Example 6.1.4:
We consider the network shown in Figure 6.2. It is similar to the example network in Example 6.1.2 but we consider a different nomination and different node potential bounds here. More precisely the differences are obtained by changing the node potential bounds in v_3 and v_7 to $\underline{\pi}_{v_3} = 90$ and $\overline{\pi}_{v_7} = 30$. Furthermore the considered nomination is adapted by setting $d_{v_3} := 7.1$ and $d_{v_7} := -7.1$. We neglect the coupling constraints $p_v|p_v| = \pi_v, v \in V$ and the pressure variables p. Recall that it holds $A_a = 0, b_a = 0$. Then relaxation (6.1.5) *writes as follows:*

$$\min \sum_{v \in V} (\Delta_v^+ + \Delta_v^-) + \sum_{a \in A'} (\Delta_a^+ + \Delta_a^-) \tag{6.1.6}$$

s. t.

$$\sum_{a \in \delta_{A'}^+(v)} (q_a - (\Delta_a^+ - \Delta_a^-)) - \sum_{a \in \delta_{A'}^-(v)} (q_a - (\Delta_a^+ - \Delta_a^-)) - (\Delta_v^+ - \Delta_v^-) = d_v \qquad \forall v \in V,$$

$$\alpha_a \, q_a|q_a| - (\pi_v - \pi_w) = 0 \qquad \forall a = (v,w) \in A', \; a \neq a_4,$$

$$\alpha_{a_4} \, q_{a_4}|q_{a_4}| - y_{a_4} - (\pi_{v_4} - \pi_{v_5}) = 0,$$

$$\pi_v \leq 100, \quad \forall v \in V, v \neq v_7$$

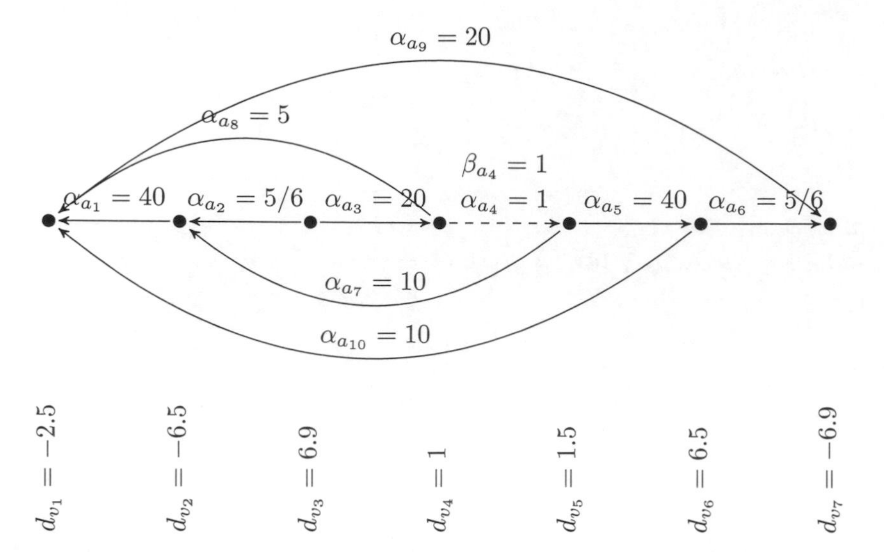

Figure 6.2: An example of a planar network. It is used in the discussion in Example 6.1.4 for showing that the flow conservation relaxation (6.1.5) is a non-convex optimization problem. The dashed arc is a control valve with constraint $\alpha_{a_4} q_{a_4}|q_{a_4}| - y_{a_4} = \pi_{v_4} - \pi_{v_5}$ and $-50 \leq y_{a_4} \leq 0$. All the other arcs $a = (v, w) \in A'$ are passive pipes, i.e., $\alpha_a q_a |q_a| = \pi_v - \pi_w$. All primal node potential values are bounded by 0 and 100, except v_3 and v_7, where upper and lower bound are $\overline{\pi}_{v_7} = 30$ and $\underline{\pi}_{v_3} = 90$, respectively. The absolute value of the arc flow is bounded by 100. The network is equal to the one shown in Figure 6.1, but the nomination and node potential bounds differ.

$$
\begin{aligned}
\pi_v &\geq 0, && \forall\, v \in V, v \neq v_3 \\
\pi_{v_3} &\geq 90, && \\
\pi_{v_7} &\leq 30, && \\
0 \leq y_a &\leq 0, && \forall\, a \in A', a \neq a_4 \\
-50 \leq y_{a_4} &\leq 0, && \\
\Delta^-_{v_7}\,(30 - \pi_{v_7}) &\leq 0, && \\
\Delta^+_{v_3}\,(\pi_{v_3} - 90) &\leq 0, && \\
\Delta^-_{v}\,(100 - \pi_v) &\leq 0 && \forall\, v \in V, v \neq v_7 \\
\Delta^+_{v}\,(\pi_v - 0) &\leq 0 && \forall\, v \in V, v \neq v_3 \\
\Delta^-_{a}\,(100 - q_a) &\leq 0 && \forall\, a \in A', \\
\Delta^+_{a}\,(q_a + 100) &\leq 0 && \forall\, a \in A', \\
\pi_v &\in \mathbb{R} && \forall\, v \in V,
\end{aligned}
$$

$$
\begin{aligned}
q_a, y_a &\in \mathbb{R} && \forall\, a \in A', \\
\Delta_v^+, \Delta_v^- &\in \mathbb{R}_{\geq 0} && \forall\, v \in V, \\
\Delta_a^+, \Delta_a^- &\in \mathbb{R}_{\geq 0} && \forall\, a \in A'.
\end{aligned}
$$

In order to show that there exist different KKT points of (6.1.6) *we consider the KKT conditions for this problem. The objective and all constraints of* (6.1.6) *are continuously differentiable. We write the Lagrange function of problem* (6.1.6) *in order to derive the KKT conditions for* (6.1.6)*:*

$$
\begin{aligned}
L(q, \pi, y, \Delta, \mu, \lambda) = &\sum_{v \in V} \left(\Delta_v^+ + \Delta_v^-\right) + \sum_{a \in A'} \left(\Delta_a^+ + \Delta_a^-\right) \\
&+ \sum_{a=(v,w) \in A'} \mu_a \left(\alpha_a\, q_a |q_a| - \beta_a y_a - (\pi_v - \pi_w)\right) \\
&+ \sum_{v \in V} \mu_v \Bigg(d_v + (\Delta_v^+ - \Delta_v^-) \\
&\qquad - \sum_{a \in \delta^+_{A'}(v)} (q_a - \Delta_a^+ + \Delta_a^-) \\
&\qquad + \sum_{a \in \delta^-_{A'}(v)} (q_a - \Delta_a^+ + \Delta_a^-) \Bigg) \\
&+ \sum_{v \in V} \left(\lambda_v^+ (\pi_v - \overline{\pi}_v) + \lambda_v^- (\underline{\pi}_v - \pi_v)\right) \\
&+ \sum_{a \in A'} \left(\lambda_a^+ (q_a - \overline{q}_a) + \lambda_a^- (\underline{q}_a - q_a)\right) \\
&+ \sum_{a \in A'} \left(\hat{\lambda}_a^+ (y_a - \overline{y}_a) + \hat{\lambda}_a^- (\underline{y}_a - y_a)\right) \\
&+ \sum_{v \in V} \left(\mu_v^+ (\overline{\pi}_v - \pi_v)\, \Delta_v^- + \mu_v^- (\pi_v - \underline{\pi}_v)\, \Delta_v^+\right) \\
&+ \sum_{a \in A'} \left(\mu_a^+ (\overline{q}_a - q_a)\, \Delta_a^- + \mu_a^- (q_a - \underline{q}_a)\, \Delta_a^+\right) \\
&- \sum_{v \in V} \left(\tilde{\lambda}_v^+ \Delta_v^+ + \tilde{\lambda}_v^- \Delta_v^-\right) - \sum_{a \in A'} \left(\tilde{\lambda}_a^+ \Delta_a^+ + \tilde{\lambda}_a^- \Delta_a^-\right).
\end{aligned}
$$

We obtain from (2.4.2a) *of the KKT conditions* (2.4.2) *of the flow conservation relaxation* (6.1.5) *that a KKT point* $(q^*, \pi^*, y^*, \Delta^*, \mu^*, \lambda^*)$ *of* (6.1.6) *is feasible for*

	a_1	a_2	a_3	a_4	a_5	a_6	a_7	a_8	a_9	a_{10}
q_a	0.5	6	1	0	0.5	6	1	2	1	1
μ_a	−1/50	−2/50	−1/50	0	−1/50	−2/50	1/50	−1/50	−1/50	1/50
$\tilde{\lambda}_a^+$	1/5	3/5	1/5	1	1/5	3/5	7/5	3/5	1/5	7/5
$\tilde{\lambda}_a^-$	9/5	7/5	9/5	1	9/5	7/5	3/5	7/5	9/5	3/5
	v_1	v_2	v_3	v_4	v_5	v_6	v_7			
π_v	50	60	90	70	70	60	30			
Δ_v^+	0	0	0.1	0	0	0	0			
Δ_v^-	0	0	0	0	0	0	0.1			
μ_v	1/5	−3/5	−1	−1/5	−1/5	3/5	1			
λ_v^+	0	0	0	0	0	0	3/50			
λ_v^-	0	0	3/50	0	0	0	0			
$\tilde{\lambda}_v^+$	6/5	2/5	0	4/5	4/5	8/5	2			
$\tilde{\lambda}_v^-$	4/5	8/5	2	6/5	6/5	2/5	0			

Table 6.3: A KKT point of the flow conservation relaxation (6.1.5) with positive objective value. Those variables not depicted have zero values, i.e., $y_a, \Delta_a^\pm, \lambda_a^\pm, \mu_a^\pm = 0, a \in A$ and $\mu_v^\pm = 0, v \in V$. The KKT point is used for a discussion in Example 6.1.4.

$$
\begin{aligned}
\frac{\partial L}{\partial q_a} = 0 \Rightarrow \quad & \mu_a 2\alpha_a |q_a| + \lambda_a^+ - \lambda_a^- = \mu_v - \mu_w && \forall\, a = (v,w) \in A', \\
\frac{\partial L}{\partial \pi_v} = 0 \Rightarrow \quad & -(\mu_v^+ \Delta_v^- - \mu_v^- \Delta_v^+) \\
& + \sum_{a \in \delta_{A'}^+(v)} \mu_a - \sum_{a \in \delta_{A'}^-(v)} \mu_a = \lambda_v^+ - \lambda_v^- && \forall\, v \in V, \\
\frac{\partial L}{\partial \Delta_v^+} = 0 \Rightarrow \quad & \mu_v + \mu_v^- (\pi_v - \underline{\pi}_v) - \tilde{\lambda}_v^+ = -1 && \forall\, v \in V, \\
\frac{\partial L}{\partial \Delta_v^-} = 0 \Rightarrow \quad & -\mu_v + \mu_v^+ (\overline{\pi}_v - \pi_v) - \tilde{\lambda}_v^- = -1 && \forall\, v \in V, \\
\frac{\partial L}{\partial \Delta_a^+} = 0 \Rightarrow \quad & \mu_v - \mu_w + \mu_a^- (q_a - \underline{q}_a) - \tilde{\lambda}_a^+ = -1 && \forall\, a = (v,w) \in A', \\
\frac{\partial L}{\partial \Delta_a^-} = 0 \Rightarrow \quad & \mu_w - \mu_v + \mu_a^+ (\overline{q}_a - q_a) - \tilde{\lambda}_a^- = -1 && \forall\, a = (v,w) \in A', \\
\frac{\partial L}{\partial y_a} = 0 \Rightarrow \quad & \hat{\lambda}_a^+ - \hat{\lambda}_a^- - \beta_a \mu_a = 0. && \forall\, a \in A'.
\end{aligned}
$$

Now it is easy to see that the solution values shown in Table 6.3 and Table 6.4 fulfill the KKT conditions (2.4.2)*, i.e., they form a feasible solution for* (6.1.6)*,* (2.4.2a) *as stated above for the considered problem and the complementarity constraints* (2.4.2e)*, and hence are KKT points of the flow conservation relaxation* (6.1.6)*. The primal part*

	a_1	a_2	a_3	a_4	a_5	a_6	a_7	a_8	a_9	a_{10}
q_a	0.18	7.11	-0.21	-2.20	-0.26	5.76	-0.43	2.98	1.13	0.46
y_a	0	0	0	-50	0	0	0	0	0	0
$\tilde{\lambda}_a^+$	1	1	1	1	1	1	1	1	1	1
$\tilde{\lambda}_a^-$	1	1	1	1	1	1	1	1	1	1
	v_1	v_2	v_3	v_4	v_5	v_6	v_7			
π_v	49.71	51.02	93.26	94.23	49.09	51.88	24.15			
$\tilde{\lambda}_v^+$	1	1	1	1	1	1	1			
$\tilde{\lambda}_v^-$	1	1	1	1	1	1	1			

Table 6.4: A KKT point of the flow conservation relaxation (6.1.5) with zero objective value. All numbers are depicted up to a precision of 2 digits. Those variables not depicted have zero values, i.e., $\mu_a, \Delta_a^\pm, \lambda_a^\pm, \mu_a^\pm = 0, a \in A$ and $\Delta_v^+, \mu_v^\pm, \mu_v, \lambda_v^\pm = 0, v \in V$. The KKT point is used for a discussion in Example 6.1.4.

shown in Table 6.4 has zero objective value while this is not the case for the primal part shown in Table 6.3. A convex combination of both primal feasible solutions is not feasible for the relaxation (6.1.6) *because of the constraints*

$$\alpha_a\, q_a |q_a|^{k_a} - \beta_a y_a - (\pi_v - \gamma_a \pi_w) = 0 \qquad \forall\, a = (v, w) \in A'.$$

We conclude that (6.1.5) *is a non-convex optimization problem. As both solutions have different objective values this shows that a nonlinear solver like* IPOPT*, which computes KKT points, cannot guarantee to compute the global optimal solution of* (6.1.5).

6.2 Detecting Infeasibility of the Active Transmission Problem by MILP

In the previous section we presented two relaxations (6.1.2) and (6.1.5) of the active transmission problem (6.1.1), which are non-convex optimization problems. Let us consider a feasible solution $(\tilde{q}, \tilde{\pi}, \tilde{p}, \tilde{\Delta}, \tilde{y})$ of these relaxations such that (q, π, p, y) is not feasible for the active transmission problem. In this section we are going to present an MILP for evaluating the infeasibility of the active transmission problem. This MILP is called *infeasibility detection MILP*. Its definition depends on the vector $(\tilde{q}, \tilde{\pi}, \tilde{p}, \tilde{y})$ and is stated in Definition 6.2.1. The main Theorem 6.2.2 of this section states that if the *infeasibility detection MILP* is infeasible or has optimal objective value zero, then the active transmission problem (6.1.1) is infeasible.

For a motivation of the definition of the infeasibility detection MILP we consider a feasible solution (q^*, π^*, p^*, y^*) for the active transmission problem (6.1.1). From constraint (6.1.1b) and (6.1.2b) and the corresponding constraint in (6.1.5) we obtain

$$\begin{aligned} \alpha_a q_a^* |q_a^*|^{k_a} - \beta_a y_a^* - (\pi_v^* - \gamma_a \pi_w^*) &= 0, \\ \alpha_a \tilde{q}_a |\tilde{q}_a|^{k_a} - \beta_a \tilde{y}_a - (\tilde{\pi}_v - \gamma_a \tilde{\pi}_w) &= 0, \end{aligned} \tag{6.2.1}$$

for every arc $a = (v, w) \in A'$. Throughout this section we use $\gamma_{r,v}$ from Definition 4.2.3 and the function $\pi'_v(\pi) = \gamma_{r,v}\pi_v$ defined in (4.2.4) which allows the relation (4.2.5), i.e.,

$$\pi'_v(\pi) - \pi'_w(\pi) = \gamma_{r,v}\,(\pi_v - \gamma_a \pi_w).$$

Using π' we derive from (6.2.1):

$$\begin{aligned} &(\pi'_v(\pi^*) - \pi'_v(\tilde{\pi})) - (\pi'_w(\pi^*) - \pi'_w(\tilde{\pi})) + \gamma_{r,v}\beta_a(y_a^* - \tilde{y}_a) \\ &= \gamma_{r,v}(\pi_v^* - \gamma_a \pi_w^* + \beta_a y_a^* - (\tilde{\pi}_v - \gamma_a \tilde{\pi}_w + \beta_a \tilde{y}_a)) \\ &= \gamma_{r,v}\, \alpha_a (q_a^* |q_a^*|^{k_a} - \tilde{q}_a |\tilde{q}_a|^{k_a}) \\ &\begin{cases} > 0 & \text{if } \alpha_a (q_a^* - \tilde{q}_a) > 0, \\ = 0 & \text{if } \alpha_a (q_a^* - \tilde{q}_a) = 0, \\ < 0 & \text{if } \alpha_a (q_a^* - \tilde{q}_a) < 0, \end{cases} \end{aligned}$$

for every arc $a = (v, w) \in A'$. We write this inequality for short as

$$s_v - s_w + s_a \begin{cases} > 0 & \text{if } \alpha_a (q_a^* - \tilde{q}_a) > 0, \\ = 0 & \text{if } \alpha_a (q_a^* - \tilde{q}_a) = 0, \\ < 0 & \text{if } \alpha_a (q_a^* - \tilde{q}_a) < 0. \end{cases} \tag{6.2.2}$$

Here s_a corresponds to $\gamma_{r,v}\beta_a(y_a^* - \tilde{y}_a)$ and s_v to $\pi'_v(\pi^*) - \pi'_v(\tilde{\pi})$.

For the node potential π' we observe after identifying x_v with $\pi'_v(\pi^*) - \pi'_v(\tilde{\pi})$ the conditions

$$\begin{aligned} x_v &\in \mathbb{R} && \text{if } \underline{\pi}_v < \tilde{\pi}_v < \overline{\pi}_v, \\ x_v &\leq 0 && \text{if } \tilde{\pi}_v = \overline{\pi}_v, \\ x_v &< 0 && \text{if } \tilde{\pi}_v > \overline{\pi}_v, \\ x_v &\geq 0 && \text{if } \underline{\pi}_v = \tilde{\pi}_v, \\ x_v &> 0 && \text{if } \underline{\pi}_v > \tilde{\pi}_v. \end{aligned} \tag{6.2.3}$$

We write these conditions (6.2.3) as a single constraint

$$x_v^+ - x_v^- - x_v - \kappa_v\, z = 0 \quad \text{with} \quad 0 \le x_v^+ \le \overline{x}_v^+, 0 \le x_v^- \le \overline{x}_v^-, z > 0 \tag{6.2.4}$$

where the variable bounds and κ are defined as

$$\overline{x}_v^+ := \begin{cases} 0 & \text{if } \tilde{\pi}_v \ge \overline{\pi}_v, \\ \infty & \text{else,} \end{cases}$$

$$\overline{x}_v^- := \begin{cases} 0 & \text{if } \tilde{\pi}_v \le \underline{\pi}_v, \\ \infty & \text{else,} \end{cases}$$

$$\kappa_v := \begin{cases} 1 & \text{if } \tilde{\pi}_v > \overline{\pi}_v, \\ -1 & \text{if } \tilde{\pi}_v < \underline{\pi}_v, \\ 0 & \text{else.} \end{cases}$$

The flow conservation constraint (6.1.1c), constraint (6.2.4) and (in)equality (6.2.2) form the main part of the infeasibility detection MILP. Keeping this idea in mind, the MILP, which contains indicator constraints, is defined as follows:

Definition 6.2.1:
Let $(\tilde{q}, \tilde{\pi}, \tilde{p}, \tilde{y})$ *be a solution of the active transmission problem* (6.1.1) *fulfilling at least constraint* (6.1.2b) *and* (6.1.2e)*. The* ***infeasibility detection MILP*** *is defined as follows:*

$$\max\ z \tag{6.2.5a}$$

$$\text{s. t.} \quad \sum_{a \in \delta^+_{A'}(v)} q_a - \sum_{a \in \delta^-_{A'}(v)} q_a = d_v \qquad \forall\, v \in V, \tag{6.2.5b}$$

$$x_v^+ - x_v^- - x_v - \kappa_v\, z = 0 \qquad \forall\, v \in V, \tag{6.2.5c}$$

$$\tilde{\kappa}_a\,(q_a - \tilde{q}_a) > 0 \Rightarrow x_v - x_w + x_a \ge 0 \qquad \forall\, a = (v,w) \in A', \tag{6.2.5d}$$

$$\tilde{\kappa}_a\,(q_a - \tilde{q}_a) = 0 \Rightarrow x_v - x_w + x_a = 0 \qquad \forall\, a = (v,w) \in A', \tag{6.2.5e}$$

$$\tilde{\kappa}_a\,(q_a - \tilde{q}_a) < 0 \Rightarrow x_v - x_w + x_a \le 0 \qquad \forall\, a = (v,w) \in A', \tag{6.2.5f}$$

$$\alpha_a\,(q_a - \tilde{q}_a) > 0 \Rightarrow s_v - s_w + s_a \ge \kappa_a z \qquad \forall\, a = (v,w) \in A', \tag{6.2.5g}$$

$$\alpha_a\,(q_a - \tilde{q}_a) = 0 \Rightarrow s_v - s_w + s_a = 0 \qquad \forall\, a = (v,w) \in A', \tag{6.2.5h}$$

$$\alpha_a\,(q_a - \tilde{q}_a) < 0 \Rightarrow s_v - s_w + s_a \le \kappa_a z \qquad \forall\, a = (v,w) \in A', \tag{6.2.5i}$$

$$\underline{q}_a \leq q_a \leq \overline{q}_a \quad \forall a \in A', \tag{6.2.5j}$$

$$\underline{s}_a \leq s_a \leq \overline{s}_a \quad \forall a \in A', \tag{6.2.5k}$$

$$\underline{x}_a \leq x_a \leq \overline{x}_a \quad \forall a \in A', \tag{6.2.5l}$$

$$x_v^+ \leq \overline{x}_v^+ \quad \forall v \in V, \tag{6.2.5m}$$

$$x_v^- \leq \overline{x}_v^- \quad \forall v \in V, \tag{6.2.5n}$$

$$x_v, s_v \in \mathbb{R} \quad \forall v \in V, \tag{6.2.5o}$$

$$x_v^+, x_v^- \in \mathbb{R}_{\geq 0} \quad \forall v \in V, \tag{6.2.5p}$$

$$x_a, s_a, q_a \in \mathbb{R} \quad \forall a \in A', \tag{6.2.5q}$$

$$z \in \mathbb{R}_{\geq 0}. \tag{6.2.5r}$$

For simplicity we do not state this problem at once as a mixed-integer nonlinear optimization problem, but give an equivalent reformulation in Remark 6.2.3.

Constraint (6.2.5c) *originates from* (6.2.4) *by expressing* $z > 0$ *as objective. Similarly* (6.2.5g)–(6.2.5i) *originate from* (6.2.2). *Constraints* (6.2.5d)–(6.2.5f) *form a weaker version of* (6.2.2).

For the definition of this MILP we make use of different constants which are defined below. Especially the bounds on s_a *for an arc* $a = (v, w) \in A'$ *originate from the previously described relation that* s_a *corresponds to* $\gamma_{r,v}\beta_a(y_a^* - \tilde{y}_a)$*:*

$$\overline{s}_a := \begin{cases} 0 & \text{if } \beta_a\tilde{y}_a = \max\{\beta_a\underline{y}_a, \beta_a\overline{y}_a\}, \\ \infty & \text{else} \end{cases} \quad \forall a \in A',\ a = (v, w), \tag{6.2.6a}$$

$$\underline{s}_a := \begin{cases} 0 & \text{if } \beta_a\tilde{y}_a = \min\{\beta_a\underline{y}_a, \beta_a\overline{y}_a\}, \\ -\infty & \text{else} \end{cases} \quad \forall a \in A',\ a = (v, w), \tag{6.2.6b}$$

$$\overline{x}_a := \begin{cases} 0 & \text{if } \beta_a\tilde{y}_a = \max\{\beta_a\underline{y}_a, \beta_a\overline{y}_a\}, \\ 0 & \text{if } \exists k : [A_a(\tilde{q}_a, \tilde{p}_v, \tilde{p}_w)]_k \geq [b_a]_k, \\ & (A_a)_{(k,1)} \geq 0, (A_a)_{(k,2)} < 0, \\ & (A_a)_{(k,3)} > 0, \\ \infty & \text{else} \end{cases} \quad \forall a \in A',\ a = (v, w), \tag{6.2.6c}$$

$$\underline{x}_a := \begin{cases} 0 & \textit{if } \beta_a \tilde{y}_a = \min\{\beta_a \underline{y}_a, \beta_a \overline{y}_a\}, \\ 0 & \textit{if } \exists k : [A_a(\tilde{q}_a, \tilde{p}_v, \tilde{p}_w)]_k \geq [b_a]_k, \\ & (A_a)_{(k,1)} \leq 0, (A_a)_{(k,2)} > 0, \\ & (A_a)_{(k,3)} < 0, \\ -\infty & \textit{else} \end{cases} \qquad \forall a \in A', \; a = (v, w), \tag{6.2.6d}$$

$$\overline{x}_v^+ := \begin{cases} 0 & \textit{if } \tilde{\pi}_v \geq \overline{\pi}_v, \\ \infty & \textit{else} \end{cases} \qquad \forall v \in V, \tag{6.2.6e}$$

$$\overline{x}_v^- := \begin{cases} 0 & \textit{if } \tilde{\pi}_v \leq \underline{\pi}_v, \\ \infty & \textit{else} \end{cases} \qquad \forall v \in V, \tag{6.2.6f}$$

$$\kappa_v := \begin{cases} 1 & \textit{if } \tilde{\pi}_v > \overline{\pi}_v, \\ -1 & \textit{if } \tilde{\pi}_v < \underline{\pi}_v, \\ 0 & \textit{else} \end{cases} \qquad \forall v \in V, \tag{6.2.6g}$$

$$\kappa_a := \begin{cases} -1 & \textit{if } \alpha_a > 0, \tilde{q}_a > \overline{q}_a, \\ 1 & \textit{if } \alpha_a > 0, \tilde{q}_a < \underline{q}_a, \\ 0 & \textit{else} \end{cases} \qquad \forall a \in A', \tag{6.2.6h}$$

$$\tilde{\kappa}_a := \begin{cases} 1 & \textit{if } \alpha_a > 0, \\ 1 & \textit{if } \exists k : [A_a(\tilde{q}_a, \tilde{p}_v, \tilde{p}_w)]_k \geq [b_a]_k, \\ & (A_a)_{(k,1)} \neq 0, \\ 0 & \textit{else} \end{cases} \qquad \forall a \in A'. \tag{6.2.6i}$$

Theorem 6.2.2:
Let $(\tilde{q}, \tilde{\pi}, \tilde{p}, \tilde{y})$ be a solution for the active transmission problem (6.1.1) *fulfilling at least constraint* (6.1.2b) *and* (6.1.2e)*. If the infeasibility detection MILP* (6.2.5) *is infeasible or has optimal objective value zero, then the active transmission problem* (6.1.1) *is infeasible.*

Proof. Let $(\tilde{q}, \tilde{\pi}, \tilde{p}, \tilde{y})$ be a solution for the active transmission problem (6.1.1) fulfilling at least constraint (6.1.2b) and (6.1.2e). If the infeasibility detection MILP (6.2.5), which depends on $(\tilde{q}, \tilde{\pi}, \tilde{p}, \tilde{y})$, is infeasible, then there does not exist a flow vector $q' \in \mathbb{R}^{A'}$ which fulfills

$$\sum_{a \in \delta^+_{A'}(v)} q_a - \sum_{a \in \delta^-_{A'}(v)} q_a = d_v \quad \forall v \in V, \qquad \underline{q}_a \le q_a \le \overline{q}_a \quad \forall a \in A'. \tag{6.2.7}$$

This can be seen as follows: Otherwise, if there exists a vector q' fulfilling (6.2.7), then $(q', 0)$ is a feasible solution for (6.2.5). We conclude that the active transmission problem (6.1.1) is infeasible if MIP (6.2.5) is infeasible.

Now assume that the MIP (6.2.5) has an optimal solution with objective value zero. We prove that this implies that the active transmission problem (6.1.1) is infeasible. Therefor we assume that the active transmission problem has a feasible solution (q^*, π^*, p^*, y^*) and show that there exists a feasible solution (q^*, x^*, s^*, z^*) to MIP (6.2.5) with positive objective, i.e., $z^* > 0$. In the following we describe how this solution (q^*, x^*, s^*, z^*) is defined. First we give the definition of s^* and z^* and show that (6.2.5g)–(6.2.5i) and (6.2.5k) are fulfilled. Then we turn to the definition of x^* and prove that (6.2.5c)–(6.2.5f) and (6.2.5l)–(6.2.5n) are fulfilled. As the flow vector q^* is feasible for the flow conservation (6.2.5b) and the bound constraints (6.2.5j) we conclude that (q^*, x^*, s^*, z^*) is feasible for MIP (6.2.5).

Recall that r was used for the definition of $\gamma_{r,v}$ for every node $v \in V$. The vector (s^*, z^*) is defined as follows:

$$\begin{aligned}
s^*_v &:= \pi'_v(\pi^*) - \pi'_v(\tilde{\pi}) && \forall v \in V,\\
s^*_a &:= \gamma_{r,v}\beta_a(y^*_a - \tilde{y}_a) && \forall a = (v, w) \in A',\\
z^* &:= \min\left\{1, \min\left\{|\gamma_{r,v}\alpha_a(q^*_a|q^*_a|^{k_a} - \tilde{q}_a|\tilde{q}_a|^{k_a})| \,\middle|\, a = (v, w) \in A' : \alpha_a q^*_a \ne \alpha_a \tilde{q}_a \right\}\right\}.
\end{aligned}$$

Let us now prove that (s^*, z^*) is feasible for (6.2.5g)–(6.2.5i) and (6.2.5k). First we prove that $s^*_a \le \overline{s}_a$ holds for every arc $a \in A'$. Let $a \in A'$. By definition (6.2.6a) we have to show $s^*_a \le 0$ if $\beta_a \tilde{y}_a = \max\{\beta_a \underline{y}_a, \beta_a \overline{y}_a\}$. This means that one of the following three cases applies:

1. $\tilde{y}_a = \overline{y}_a, \beta_a > 0 \Rightarrow y^*_a \le \tilde{y}_a \Rightarrow s^*_a \le 0,$

2. $\tilde{y}_a = \underline{y}_a, \beta_a < 0 \Rightarrow y^*_a \ge \tilde{y}_a \Rightarrow s^*_a \le 0,$

3. $\beta_a = 0 \Rightarrow s^*_a = 0.$

Hence $s_a^* \leq 0$ if $\beta_a \tilde{y}_a = \max\{\beta_a \underline{y}_a, \beta_a \overline{y}_a\}$. Similarly we prove that $s_a^* \geq \underline{s}_a$ holds for every arc $a \in A'$. We conclude that s^* is feasible for (6.2.5k). Now we turn to the constraints (6.2.5g)–(6.2.5i). We consider an arc $a = (v, w) \in A'$ and obtain:

$$\begin{aligned} s_v^* - s_w^* + s_a^* &= \pi_v'(\pi^*) - \pi_w'(\pi^*) + \gamma_{r,v}\beta_a y_a^* - (\pi_v'(\tilde{\pi}) - \pi_w'(\tilde{\pi}) + \gamma_{r,v}\beta_a \tilde{y}_a) \\ &= \gamma_{r,v}\alpha_a(q_a^*|q_a^*|^{k_a} - \tilde{q}_a|\tilde{q}_a|^{k_a}) \\ &\begin{cases} > 0 & \text{if } \alpha_a(q_a^* - \tilde{q}_a) > 0, \\ = 0 & \text{if } \alpha_a(q_a^* - \tilde{q}_a) = 0, \\ < 0 & \text{if } \alpha_a(q_a^* - \tilde{q}_a) < 0. \end{cases} \end{aligned}$$

We conclude that (s^*, z^*) is feasible for (6.2.5g)–(6.2.5i) and (6.2.5k).

By Lemma 6.2.5 there exists a vector x^* for z^* such that (x^*, z^*) is feasible for (6.2.5c)–(6.2.5f) and (6.2.5l)–(6.2.5n). Furthermore the flow conservation constraint (6.2.5b) and the bound constraints (6.2.5j) are fulfilled by q^* as (q^*, π^*, p^*, y^*) is a feasible solution for the active transmission problem (6.1.1). Hence (q^*, x^*, s^*, z^*) is a feasible solution for MIP (6.2.5).

We finally show that $z^* > 0$. Because of $\gamma_{r,v} > 0$ for all nodes $v \in V$ we have that $\alpha_a q_a^* \neq \alpha_a \tilde{q}_a$ for an arc $a = (v, w) \in A'$ implies $\gamma_{r,v}\alpha_a(q_a^*|q_a^*|^{k_a} - \tilde{q}_a|\tilde{q}_a|^{k_a}) \neq 0$. This proves $z^* > 0$. □

Remark 6.2.3:
As discussed below in Section 6.4 we will consider the infeasibility detection MILP (6.2.5) *only for a solution* $(\tilde{q}, \tilde{\pi}, \tilde{p}, \tilde{y})$ *of* (6.1.1) *which does not violate the flow conservation constraint* (6.1.2c). *We roughly describe how the* MILP *formulation of* (6.2.5) *is obtained for this case. At first we replace* $(q_a - \tilde{q}_a)$ *by* Δ_a *for every arc* $a \in A'$. *As* $\tilde{q}$ *fulfills the flow conservation we obtain that* Δ *is a circulation, i.e., we replace the flow conservation* (6.2.5b) *by*

$$\sum_{a \in \delta_{A'}^+(v)} \Delta_a - \sum_{a \in \delta_{A'}^-(v)} \Delta_a = 0 \qquad \forall v \in V.$$

We define bounds for Δ *by*

$$\overline{\Delta}_a := \begin{cases} \infty & \text{if } \tilde{q}_a < \overline{q}_a, \\ 0 & \text{if } \tilde{q}_a = \overline{q}_a, \end{cases} \qquad \underline{\Delta}_a := \begin{cases} 0 & \text{if } \tilde{q}_a = \underline{q}_a, \\ -\infty & \text{if } \tilde{q}_a > \underline{q}_a, \end{cases}$$

and replace (6.2.5j) *by* $\underline{\Delta} \leq \Delta \leq \overline{\Delta}$. *Then it is easy to see, as* Δ *is a circulation, that* Δ *can be chosen such that either* $\Delta_a = 0$ *or* $|\Delta_a| \geq 1$ *holds. We introduce binary variables* $x_a^{FW}, x_a^{BW}, s_a^{FW}, s_a^{BW} \in \{0,1\}$ *in combination with indicator constraints as follows:*

$$\begin{aligned}
x_a^{FW} &= 1 \Rightarrow \tilde{\kappa}_a \Delta_a \geq 1 && \forall a \in A', \\
x_a^{FW} &= 0 \Rightarrow \tilde{\kappa}_a \Delta_a \leq 0 && \forall a \in A', \\
x_a^{BW} &= 1 \Rightarrow \tilde{\kappa}_a \Delta_a \leq -1 && \forall a \in A', \\
x_a^{BW} &= 0 \Rightarrow \tilde{\kappa}_a \Delta_a \geq 0 && \forall a \in A', \\
s_a^{FW} &= 1 \Rightarrow \alpha_a \Delta_a \geq 1 && \forall a \in A', \\
s_a^{FW} &= 0 \Rightarrow \alpha_a \Delta_a \leq 0 && \forall a \in A', \\
s_a^{BW} &= 1 \Rightarrow \alpha_a \Delta_a \leq -1 && \forall a \in A', \\
s_a^{BW} &= 0 \Rightarrow \alpha_a \Delta_a \geq 0 && \forall a \in A'.
\end{aligned}$$

Then (6.2.5d)–(6.2.5i) *are replaced by*

$$\begin{aligned}
x_a^{FW} &= 1 \Rightarrow x_v - x_w + x_a \geq 0 && \forall a = (v,w) \in A', \\
x_a^{FW} = 0, x_a^{BW} &= 0 \Rightarrow x_v - x_w + x_a = 0 && \forall a = (v,w) \in A', \\
x_a^{BW} &= 1 \Rightarrow x_v - x_w + x_a \leq 0 && \forall a = (v,w) \in A', \\
s_a^{FW} &= 1 \Rightarrow s_v - s_w + s_a \geq \kappa_a z && \forall a = (v,w) \in A', \\
s_a^{FW} = 0, s_a^{BW} &= 0 \Rightarrow s_v - s_w + s_a = 0 && \forall a = (v,w) \in A', \\
s_a^{BW} &= 1 \Rightarrow s_v - s_w + s_a \leq \kappa_a z && \forall a = (v,w) \in A'.
\end{aligned}$$

All these reformulations yield an MILP *with indicator constraints which is equivalent to* (6.2.5).

In the remaining part of this section we prove Lemma 6.2.5 which was used in the previous proof of Theorem 6.2.2. Therefor we prove an auxiliary lemma.

Lemma 6.2.4:

Let $(\tilde{q}, \tilde{\pi}, \tilde{p}, \tilde{y})$ *be a solution for the active transmission problem* (6.1.1) *fulfilling at least constraints* (6.1.2b) *and* (6.1.2e). *Furthermore let* (q^*, π^*, p^*, y^*) *be a feasible solution for* (6.1.1). *There exists a partition of the node set* $V = M_1 \dot\cup M_2 \dot\cup M_3$ *fulfilling the following conditions:*

- $\forall\, a = (v, w) \in \delta^+_{A'}(M_1) \cup \delta^-_{A'}(M_3)$:

$$\begin{gathered}(\pi'_v(\pi^*) - \pi'_w(\pi^*)) < (\pi'_v(\tilde{\pi}) - \pi'_w(\tilde{\pi})),\\ \nexists k : [A_a\,(\tilde{q}_a, \tilde{p}_v, \tilde{p}_w)]_k \geq [b_a]_k, (A_a)_{(k,1)}(q^*_a - \tilde{q}_a) \geq 0,\\ (A_a)_{(k,2)} < 0, (A_a)_{(k,3)} > 0,\end{gathered}$$

- $\forall\, a = (v, w) \in \delta^-_{A'}(M_1) \cup \delta^+_{A'}(M_3)$:

$$\begin{gathered}(\pi'_v(\pi^*) - \pi'_w(\pi^*)) > (\pi'_v(\tilde{\pi}) - \pi'_w(\tilde{\pi})),\\ \nexists k : [A_a\,(\tilde{q}_a, \tilde{p}_v, \tilde{p}_w)]_k \geq [b_a]_k, (A_a)_{(k,1)}(q^*_a - \tilde{q}_a) \geq 0,\\ (A_a)_{(k,2)} > 0, (A_a)_{(k,3)} < 0.\end{gathered}$$

Furthermore it holds

$$\begin{aligned}\{v \in V \mid \tilde{\pi}_v > \overline{\pi}_v\} &\subseteq M_1, & \{v \in V \mid \tilde{\pi}_v \leq \underline{\pi}_v\} \cap M_1 &= \varnothing,\\ \{v \in V \mid \tilde{\pi}_v < \underline{\pi}_v\} &\subseteq M_3, & \{v \in V \mid \tilde{\pi}_v \geq \overline{\pi}_v\} \cap M_3 &= \varnothing.\end{aligned}$$

Proof. Let $(\tilde{q}, \tilde{\pi}, \tilde{p}, \tilde{y})$ be a solution for the active transmission problem (6.1.1) fulfilling at least constraint (6.1.2b) and (6.1.2e). Furthermore let (q^*, π^*, p^*, y^*) be a feasible solution for (6.1.1). We iteratively construct the sets M_1, M_2, M_3 with $V = M_1 \mathbin{\dot{\cup}} M_2 \mathbin{\dot{\cup}} M_3$ as follows:

1. Initially we set $M_1 := \{v \in V \mid \tilde{\pi}_v > \overline{\pi}_v\}$. Then we iteratively extend this set by considering every arc $a = (v, w) \in \delta^+_{A'}(M_1) \cup \delta^-_{A'}(M_1)$. If this arc does not fulfill one of the following cases, then we either add v to M_1 if $v \notin M_1$ and set the predecessor $p(v) := w$ or we add w to M_1 if $w \notin M_1$ and set $p(w) := v$.

   Case $a = (v, w) \in \delta^+_{A'}(M_1)$:

$$(\pi'_v(\pi^*) - \pi'_w(\pi^*)) < (\pi'_v(\tilde{\pi}) - \pi'_w(\tilde{\pi})), \tag{6.2.8a}$$

$$\begin{aligned}&\nexists k \in \{1, \ldots \nu_a\} : [A_a\,(\tilde{q}_a, \tilde{p}_v, \tilde{p}_w)]_k \geq [b_a]_k, \text{ with}\\ &\qquad (A_a)_{(k,1)}(q^*_a - \tilde{q}_a) \geq 0, (A_a)_{(k,2)} < 0,\\ &\qquad (A_a)_{(k,3)} > 0.\end{aligned} \tag{6.2.8b}$$

Case $a = (v, w) \in \delta^-_{A'}(M_1)$:

$$(\pi'_v(\pi^*) - \pi'_w(\pi^*)) > (\pi'_v(\tilde{\pi}) - \pi'_w(\tilde{\pi})), \tag{6.2.9a}$$

$$\begin{aligned} \nexists k \in \{1, \dots \nu_a\} : [A_a\,(\tilde{q}_a, \tilde{p}_v, \tilde{p}_w)]_k \geq [b_a]_k, \text{ with} \\ (A_a)_{(k,1)}(q^*_a - \tilde{q}_a) \geq 0, (A_a)_{(k,2)} > 0, \\ (A_a)_{(k,3)} < 0. \end{aligned} \tag{6.2.9b}$$

This way we obtain the node set M_1 such that every arc $a = (v, w) \in \delta^+_{A'}(M_1)$ fulfills (6.2.8) and every arc $a = (v, w) \in \delta^-_{A'}(M_1)$ fulfills (6.2.9). Furthermore it holds

$$M_1 \cap \{v \in V \mid \tilde{\pi}_v \leq \underline{\pi}_v\} = \varnothing.$$

This can be seen as follows: If $\tilde{\pi}_v \leq \overline{\pi}_v$ holds for every node $v \in V$, then $M_1 = \varnothing$ by construction. Assume that M_1 contains a node t with $\tilde{\pi}_t \leq \underline{\pi}_t$. Then we consider the nodes t, $p(t)$, $p(p(t))$, …, s where $s \in M_1$ has no predecessor. These nodes define the nodes of an edge-disjoint s-t-path P in the undirected graph $(M_1, E'[M_1])$ which originates from $(M_1, A'[M_1])$ by removing the orientation of each arc $a \in A'[M_1]$. Note that $(M_1, E'[M_1])$ might contain multiple parallel edges. This way each arc $a \in A'[M_1]$ corresponds uniquely to an edge $e \in E'[M_1]$ and vice versa. Let $v_1, \dots, v_{n+1}$ be the nodes and $e_1, \dots, e_n$ with $e_i = \{v_i, v_{i+1}\}$ be the ordered edges of P and $a_1, \dots, a_n$ be the corresponding arcs in $(V, A'[M_1])$. We have that for every arc $a_i, i = 1, \dots, n$ neither (6.2.8) nor (6.2.9) applies because otherwise t would not be contained in M_1 by construction of M_1. This means that one of the following cases holds for every arc $a_i, i = 1, \dots, n$:

Case (6.2.8a) **and** (6.2.9a) **do not apply:** Node potential loss estimation derived as (6.2.8a) and (6.2.9a) do not apply, hence:

$$\pi'_{v_i}(\pi^*) - \pi'_{v_{i+1}}(\pi^*) \geq \pi'_{v_i}(\tilde{\pi}) - \pi'_{v_{i+1}}(\tilde{\pi}).$$

Case (6.2.8b) **and** (6.2.9b) **do not apply:** In this case we differ between the orientation of arc a_i.

- If arc $a_i = (v_i, v_{i+1})$ then, as (6.2.8b) does not apply, there exists an index $k \in \{1, \dots, \nu_{a_i}\}$ such that $[A_{a_i}(\tilde{q}_{a_i}, \tilde{p}_{v_i}, \tilde{p}_{v_{i+1}})]_k \geq [b_{a_i}]_k$ holds with $(A_{a_i})_{(k,1)}(q^*_{a_i} - \tilde{q}_{a_i}) \geq 0$, $(A_{a_i})_{(k,2)} < 0$, $(A_{a_i})_{(k,3)} > 0$. We rewrite this inequality as $a_1\tilde{q}_{a_i} - a_3 \geq \tilde{p}_{v_i} - a_2\tilde{p}_{v_{i+1}}$ with $a_2 \in \mathbb{R}_{\geq 0}$ and $a_1, a_3 \in \mathbb{R}$. Then we derive the estimation

$$\tilde{p}_{v_i} - a_2\tilde{p}_{v_{i+1}} \leq a_1\tilde{q}_{a_i} - a_3 \leq a_1 q^*_{a_i} - a_3 \leq p^*_{v_i} - a_2 p^*_{v_{i+1}}.$$

- If arc $a_i = (v_{i+1}, v_i)$ then, as (6.2.9b) does not apply, there exists an index $k \in \{1, \ldots, \nu_{a_i}\}$ such that $[A_{a_i}(\tilde{q}_{a_i}, \tilde{p}_{v_{i+1}}, \tilde{p}_{v_i})]_k \geq [b_{a_i}]_k$ holds with $(A_{a_i})_{(k,1)}(q^*_{a_i} - \tilde{q}_{a_i}) \geq 0$, $(A_{a_i})_{(k,2)} > 0$, $(A_{a_i})_{(k,3)} < 0$. We rewrite this inequality as $a_1\tilde{q}_{a_i} - a_3 \geq \tilde{p}_{v_i} - a_2\tilde{p}_{v_{i+1}}$ with $a_2 \in \mathbb{R}_{\geq 0}$ and $a_1, a_3 \in \mathbb{R}$. Again we derive the estimation

$$\tilde{p}_{v_i} - a_2\tilde{p}_{v_{i+1}} \leq a_1\tilde{q}_{a_i} - a_3 \leq a_1 q^*_{a_i} - a_3 \leq p^*_{v_i} - a_2 p^*_{v_{i+1}}.$$

Because of the coupling constraint $\tilde{p}_v|\tilde{p}_v| = \tilde{\pi}_v$ relating the pressure and node potential variables for each node $v \in V$, by using the previous estimations, we obtain:

$$\begin{aligned} \tilde{\pi}_{v_1} &> \pi^*_{v_1} &&\Rightarrow \tilde{\pi}_{v_2} > \pi^*_{v_2}, \ldots, \tilde{\pi}_{v_{n+1}} > \pi^*_{v_{n+1}}, \\ \tilde{\pi}_{v_{n+1}} &< \pi^*_{v_{n+1}} &&\Rightarrow \tilde{\pi}_{v_n} < \pi^*_{v_n}, \ldots, \quad \tilde{\pi}_{v_1} < \pi^*_{v_1}. \end{aligned}$$

The path P is chosen such that the start node v_1 violates its upper node potential bound, i.e., $\tilde{\pi}_{v_1} > \overline{\pi}_{v_1}$ and for the end node v_{n+1} it holds $\tilde{\pi}_{v_{n+1}} \leq \underline{\pi}_{v_{n+1}}$. Hence the first of the above cases applies. We conclude that π^* violates a node potential bound in v_{n+1} which is a contradiction to the assumption that (q^*, π^*, p^*, y^*) is feasible for the active transmission problem.

2. In a second step we initially set $M_3 := \{v \in V \mid \pi_v < \underline{\pi}_v\}$. We now concentrate on the graph $(V \setminus M_1, A'[V \setminus M_1])$. Again we iteratively consider each arc $a = (v, w) \in \delta^+_{A'}(M_3) \cup \delta^-_{A'}(M_3)$. If neither (6.2.8) applies for an ingoing arc nor (6.2.9) applies for an outgoing arc, then we add v to M_3 if $v \notin M_3$ and w if $w \notin M_3$. By definition it follows $M_1 \cap M_3 = \varnothing$. By a similar reasoning as in the previous item we conclude

$$M_3 \cap \{v \in V \mid \tilde{\pi}_v \geq \overline{\pi}_v\} = \varnothing.$$

3. In a third step we define $M_2 := \{v \in V \mid v \notin M_1 \cup M_3\}$. The previously defined sets then have the property $V = M_1 \,\dot{\cup}\, M_2 \,\dot{\cup}\, M_3$.

□

Lemma 6.2.5:
Let $(\tilde{q}, \tilde{\pi}, \tilde{p}, \tilde{y})$ *be a solution for the active transmission problem* (6.1.1) *fulfilling at least constraint* (6.1.2b) *and* (6.1.2e). *Furthermore let* (q^*, π^*, p^*, y^*) *be a feasible solution for* (6.1.1) *and* $0 \leq z^* \leq 1$. *There exists a vector* $x^* = (x_v^+, x_v^-, x_v, x_a)^*_{v \in V, a \in A'}$ *with* $x_v^{*+}, x_v^{*-} \in \mathbb{R}_{\geq 0}, x_v^*, x_a^* \in \mathbb{R}$ *for* $v \in V$ *and* $a \in A'$ *with* $x^* \neq 0$ *which is feasible for* (6.2.5c)–(6.2.5f) *and* (6.2.5l)–(6.2.5n).

Proof. Let $(\tilde{q}, \tilde{\pi}, \tilde{p}, \tilde{y})$ be a solution for the active transmission problem (6.1.1) fulfilling at least constraint (6.1.2b) and (6.1.2e). Furthermore let (q^*, π^*, p^*, y^*) be a feasible solution for the active transmission problem (6.1.1). For $0 \leq z^* \leq 1$ we describe how to define $x^* = (x_v, x_a)^*_{v \in V, a \in A'}$ which is feasible for (6.2.5d)–(6.2.5f) and (6.2.5l).

By Lemma 6.2.4 there exists a partition $V = M_1 \dot{\cup} M_2 \dot{\cup} M_3$ such that the following holds:

- $\forall\, a = (v, w) \in \delta^+_{A'}(M_1) \cup \delta^-_{A'}(M_3)$:

$$\begin{aligned}
&(\pi'_v(\pi^*) - \pi'_w(\pi^*)) < (\pi'_v(\tilde{\pi}) - \pi'_w(\tilde{\pi})), \\
&\nexists k \in \{1, \ldots, \nu_a\} : [A_a\,(\tilde{q}_a, \tilde{p}_v, \tilde{p}_w)]_k \geq [b_a]_k, \text{ with} \\
&\qquad (A_a)_{(k,1)}(q_a^* - \tilde{q}_a) \geq 0, (A_a)_{(k,2)} < 0, (A_a)_{(k,3)} > 0.
\end{aligned} \tag{6.2.10}$$

- $\forall\, a = (v, w) \in \delta^-_{A'}(M_1) \cup \delta^+_{A'}(M_3)$:

$$\begin{aligned}
&(\pi'_v(\pi^*) - \pi'_w(\pi^*)) > (\pi'_v(\tilde{\pi}) - \pi'_w(\tilde{\pi})), \\
&\nexists k \in \{1, \ldots, \nu_a\} : [A_a\,(\tilde{q}_a, \tilde{p}_v, \tilde{p}_w)]_k \geq [b_a]_k, \text{ with} \\
&\qquad (A_a)_{(k,1)}(q_a^* - \tilde{q}_a) \geq 0, (A_a)_{(k,2)} > 0, (A_a)_{(k,3)} < 0.
\end{aligned} \tag{6.2.11}$$

From this we obtain an estimation which is needed in the following:

$$\begin{aligned}
&a = (v, w) \in \delta^+(M_1) \cup \delta^-_{A'}(M_3) \\
&\Rightarrow \alpha_a q_a^* |q_a^*|^{k_a} - \beta_a y_a^* = \gamma_{r,v}^{-1}(\pi'_v(\pi^*) - \pi'_w(\pi^*)) \\
&\qquad < \gamma_{r,v}^{-1}(\pi'_v(\tilde{\pi}) - \pi'_w(\tilde{\pi})) = \alpha_a \tilde{q}_a |\tilde{q}_a|^{k_a} - \beta_a \tilde{y}_a
\end{aligned} \tag{6.2.12a}$$

$$\begin{aligned}
&a = (v, w) \in \delta^-(M_1) \cup \delta^+_{A'}(M_3) \\
&\Rightarrow \alpha_a q_a^* |q_a^*|^{k_a} - \beta_a y_a^* = \gamma_{r,v}^{-1}(\pi'_v(\pi^*) - \pi'_w(\pi^*)) \\
&\qquad > \gamma_{r,v}^{-1}(\pi'_v(\tilde{\pi}) - \pi'_w(\tilde{\pi})) = \alpha_a \tilde{q}_a |\tilde{q}_a|^{k_a} - \beta_a \tilde{y}_a
\end{aligned} \tag{6.2.12b}$$

We use the sets M_1, M_2, M_3 to define the values $x_v^*, v \in V$ and $x_a^*, a = (v,w) \in A'$ as follows:

$$x_v^* := \begin{cases} -1 & \text{if } v \in M_1, \\ 0 & \text{if } v \in M_2, \\ 1 & \text{if } v \in M_3, \end{cases} \tag{6.2.13a}$$

$$x_a^* := \begin{cases} 0 & \text{if } a \in \delta^+_{A'}(M_1) \cup \delta^-_{A'}(M_3) : \tilde{\kappa}_a(q_a^* - \tilde{q}_a) < 0, \\ 0 & \text{if } a \in \delta^-_{A'}(M_1) \cup \delta^+_{A'}(M_3) : \tilde{\kappa}_a(q_a^* - \tilde{q}_a) > 0, \\ x_w^* - x_v^* & \text{else.} \end{cases} \tag{6.2.13b}$$

We proceed by showing that this definition is feasible for constraints (6.2.5d)–(6.2.5f) and (6.2.5l). We have $x_v^* = x_w^*$ and $x_a^* = 0$ for all arcs $a = (v,w) \in A'(M_1 : M_1) \cup A'(M_2 : M_2) \cup A'(M_3 : M_3)$. Thus (6.2.5d)–(6.2.5f) and (6.2.5l) are fulfilled for these arcs. Let us now turn to the remaining arcs. At first we observe:

- For every arc $a \in A'(M_1 : M_2) \cup A'(M_2 : M_3) \cup A'(M_1 : M_3)$ it holds $a \in \delta^+_{A'}(M_1) \cup \delta^-_{A'}(M_3)$.
- For every arc $a \in A'(M_2 : M_1) \cup A'(M_3 : M_1) \cup A'(M_3 : M_2)$ it holds $a \in \delta^-_{A'}(M_1) \cup \delta^+_{A'}(M_3)$.

These are the two cases that we distinguish in the following:

Case $q_a^* < \tilde{q}_a$ **:** We distinguish two cases.

Case $a = (v,w) \in \delta^+_{A'}(M_1) \cup \delta^-_{A'}(M_3)$ **:** In this case we have $x_v^* - x_w^* \leq -1$.

- By (6.2.5f) $\tilde{\kappa}_a \neq 0$ means $x_v - x_w + x_a \leq 0$ must be fulfilled by x^*. We have $\tilde{\kappa}_a \neq 0 \Rightarrow \tilde{\kappa}_a > 0 \Rightarrow \tilde{\kappa}_a(q_a^* - \tilde{q}_a) < 0 \Rightarrow x_a^* = 0$ by (6.2.13b). Hence it holds $x_v^* - x_w^* + x_a^* \leq 0$ and $\underline{x}_a \leq x_a^* \leq \overline{x}_a$.
- By (6.2.5e) $\tilde{\kappa}_a = 0$ means $x_v - x_w + x_a = 0$ must be fulfilled by x^*. We have $x_a^* = -(x_v^* - x_w^*) > 0$ by (6.2.13b). Hence it holds $x_v^* - x_w^* + x_a^* = 0$. $\tilde{\kappa}_a = 0$ means $\alpha_a = 0$ by (6.2.6i). This in combination with (6.2.12a) implies $\beta_a y_a^* > \beta_a \tilde{y}_a$. By (6.2.10) there exists no index k so that $[A_a(\tilde{q}_a, \tilde{p}_v, \tilde{p}_w)]_k \geq [b_a]_k$, with $(A_a)_{(k,1)}(q_a^* - \tilde{q}_a) \geq 0$,$(A_a)_{(k,2)} < 0$, $(A_a)_{(k,3)} > 0$. This and the conclusions that there exists no index k such that $[A_a(\tilde{q}_a, \tilde{p}_v, \tilde{p}_w)]_k \geq [b_a]_k$, with $(A_a)_{(k,1)} \neq 0$ (because of $\tilde{\kappa}_a = 0$) especially implies that there exists no index k such that $[A_a(\tilde{q}_a, \tilde{p}_v, \tilde{p}_w)]_k \geq [b_a]_k$, with $(A_a)_{(k,1)} \geq 0$,$(A_a)_{(k,2)} < 0$ and $(A_a)_{(k,3)} > 0$. From this we conclude $\overline{x}_a = \infty$ by (6.2.6c). This yields $\underline{x}_a \leq 0 < x_a^* \leq \overline{x}_a$.

Case $a = (v, w) \in \delta^-_{A'}(M_1) \cup \delta^+_{A'}(M_3)$: In this case we have $x^*_v - x^*_w \geq 1$.

- From $\tilde{\kappa}_a \geq 0$ it follows $\tilde{\kappa}_a(q^*_a - \tilde{q}_a) \leq 0$. Hence we obtain from (6.2.13b) that $x^*_a = -(x^*_v - x^*_w) \leq -1 < 0$ holds. By (6.2.5e) and (6.2.5f) $x_v - x_w + x_a \leq 0$ or $x_v - x_w + x_a = 0$ must be fulfilled by x^*, which is obviously true. By (6.2.12b) we have $\beta_a y^*_a < \beta_a \tilde{y}_a$. By (6.2.11) there exists no index k such that $[A_a(\tilde{q}_a, \tilde{p}_v, \tilde{p}_w)]_k \geq [b_a]_k$, with $(A_a)_{(k,1)}(q^*_a - \tilde{q}_a) \geq 0$, $(A_a)_{(k,2)} > 0$, $(A_a)_{(k,3)} < 0$. This especially means that there exists no index k such that $[A_a(\tilde{q}_a, \tilde{p}_v, \tilde{p}_w)]_k \geq [b_a]_k$, with $(A_a)_{(k,1)} \leq 0$, $(A_a)_{(k,2)} > 0, (A_a)_{(k,3)} < 0$. Hence we have $\underline{x}_a = -\infty$ by (6.2.6d). This yields $\underline{x}_a < x^*_a \leq 0 \leq \overline{x}_a$.

Case $q^*_a = \tilde{q}_a$: We distinguish two cases.

Case $a = (v, w) \in \delta^+_{A'}(M_1) \cup \delta^-_{A'}(M_3)$: In this case we have $x^*_v - x^*_w \leq -1$. By (6.2.12a) we have $\beta_a y^*_a > \beta_a \tilde{y}_a$.

- By (6.2.5e) $x_v - x_w + x_a = 0$ must be fulfilled by x^*. Because of $\tilde{\kappa}_a(q^*_a - \tilde{q}_a) = 0$ we have $x^*_a = -(x^*_v - x^*_w) \geq 1 > 0$ by (6.2.13b). Hence $x^*_v - x^*_w + x^*_a = 0$. By (6.2.10) there exists no index k such that $[A_a(\tilde{q}_a, \tilde{p}_v, \tilde{p}_w)]_k \geq [b_a]_k$, with $(A_a)_{(k,1)}(q^*_a - \tilde{q}_a) \geq 0$, $(A_a)_{(k,2)} < 0$, $(A_a)_{(k,3)} > 0$. This especially implies that there exists no index k such that $[A_a(\tilde{q}_a, \tilde{p}_v, \tilde{p}_w)]_k \geq [b_a]_k$, with $(A_a)_{(k,1)} \geq 0$, $(A_a)_{(k,2)} < 0$, $(A_a)_{(k,3)} > 0$. Hence $\overline{x}_a = \infty$ by (6.2.6c). So we obtain $\underline{x}_a \leq 0 < x^*_a < \overline{x}_a$.

Case $a = (v, w) \in \delta^-_{A'}(M_1) \cup \delta^+_{A'}(M_3)$: In this case we have $x^*_v - x^*_w \geq 1$. By (6.2.12b) we have $\beta_a y^*_a < \beta_a \tilde{y}_a$.

- By (6.2.5e) $x_v - x_w + x_a = 0$ must be fulfilled by x^*. Because of $\tilde{\kappa}_a(q^*_a - \tilde{q}_a) = 0$ we have $x^*_a = -(x^*_v - x^*_w) \leq -1 < 0$ by (6.2.13b). Hence $x^*_v - x^*_w + x^*_a = 0$. By (6.2.11) there exists no index k such that $[A_a(\tilde{q}_a, \tilde{p}_v, \tilde{p}_w)]_k \geq [b_a]_k$, with $(A_a)_{(k,1)}(q^*_a - \tilde{q}_a) \geq 0$, $(A_a)_{(k,2)} > 0$, $(A_a)_{(k,3)} < 0$. This especially implies that there exists no index k such that $[A_a(\tilde{q}_a, \tilde{p}_v, \tilde{p}_w)]_k \geq [b_a]_k$, with $(A_a)_{(k,1)} \leq 0$, $(A_a)_{(k,2)} > 0$, $(A_a)_{(k,3)} < 0$. Hence $\underline{x}_a = -\infty$ by (6.2.6d). This yields $\underline{x}_a < x^*_a < 0 \leq \overline{x}_a$.

Case $q^*_a > \tilde{q}_a$: We distinguish two cases:

Case $a = (v, w) \in \delta^+_{A'}(M_1) \cup \delta^-_{A'}(M_3)$: In this case we have $x^*_v - x^*_w \leq -1$. By (6.2.12a) we have $\beta_a y^*_a > \beta_a \tilde{y}_a$.

- From $\tilde{\kappa}_a \geq 0$ it follows $\tilde{\kappa}_a(q^*_a - \tilde{q}_a) \geq 0$. Hence we obtain from (6.2.13b) that $x^*_a = -(x^*_v - x^*_w) \geq 1 > 0$ holds. By (6.2.5d) and (6.2.5e) $x_v - x_w + x_a \geq 0$ or $x_v - x_w + x_a = 0$ must be fulfilled by x^*, which is obviously true. By (6.2.10) there exists no index k such that $[A_a(\tilde{q}_a, \tilde{p}_v, \tilde{p}_w)]_k \geq [b_a]_k$, with $(A_a)_{(k,1)}(q_a^* - \tilde{q}_a) \geq 0$, $(A_a)_{(k,2)} < 0$, $(A_a)_{(k,3)} > 0$. This especially implies that there exists no index k such that $[A_a(\tilde{q}_a, \tilde{p}_v, \tilde{p}_w)]_k \geq [b_a]_k$, with $(A_a)_{(k,1)} \geq 0$, $(A_a)_{(k,2)} < 0$, $(A_a)_{(k,3)} > 0$. Hence $\overline{x}_a = \infty$ by (6.2.6c). This yields $\underline{x}_a \leq 0 < x_a^* < \infty = \overline{x}_a$.

Case $a = (v,w) \in \delta^-_{A'}(M_1) \cup \delta^+_{A'}(M_3)$: In this case we have $x_v^* - x_w^* \geq 1$.

- By (6.2.5d) $\tilde{\kappa}_a \neq 0$ means $x_v - x_w + x_a \geq 0$ must be fulfilled by x^*. We have $\tilde{\kappa}_a \neq 0 \Rightarrow \tilde{\kappa}_a > 0 \Rightarrow \tilde{\kappa}_a(q_a^* - \tilde{q}_a) > 0 \Rightarrow x_a^* = 0$ by (6.2.13b). Hence it holds $x_v^* - x_w^* + x_a^* = x_v^* - x_w^* \geq 1 \geq 0$ and $\underline{x}_a \leq x_a^* \leq \overline{x}_a$.
- By (6.2.5e) $\tilde{\kappa}_a = 0$ means $x_v - x_w + x_a = 0$ must be fulfilled by x^*. By (6.2.13b) we have $x_a^* = -(x_v^* - x_w^*) < 0$ which implies $x_v^* - x_w^* + x_a^* = 0$. $\tilde{\kappa}_a = 0$ means $\alpha_a = 0$. This implies in combination with (6.2.12b) that $\beta_a y_a^* < \beta_a \tilde{y}_a$ holds. By (6.2.11) there exists no index k such that $[A_a(\tilde{q}_a, \tilde{p}_v, \tilde{p}_w)]_k \geq [b_a]_k$, with $(A_a)_{(k,1)}(q_a^* - \tilde{q}_a) \geq 0$, $(A_a)_{(k,2)} > 0$, $(A_a)_{(k,3)} < 0$. Additionally $\tilde{\kappa}_a = 0$ yields that $\nexists k : [A_a(\tilde{q}_a, \tilde{p}_v, \tilde{p}_w)]_k \geq [b_a]_k$ with $(A_a)_{(k,1)} \neq 0$ by (6.2.6i). This especially implies that there exists no index k such that $[A_a(\tilde{q}_a, \tilde{p}_v, \tilde{p}_w)]_k \geq [b_a]_k$, with $(A_a)_{(k,1)} \leq 0$, $(A_a)_{(k,2)} > 0$, $(A_a)_{(k,3)} < 0$. Hence $\underline{x}_a = -\infty$ by (6.2.6d). This yields $\underline{x}_a < x_a^* < 0 \leq \overline{x}_a$.

This case discussion proves that $(x_v^*, x_a^*)_{v \in V, a \in A'}$ is feasible for (6.2.5d)–(6.2.5f) and (6.2.5l). We set

$$\begin{aligned} x_v^{*+} &:= \max\{0, x_v^* + \kappa_v z^*\} && \forall v \in V, \\ x_v^{*-} &:= \max\{0, -x_v^* - \kappa_v z^*\} && \forall v \in V. \end{aligned} \tag{6.2.14}$$

From this definition it follows that (6.2.5c) is fulfilled. We prove that this definition is feasible for (6.2.5m) and (6.2.5n). Therefor we make use of

$$\{v \in V \mid \tilde{\pi}_v > \overline{\pi}_v\} \subseteq M_1 \qquad \{v \in V \mid \tilde{\pi}_v \leq \underline{\pi}_v\} \cap M_1 = \varnothing, \tag{6.2.15a}$$

$$\{v \in V \mid \tilde{\pi}_v < \underline{\pi}_v\} \subseteq M_3 \qquad \{v \in V \mid \tilde{\pi}_v \geq \overline{\pi}_v\} \cap M_3 = \varnothing, \tag{6.2.15b}$$

which holds by Lemma 6.2.4. We consider the bound constraints (6.2.5m) and (6.2.5n) separately:

- It holds $x_v^{*+} \leq \overline{x}_v^+$ for every node $v \in V$: Let $v \in V$. According to (6.2.6e) we have to show $x_v^{*+} = 0$ if $\tilde{\pi}_v \geq \overline{\pi}_v$. In this case it holds $v \in M_1 \cup M_2$ by (6.2.15b). We distinguish two cases:

$$\tilde{\pi}_v = \overline{\pi}_v \overset{(6.2.6g)}{\Rightarrow} \kappa_v = 0$$

$$\Rightarrow x_v^* + \kappa_v z^* = x_v^* \overset{(6.2.13a),(6.2.15b)}{\leq} 0 \Rightarrow x_v^{*+} = \max\{0, x_v^* + \kappa_v z_v^*\} = 0.$$

$$\tilde{\pi}_v > \overline{\pi}_v \overset{(6.2.6g),(6.2.15a),(6.2.13a)}{\Rightarrow} \kappa_v = 1, x_v^* = -1$$
$$\Rightarrow x_v^* + \kappa_v z^* = -1 + z^* \overset{z^* \in [0,1]}{\leq} 0 \Rightarrow x_v^{*+} = \max\{0, x_v^* + \kappa_v z^*\} = 0.$$

- It holds $x_v^{*-} \leq \overline{x}_v^-$ for every node $v \in V$: Let $v \in V$. According to (6.2.6f) we have to show $x_v^{*-} = 0$ if $\tilde{\pi}_v \leq \underline{\pi}_v$. In this case it holds $v \in M_3 \cup M_2$ by (6.2.15a). We distinguish two cases:

$$\tilde{\pi}_v = \underline{\pi}_v \overset{(6.2.6g)}{\Rightarrow} \kappa_v = 0$$
$$\Rightarrow x_v^* + \kappa_v z^* = x_v^* \overset{(6.2.13a),(6.2.15a)}{\geq} 0 \Rightarrow x_v^{*-} = \min\{0, x_v^* + \kappa_v z^*\} = 0.$$

$$\tilde{\pi}_v < \underline{\pi}_v \overset{(6.2.6g),(6.2.15b),(6.2.13a)}{\Rightarrow} \kappa_v = -1, x_v^* = 1$$
$$\Rightarrow x_v^* + \kappa_v z^* = 1 - z^* \overset{z^* \in [0,1]}{\geq} 0 \Rightarrow x_v^{*-} = \min\{0, x_v^* + \kappa_v z^*\} = 0.$$

We conclude that the definition (6.2.13a), (6.2.13b), and (6.2.14) yield a vector x^* which is feasible for (6.2.5c)–(6.2.5f) and (6.2.5l)–(6.2.5n). □

6.3 Interpretation of the Infeasibility Detection MILP

In this section we present the initial idea which led to the formulation of the infeasibility detection MILP (6.2.5). The concept becomes visible when looking at the dual problem of (6.2.5) for a fixed flow vector $q' \in \mathbb{R}^{A'}$ which fulfills the flow conservation and bound constraints (6.2.7), i.e.,

$$\sum_{a \in \delta^+_{A'}(v)} q'_a - \sum_{a \in \delta^-_{A'}(v)} q'_a = d_v \quad \forall v \in V, \qquad \underline{q}_a \leq q'_a \leq \overline{q}_a \quad \forall a \in A'.$$

Let $(\tilde{q}, \tilde{\pi}, \tilde{p}, \tilde{y})$ be a solution of the active transmission problem (6.1.1) fulfilling at least constraint (6.1.2b) and (6.1.2e). Assume that the MILP (6.2.5) has optimal objective value zero. By Theorem 6.2.2 we conclude that there do not exist vectors $\pi' \in \mathbb{R}^V$, $p' \in \mathbb{R}^V$, $y' \in \mathbb{R}^{A'}$ such that (q', π', p', y') is feasible for the active transmission problem (6.1.1). In the following we show an example demonstrating that especially $\pi' \in \mathbb{R}^V$ with $\underline{\pi} \leq \pi' \leq \overline{\pi}$ cannot exist. Therefor we assume that (q', π', p', y') is a feasible solution for (6.1.1) and derive a contradiction by comparing (q', π', p', y') and $(\tilde{q}, \tilde{\pi}, \tilde{p}, \tilde{y})$.

As the MILP (6.2.5) has optimal objective value zero it is easy to see that the following linear optimization problem is bounded (recall that q' and $\tilde{q}$ are fixed):

$$\max z \tag{6.3.1}$$

$$\begin{aligned}
[\lambda_v] \quad \text{s.t.} \quad & x_v^+ - x_v^- - x_v - \kappa_v\, z = 0 && \forall v \in V,\\
[\mu_a] \quad & x_v - x_w + x_a - x_a^+\, \tilde{\kappa}_a\, (q_a' - \tilde{q}_a) = 0 && \forall a = (v,w) \in A',\\
[\nu_a] \quad & s_v - s_w + s_a - s_a^+\, \alpha_a\, (q_a' - \tilde{q}_a) - \kappa_a z = 0 && \forall a = (v,w) \in A',\\
& \underline{s}_a \le s_a \le \overline{s}_a && \forall a \in A',\\
& \underline{x}_a \le x_a \le \overline{x}_a && \forall a \in A',\\
& x_v^+ \le \overline{x}_v^+ && \forall v \in V,\\
& x_v^- \le \overline{x}_v^- && \forall v \in V,\\
& x_v, s_v \in \mathbb{R} && \forall v \in V,\\
& x_v^+, x_v^- \in \mathbb{R}_{\ge 0} && \forall v \in V,\\
& x_a, s_a \in \mathbb{R} && \forall a \in A',\\
& x_a^+, s_a^+ \in \mathbb{R}_{\ge 0} && \forall a \in A',\\
& z \in \mathbb{R}_{\ge 0}.
\end{aligned}$$

We associated dual variables λ_v for each node $v \in V$ and μ_a, ν_a for each arc $a \in A'$. As (6.3.1) is bounded it follows that its dual is feasible. This dual is as follows:

$$\exists\, \lambda, \mu, \nu \tag{6.3.2a}$$

$$\text{s.t.} \quad \sum_{a \in \delta_{A'}^+(v)} \nu_a - \sum_{a \in \delta_{A'}^-(v)} \nu_a = 0 \quad \forall v \in V, \tag{6.3.2b}$$

$$\sum_{a \in \delta_{A'}^+(v)} \mu_a - \sum_{a \in \delta_{A'}^-(v)} \mu_a - \lambda_v = 0 \quad \forall v \in V, \tag{6.3.2c}$$

$$\sum_{v \in V : \tilde{\pi}_v > \overline{\pi}_v} \lambda_v - \sum_{v \in V_\pi : \tilde{\pi}_v < \underline{\pi}_v} \lambda_v + \sum_{a \in A'} \kappa_a\, \nu_a \ge 1, \tag{6.3.2d}$$

$$\tilde{\kappa}_a (q_a' - \tilde{q}_a)\, \mu_a \ge 0 \quad \forall a \in A', \tag{6.3.2e}$$

$$\alpha_a (q_a' - \tilde{q}_a)\, \nu_a \ge 0 \quad \forall a \in A', \tag{6.3.2f}$$

$$\underline{\lambda}_v \le \lambda_v \le \overline{\lambda}_v \quad \forall v \in V, \tag{6.3.2g}$$

$$\underline{\mu}_a \le \mu_a \le \overline{\mu}_a \quad \forall a \in A', \tag{6.3.2h}$$

$$\underline{\nu}_a \le \nu_a \le \overline{\nu}_a \quad \forall a \in A', \tag{6.3.2i}$$

$$\mu_a, \nu_a \in \mathbb{R} \quad \forall a \in A', \tag{6.3.2j}$$

$$\lambda_v \in \mathbb{R} \quad \forall v \in V. \tag{6.3.2k}$$

Here the variable bounds are defined as

$$\overline{\lambda}_v := \begin{cases} \infty & \text{if } \tilde{\pi}_v \geq \overline{\pi}_v, \\ 0 & \text{else} \end{cases} \qquad \forall v \in V, \tag{6.3.3a}$$

$$\underline{\lambda}_v := \begin{cases} -\infty & \text{if } \tilde{\pi}_v \leq \underline{\pi}_v, \\ 0 & \text{else} \end{cases} \qquad \forall v \in V, \tag{6.3.3b}$$

$$\overline{\nu}_a := \begin{cases} \infty & \text{if } \beta_a \tilde{y}_a = \max\{\beta_a \underline{y}_a, \beta_a \overline{y}_a\} \\ 0 & \text{else} \end{cases} \qquad \begin{array}{r} \forall a \in A', \\ a = (v, w), \end{array} \tag{6.3.3c}$$

$$\underline{\nu}_a := \begin{cases} -\infty & \text{if } \beta_a \tilde{y}_a = \min\{\beta_a \underline{y}_a, \beta_a \overline{y}_a\} \\ 0 & \text{else} \end{cases} \qquad \begin{array}{r} \forall a \in A', \\ a = (v, w). \end{array} \tag{6.3.3d}$$

$$\overline{\mu}_a := \begin{cases} \infty & \text{if } \beta_a \tilde{y}_a = \max\{\beta_a \underline{y}_a, \beta_a \overline{y}_a\} \\ \infty & \text{if } \exists k : [A_a\,(\tilde{q}_a, \tilde{p}_v, \tilde{p}_w)]_k \geq [b_a]_k, \\ & (A_a)_{(k,1)} \geq 0, (A_a)_{(k,2)} < 0, \\ & (A_a)_{(k,3)} > 0, \\ 0 & \text{else} \end{cases} \qquad \begin{array}{r} \forall a \in A', \\ a = (v, w), \end{array} \tag{6.3.3e}$$

$$\underline{\mu}_a := \begin{cases} -\infty & \text{if } \beta_a \tilde{y}_a = \min\{\beta_a \underline{y}_a, \beta_a \overline{y}_a\} \\ -\infty & \text{if } \exists k : [A_a\,(\tilde{q}_a, \tilde{p}_v, \tilde{p}_w)]_k \geq [b_a]_k, \\ & (A_a)_{(k,1)} \leq 0, (A_a)_{(k,2)} > 0, \\ & (A_a)_{(k,3)} < 0, \\ 0 & \text{else} \end{cases} \qquad \begin{array}{r} \forall a \in A', \\ a = (v, w), \end{array} \tag{6.3.3f}$$

Now let $(\lambda^*, \mu^*, \nu^*)$ be a feasible solution for (6.3.2). The vectors ν^* and μ^* form a network flow by constraints (6.3.2b) and (6.3.2c). For the following discussion we focus on the case that either $\mu^* \geq 0$ if $\lambda^* \neq 0$ or $\nu^* \geq 0$ holds. The case where these assumptions are not fulfilled can be led back to the case fulfilling the assumptions by changing the orientation of some arcs. Our initial motivation for the definition of MILP (6.2.5) was to look either for a path or a circuit as discussed in the following two cases:

Case $\lambda^* \neq 0$**:** We split the network flow μ^* into sets of flow along paths $P_1, \ldots, P_m$ and flow along circuits $C_1, \ldots, C_n$, see Theorem 4.2.5. This way we obtain from $\mu^* \geq 0$ that there exist flow values $\mu_{P_i} > 0, i = 1, \ldots, m$ and $\mu_{C_i} > 0, i = 1, \ldots, n$ such that

$$\mu^*_a = \sum_{\substack{i=1,\ldots,m:\\ a \in A'(P_i)}} \mu_{P_i} + \sum_{\substack{i=1,\ldots,n:\\ a \in A'(C_i)}} \mu_{C_i} \qquad \forall\, a \in A'.$$

Consider a path P_ℓ that starts in node v and ends in node w. Because of constraint (6.3.2d) the index ℓ can be chosen such that either $\tilde{\pi}_v > \overline{\pi}_v$ and $\tilde{\pi}_w \leq \underline{\pi}_w$ or $\tilde{\pi}_v \geq \overline{\pi}_v$ and $\tilde{\pi}_w < \underline{\pi}_w$ holds. Let the nodes of P_ℓ be given by $v_1, \ldots, v_{n+1}$ where $v_1 = v$ and $v_{n+1} = w$ and connecting arcs by $a_1, \ldots, a_n$.

In order to show that (q', π', p', y') is not feasible for the active transmission problem (6.1.1) we distinguish two cases for each arc a_i of the path P_ℓ:

1. In the case that $\beta_{a_i}\tilde{y}_{a_i} = \max\{\beta_{a_i}\underline{y}_{a_i}, \beta_{a_i}\overline{y}_{a_i}\}$ we obtain the following estimation from $\tilde{q}_{a_i} \leq q'_{a_i}$ if $\alpha_{a_i} \neq 0$ (by (6.3.2e)) and $\beta_{a_i}y'_{a_i} \leq \beta_{a_i}\tilde{y}_{a_i}$:

$$\begin{aligned}\tilde{\pi}_{v_i} - \gamma_{a_i}\tilde{\pi}_{v_{i+1}} &= \alpha_{a_i}\,\tilde{q}_{a_i}|\tilde{q}_{a_i}|^{k_{a_i}} - \beta_{a_i}\tilde{y}_{a_i}\\ &\leq \alpha_{a_i}\,q'_{a_i}|q'_{a_i}|^{k_{a_i}} - \beta_{a_i}y'_{a_i} = \pi'_{v_i} - \gamma_{a_i}\pi'_{v_{i+1}}.\end{aligned}$$

2. In the case that $[A_{a_i}(\tilde{q}_{a_i}, \tilde{p}_{v_i}, \tilde{p}_{v_{i+1}})]_k \geq [b_{a_i}]_k$ holds for an index k with $(A_{a_i})_{(k,1)} \geq 0, (A_{a_i})_{(k,2)} < 0, (A_{a_i})_{(k,3)} > 0$ we rewrite this inequality as $a_1\tilde{q}_{a_i} - a_3 \geq \tilde{p}_{v_i} - a_2\tilde{p}_{v_{i+1}}$ with $a_1 \in \mathbb{R}_{\geq 0}$, $a_2 \in \mathbb{R}_{>0}$ and $a_3 \in \mathbb{R}$. Then we derive the estimation (using (6.3.2e) and (6.2.6i), which yields $a_1 > 0 \Rightarrow \tilde{q}_{a_i} \leq q'_{a_i}$):

$$\tilde{p}_{v_i} - a_2\tilde{p}_{v_{i+1}} \leq a_1\tilde{q}_{a_i} - a_3 \leq a_1 q'_{a_i} - a_3 \leq p'_{v_i} - a_2 p'_{v_{i+1}}.$$

We note that at least one of the previous cases applies because of $0 < \mu^*_{a_i} \leq \overline{\mu}_{a_i}$ and (6.3.3e). Because of the coupling constraint $\tilde{p}_v|\tilde{p}_v| = \tilde{\pi}_v$ relating the pressure and node potential variables for each node $v \in V$, we obtain:

$$\tilde{\pi}_{v_1} > \pi'_{v_1} \quad \Rightarrow \tilde{\pi}_{v_2} > \pi'_{v_2}, \ \ldots, \tilde{\pi}_{v_{n+1}} > \pi'_{v_{n+1}}, \tag{6.3.4a}$$

$$\tilde{\pi}_{v_{n+1}} < \pi'_{v_{n+1}} \Rightarrow \tilde{\pi}_{v_n} < \pi'_{v_n}, \ \ldots, \quad \tilde{\pi}_{v_1} < \pi'_{v_1}. \tag{6.3.4b}$$

The path P_ℓ is chosen such that either the start node v_1 or the end node v_{n+1} of path P_ℓ violates its node potential bound, i.e., one of the following cases applies:

$$\tilde{\pi}_{v_1} > \overline{\pi}_{v_1} \text{ and } \tilde{\pi}_{v_{n+1}} \leq \underline{\pi}_{v_{n+1}} \quad \text{and } \overline{\pi}_{v_1} \geq \pi'_{v_1} \quad \overset{(6.3.4a)}{\Rightarrow} \pi'_{v_{n+1}} < \underline{\pi}_{v_{n+1}},$$
$$\tilde{\pi}_{v_1} \geq \overline{\pi}_{v_1} \text{ and } \tilde{\pi}_{v_{n+1}} < \underline{\pi}_{v_{n+1}} \text{ and } \underline{\pi}_{v_{n+1}} \leq \pi'_{v_{n+1}} \quad \overset{(6.3.4b)}{\Rightarrow} \quad \pi'_{v_1} > \overline{\pi}_{v_1}.$$

Hence π' violates the node potential bounds which implies that (q', π', p', y') is not feasible for the active transmission problem (6.1.1).

Case $\nu^* \neq 0, \lambda^* = 0$**:** Similar as in the previous case we split the network flow ν^* into sets of flow along circuits $C_1, \ldots, C_n$, see Theorem 4.2.5. By constraint (6.3.2d) there exists an arc $a \in A'$ with $\nu^*_a \neq 0$ and $\alpha_a > 0$ by (6.2.6h). From our assumption we obtain $\nu^*_a > 0$ for this arc. Let ℓ be chosen such that C_ℓ contains this arc. Let the nodes of C_ℓ be given by $v_1, \ldots, v_{n+1}$ where $v_1 = v_{n+1}$ and connecting arcs by $a_1, \ldots, a_n$. We note that $\beta_{a_i}\tilde{y}_{a_i} = \max\{\beta_{a_i}\underline{y}_{a_i}, \beta_{a_i}\overline{y}_{a_i}\}$ holds because of $0 < \mu^*_{a_i} \leq \overline{\mu}_{a_i}$ and (6.3.3c). From this observation we derive the following contradiction from $\tilde{q}_{a_i} \leq q'_{a_i}$ if $\alpha_{a_i} \neq 0$ (by (6.3.2f)):

$$\begin{aligned}
0 &= \sum_{i=1}^{n} \left(\prod_{j=1}^{i-1} \gamma_{a_j} \right) (\tilde{\pi}_{v_i} - \gamma_{a_i} \tilde{\pi}_{v_{i+1}}) \\
&= \sum_{i=1}^{n} \left(\prod_{j=1}^{i-1} \gamma_{a_j} \right) (\alpha_{a_i} \tilde{q}_{a_i} |\tilde{q}_{a_i}|^{k_{a_i}} - \beta_{a_i} \tilde{y}_{a_i}) \\
&< \sum_{i=1}^{n} \left(\prod_{j=1}^{i-1} \gamma_{a_j} \right) (\alpha_{a_i} q'_{a_i} |q'_{a_i}|^{k_{a_i}} - \beta_{a_i} y'_{a_i}) \\
&= \sum_{i=1}^{n} \left(\prod_{j=1}^{i-1} \gamma_{a_j} \right) (\pi'_{v_i} - \gamma_{a_i} \pi'_{v_{i+1}}) = 0.
\end{aligned}$$

The inequality is strict because $\kappa_a \nu_a > 0$ by (6.3.2d) and $\tilde{q}_a < \overline{q}_a \leq q'_a$ by (6.2.6h) and the feasibility of q'. This contradiction implies that our assumption was wrong and hence the solution (q', π', p', y') is not feasible for the active transmission problem (6.1.1).

We note that at least one of the above cases applies because of constraint (6.3.2d). This contradicts our assumption that (q', π', p', y') is feasible for the active transmission problem (6.1.1). This shows that this assumption was wrong.

6.4 Integration and Computational Results

In this chapter we focused on the topology optimization problem (3.2.1) for the third type of network considered in this thesis. Recall that every gas network is associated with this type, i.e., especially compressors and control valves might be contained.

The strategy for solving the topology optimization problem is as follows: We solve the model (3.2.1) by SCIP as described in Section 2.2. Furthermore we combine the methods of the previous sections for solving the active transmission problem (6.1.1) as follows: First we compute a feasible solution of one of the relaxations (6.1.2) or (6.1.5). We either obtain a feasible solution for (6.1.1) directly, if the slack value equals zero. Or we solve the infeasibility detection MILP (6.2.5). If this problem turns out to be infeasible or has optimal objective value zero, then we conclude that the active transmission problem (6.1.1) is infeasible by Theorem 6.2.2.

In Chapter 4 we computationally showed that the domain relaxation (4.2.1), which is based on bound relaxations, has lower computation times than the relaxation (4.3.1), which relaxes the flow conservation constraint. Let us proceed similarly for a comparison of the two non-convex problems (6.1.2) and (6.1.5). We implemented both relaxations in C and use the computational setup described in Section 3.5. A scatter plot comparing the run times of both relaxations is shown in Figure 6.3 for different scenarios and the network `net6`. It turns out that the domain relaxation (6.1.2) has better solving performance. Thus we use this relaxation in our solution approach and will not consider the flow conservation relaxation (6.1.5) any longer.

The new solution approach that we follow for solving the topology optimization problem (3.2.1) is as follows: At the time during the solving process when a node of the branching tree is considered and where all binary decisions x are fixed, we consider the corresponding active transmission problem and solve the domain relaxation (6.1.2). If we obtain a solution with zero slack, then this yields a feasible solution for the active transmission problem. In this sense, solving the domain relaxation (6.1.2) is a primal heuristic for the active transmission problem. Otherwise, if we obtain a solution with positive slack, then the solution is infeasible for the active transmission problem and violates at most the constraints (6.1.1d) and (6.1.1f)-(6.1.1i). Now we apply Theorem 6.2.2 and solve the infeasibility detection MILP (6.2.5). If (6.2.5) is infeasible or has optimal objective value zero, then the active transmission problem is infeasible by Theorem 6.2.2. In this case we prune the corresponding node of the branch-and-bound tree. Otherwise, if we cannot decide that the current active transmission problem is infeasible, then we continue with branching.

We implemented the algorithm above in C, i.e., solving the domain relaxation (6.1.2) and the infeasibility detection MILP (6.2.5). The MILP is expressed using

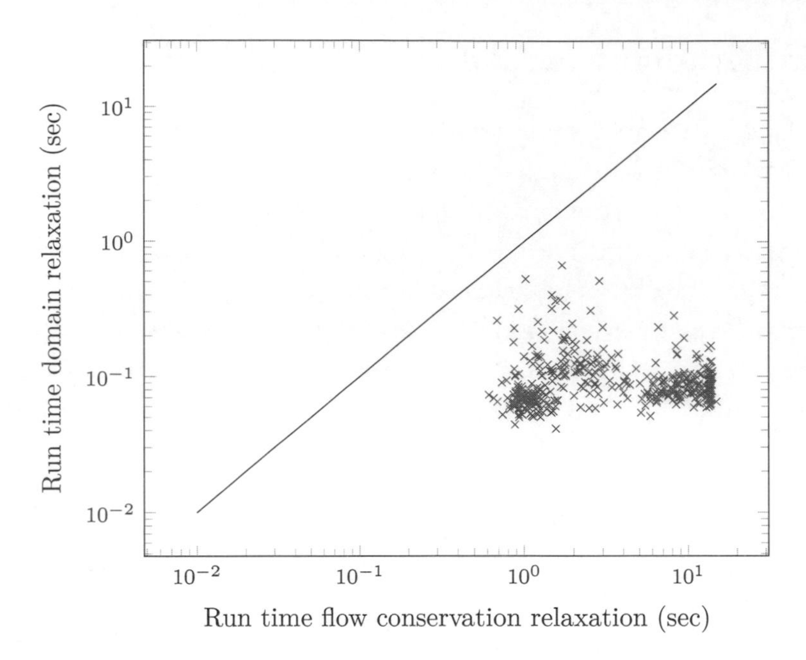

Figure 6.3: Run time comparison for the domain relaxation (6.1.2) and the flow conservation relaxation (6.1.5) on instances of network `net6`.

indicator constraints as described in Remark 6.2.3. For the computational studies we use the setup described in Section 3.5. We compare three strategies for solving the topology optimization problem (3.2.1).

1. The first strategy is to use SCIP without any adaptations on the solver settings. All branching decisions are up to the solver, and the topology optimization problem (3.2.1) is basically solved by branch-and-bound, separation and spatial branching as described in Section 2.2.

2. The second strategy is to enforce a certain branching priority rule, so that SCIP first branches on binary and discrete decision variables. Only after all discrete variables are fixed it is allowed to perform spatial branching on continuous variables.

3. The third strategy implements the domain relaxation strategy presented in this chapter for solving the active transmission problem (6.1.1). We use the nonlinear solver IPOPT for solving the domain relaxation (6.1.2) and SCIP for solving the infeasibility detection MILP (6.2.5) in advance. For solving this MILP we set a time limit of 15 s. Additionally we set branching priorities according to the second strategy.

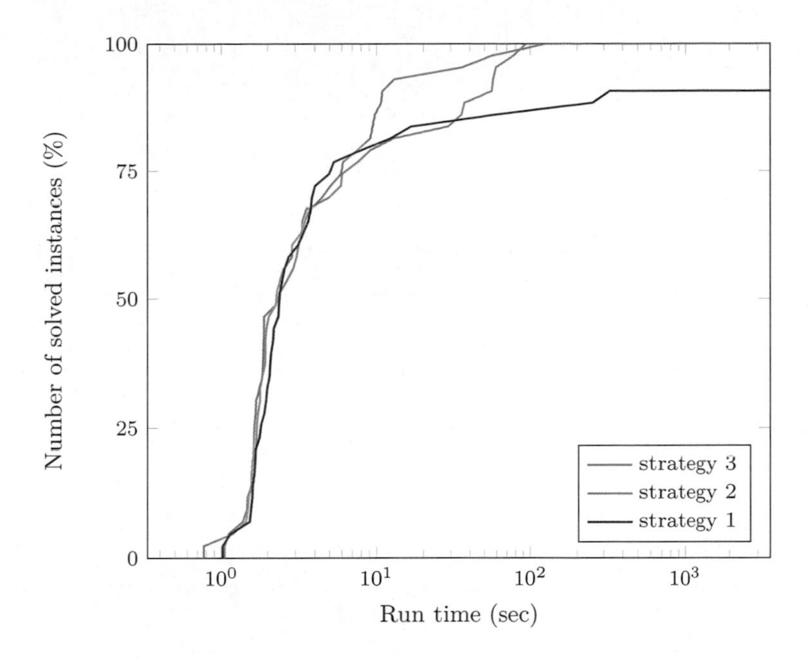

Figure 6.4: Performance plot for different nominations on the network `net6` and a time limit of 3600 s. The different strategies are described in Section 6.4. Strategy 1 and 2 mainly consist of SCIP. Strategy 3 also corresponds to SCIP together with our elaborated solution method presented in this chapter. The underlying data are available in Table A.10.

Computational Results

We consider the real-world network `net6` shown in Figure 3.10. This network contains active elements, i.e., valves, compressors and control valves and is in industrial use. We solve the nomination validation problem which is a feasibility problem. The 43 instances are nominations from our industrial partner. Figure 6.4 shows a performance plot of the instances that could be solved within a given time limit of 3600 s. From Figure 6.4 we conclude that the first strategy clearly performs worse than the second and third one. The third strategy keeps up with the second one but performs slightly worse.

Now we turn to the networks `net3`, `net4`, and `net5`. The instances are topology expansion problems, i.e., we optimize over a set of extensions and the objective function is nonzero. For all these instances we set a time limit of 39 600 s. We consider an additional fourth strategy:

4. The fourth strategy implements the domain relaxation (6.1.2) for computing a primal feasible solution. In comparison to the third strategy we do not check infeasibility conditions of the active transmission problem by solving (6.2.5).

strategy	1	2	3	4	all
solved instances	43	48	63	50	63

Table 6.5: Summary of the Tables A.11-A.13 showing the globally solved instances out of 76 nominations in total. The third strategy globally solves all instances which are solved to global optimality by the second and the first strategy.

	(A,B) = (2,3)			(A,B) = (2,4)		
	solved(48)		incomp.(3)	solved(46)		incomp.(15)
	time [s]	nodes	gap [%]	time [s]	nodes	gap [%]
strategy A	50.4	2,565	28	42.4	2,215	31
strategy B	62.3	2,021	17	59.4	2,280	40
shifted geom. mean	23 %	−21 %	−40 %	40 %	3 %	29 %

Table 6.6: Run time, number of branch-and-bound nodes and gap comparison for the strategies 2 and 3 and additionally 2 and 4 (aggregated results). The columns "solved" contain mean values for those instances globally solved by both strategies A and B. The columns "incomplete" show mean values for those instances having a primal feasible solution available but were not globally solved by both strategies A and B. The underlying data are available in Tables A.11-A.13.

So we do not detect infeasibility of the active transmission problem and thus do not manually prune any node of the branch-and-bound tree.

The computational results are shown in Tables A.11-A.13 and a summary is available in Table 6.5 and 6.6. We use the geometric mean of run time, number of branch-and-bound nodes and gap as described in Section 3.5. The third strategy clearly outperforms the other strategies in terms of number of solved instances. All instances globally solved by strategies 1 or 2 or 4 are also solved by the third strategy. Approximately 20 % more instances of the test set (15 out of 76) are solved by the third strategy compared to the second one. The second strategy performs better than the first one because it solves 5 more instances within the time limit. We conclude that branching priorities as set by the second strategy are a first step to improve the solving performance of SCIP. This is consistent with our observations from Chapter 4.

Approximately 63 % of the instances of the test set (48 out of 76) are globally solved by the second strategy. Here the run time increases by approximately 23 % on average following strategy 3 while the number of nodes is decreased by approximately 21 %. Clearly the reduction does not pay off due to the increasement in run time. We conclude that the domain relaxation (6.1.2) together with the infeasibility detection MILP (6.2.5) consume more time than spatial branching on those instances globally solved by SCIP in combination with branching priorities.

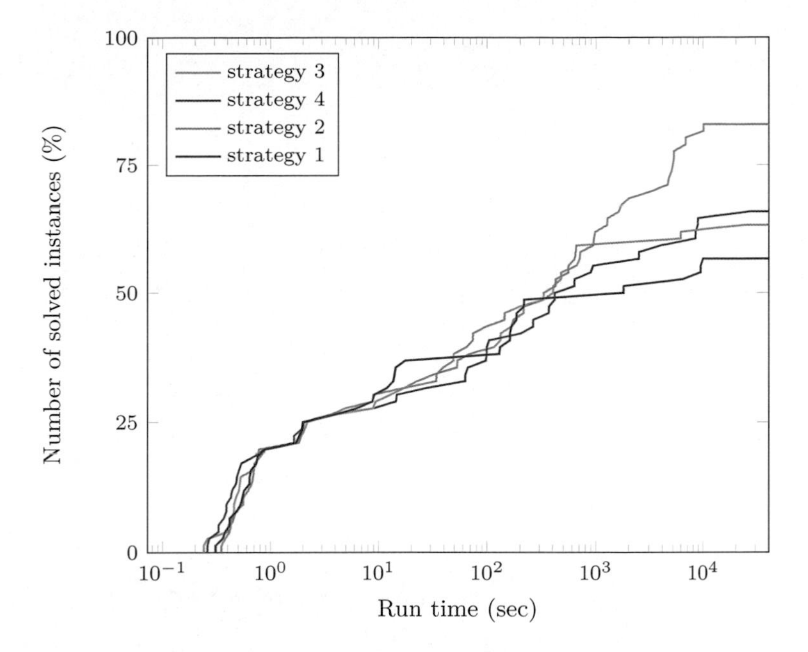

Figure 6.5: Performance plot for different nominations on the networks `net3`, `net4`, and `net5` and a time limit of 39 600 s. The different strategies are described in Section 6.4. Strategy 1 and 2 mainly consist of SCIP. Strategy 3 also corresponds to SCIP together with our elaborated solution method presented in this chapter. Strategy 4 is a weaker version of strategy 3 and only necessary for evaluation of the computational results. The underlying data are available in Tables A.11-A.13.

The fourth strategy allows to solve 65 % of the instances of the test set (50 out of 76) and hence performs nearly similar as the second strategy in terms of number of globally solved instances. But the run time increases by approximately 40 % and the number of nodes by approximately 3 %. We conclude that the primal heuristic (solving the domain relaxation (6.1.2)) as well as the verification of the infeasibility conditions (represented by the infeasibility detection MILP (6.2.5)) are both important for the performance of the third strategy.

A performance plot of all instances is shown in Figure 6.5. We conclude that the third strategy yields the best results and use it in our practical application. Here we accept an increase in run time because more instances are solved by this strategy.

In a last step we focused on the network `net7`. Here the results do not differ much in comparison to the initial computational study in Section 1.4. Thus we do not report them. Instead we present a primal heuristic for these instances in the next chapter. We also apply this heuristic to the network `net5` because very few feasible solutions are computed by the afore presented approach.

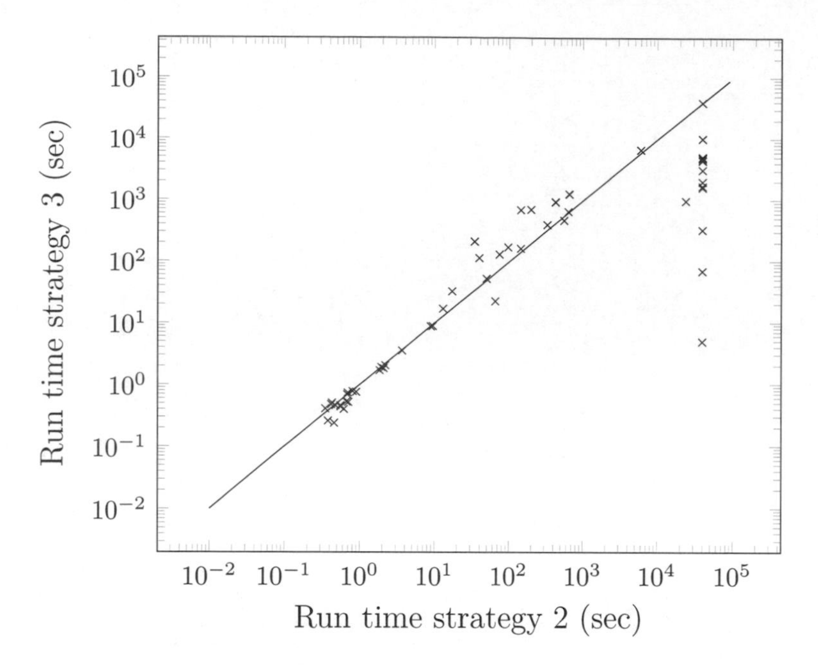

Figure 6.6: Run time comparison on different nominations on the networks `net3`, `net4`, and `net5`. Each cross (×) corresponds to a single instance of the test set. Note that multiple crosses are drawn in the upper right corner of the plots and cannot be differed. They represent those instances that ran into the time limit by both strategies. The underlying data are available in Tables A.11-A.13.

Summary

We presented a new method for proving infeasibility of the active transmission problem (6.1.1) when given a feasible solution for the relaxations (6.1.2) or (6.1.5). This allows the following solution approach for the topology optimization problem (3.2.1): We use SCIP and therein solve the active transmission problem (6.1.1) by computing a local optimal solution of the domain relaxation (6.1.2), which is a non-convex optimization problem. In advance we analyze this solution by the infeasibility detection MILP (6.2.5) and possibly conclude that the active transmission problem is infeasible. This clearly outperforms SCIP used without any adaptations on the implementation. Approximately 20 % more instances of the test set are solved to global optimality compared to the solver SCIP. The test set is formed by topology expansion instances on the third type of network considered in this thesis, i.e., networks that contain control valves and compressors.

Chapter 7

A Primal Heuristic based on Dual Information

In this chapter we present a primal heuristic for the topology optimization problem (3.2.1). In contrast to the previous Chapters 4-6 we do not focus on a special class of gas networks, i.e., we do not restrict to networks that contain only certain elements.

Recall that the subproblem which results from fixing all binary variables x in (3.2.1) yields the active transmission problem (6.1.1). It is a continuous nonlinear optimization problem and the constraints are continuously differentiable functions. The outline of the heuristic is as follows: We consider a feasible relaxation of the active transmission problem and compute a KKT point. If we obtain a primal solution with zero slack, then we derive a feasible solution for (3.2.1) by neglecting the slack variables. Otherwise we make use of the dual solution of the KKT point and apply parametric sensitivity analysis to identify constraints of the subproblem which we adapt in a next step. For these constraints we consider the associated binary variables of the MINLP and switch them to different values. The new binary values lead to another subproblem of the original MINLP which contains continuous variables only. As before, we consider a relaxation of the new active transmission problem and compute a KKT point. On the basis of the choice of the switching variables we have good chances to obtain a solution that has less slack than the solution that we computed before. We repeat this process for a predefined number of iterations. Furthermore we integrate this method in a branch-and-prune search in order to avoid cycles. Our aim is to come up with a certain active transmission problem together with a feasible solution, i.e., a feasible solution with zero slack for the corresponding relaxation.

Recall that the topology optimization problem (3.2.1) is solved by branch-and-bound, separation, and spatial branching. Within this approach the heuristic is invoked at those nodes in the branch-and-bound tree where all binary decision variables have binary values, either by branching or by the solution of the LP relaxation. Fixing all binary variables of the topology optimization problem yields the active transmission problem (6.1.1). By Lemma 6.1.1 a feasible relaxation is given by the domain relaxation (6.1.2). By using the dual solution of a KKT point of this relaxation we identify binary variables that we change in each iteration. This selection will be improved further. We make use of the problem structure of the topology optimization problem by switching selectively between the different modes of a valve, a control valve or a compressor.

We apply the heuristic to the real-world network `net7`. Note that `net7` remains with very few primal feasible solutions available compared to the computational results of the previous chapters. The associated problem for `net7` is the nomination validation problem and so we solve it only for feasibility. Our initial computational study showed that the MINLP solvers ANTIGONE, BARON and SCIP were not able to compute a feasible solution for more than one instance of the nomination validation problems on the final version of network `net7` that we were given from our cooperation partner, see Section 1.4. Using the heuristic we are able to solve 18 out of 30 instances which corresponds to approximately 60 % of all these instances. When taking all the different versions of the network into account, the heuristic allows to solve globally approximately 61 % more instances of the test set compared to SCIP without any adaptations. We also apply the heuristic to the network `net5`. Similar to `net7`, this network remains with very few primal feasible solutions available compared to the results of the previous chapters. Here we combine the heuristic with the solution approach presented in the Chapter 6.

The outline of this chapter is as follows: In Section 7.1 we describe the MINLP that we consider in this chapter. The heuristic is presented in Section 7.2. In Section 7.3 we show an adaptation of the heuristic to the topology optimization problem (3.2.1). Afterwards we give some information of the implementation details in Section 7.4. Computational results are given in Section 7.5.

7.1 A Relaxation of the MINLP

In this chapter we consider a mixed-integer nonlinear optimization problem with indicator constraints and variables x, z of the following form:

$$\min \quad f(x,z) \tag{7.1.1a}$$

$$\text{s.t.} \quad g(x,z) \leq 0, \tag{7.1.1b}$$

$$
\begin{aligned}
h_i(z) - \Delta_i \le 0 \qquad & \forall i \in I, && (7.1.1c)\\
x_i = 1 \Rightarrow \Delta_i = 0 \qquad & \forall i \in I, && (7.1.1d)\\
\underline{z} \le z \le \overline{z}, & && (7.1.1e)\\
z \in \mathbb{R}^n, & && (7.1.1f)\\
\Delta \in \mathbb{R}^I_{\ge 0}, & && (7.1.1g)\\
x \in \{0,1\}^I. & && (7.1.1h)
\end{aligned}
$$

Here $n \in \mathbb{N}$ are the dimensions of the real variables. Furthermore $I := \{1, 2, \dots\}$ with cardinality $|I|$ is a nonempty index set. The functions $f, g : \{0,1\}^I \times \mathbb{R}^n \to \mathbb{R}$ and $h_i : \mathbb{R}^n \to \mathbb{R}$ (for $i \in I$) are $\mathcal{C}^2$ (twice continuously differentiable). The binary variable x_i represents the decision whether the constraint $h_i(z) \le 0$ is active (for $x_i = 1$) or inactive (for $x_i = 0$). A constraint h_i is inactive, if and only if a slack variable Δ_i is positive, where $\Delta = (\Delta_1, \dots, \Delta_{|I|})$ denotes the vector of continuous slack variables. The constraints (7.1.1d) are the indicator constraints of the problem. For the ease of notation we first focus on MINLP (7.1.1) in a general form before turning to the problem of optimizing the topology of gas networks in Section 7.3.

Now we turn to a relaxation of a subproblem of (7.1.1) which forms the basis for our heuristic and is obtained by fixing all binary variables. Assume we are given binary values x^* for the binary variables x. Then the subproblem is obtained by fixing $x = x^*$. Note that it is a nonlinear optimization problem which might be infeasible. To circumvent its infeasibility we consider a relaxation of this subproblem of (7.1.1). This relaxation is obtained as follows: Let $I' \subseteq I$ contain those indices such that the binary variables x_i^* are 1 for all $i \in I'$. We define $I'' \subseteq I'$ and relax all constraints $h_i(z) \le 0$ with $i \in I''$. Furthermore we introduce a subset $J \subseteq \{1, \dots, n\}$ and relax the variable bounds of x_j for all $j \in J$. Therefor we introduce two vectors of slack variables $s^+, s^- \in \mathbb{R}^J_{\ge 0}$. Then the relaxation writes as follows for $p = 0 \in \mathbb{R}^{I'}$:

$$
\min \sigma f(x^*, z) + (1-\sigma) \sum_{j \in J} (s_j^+ + s_j^-) + (1-\sigma) \sum_{i \in I''} \Delta_i \qquad (7.1.2a)
$$

$$
\begin{aligned}
\text{s.t.} \quad g(x^*, z) \le 0, & && (7.1.2b)\\
h_i(z) \le p_i \quad & \forall i \in I' \setminus I'', && (7.1.2c)\\
h_i(z) - \Delta_i \le p_i \quad & \forall i \in I'', && (7.1.2d)\\
z_j - s_j^+ \le \overline{z}_j \quad & \forall j \in J, && (7.1.2c)\\
z_j + s_j^- \ge \underline{z}_j \quad & \forall j \in J, && (7.1.2f)\\
z \in \mathbb{R}^n, & && (7.1.2g)
\end{aligned}
$$

$$\Delta \in \mathbb{R}_+^{I''}, \tag{7.1.2h}$$
$$s^\pm \in \mathbb{R}_+^J. \tag{7.1.2i}$$

We assume a definition of the index sets I'' and J in such a way that the relaxation (7.1.2) is feasible. The parameter $\sigma \in [0,1]$ in the objective function (7.1.2a) controls the compromise between improving feasibility and optimality: For $\sigma = 0$ we aim to compute a primal feasible solution for MINLP (7.1.1) while for $\sigma = 1$ our goal is to compute a globally optimal solution for (7.1.1). Hence a lower value of σ puts a higher emphasis on feasibility. The term

$$(1-\sigma)\sum_{j\in J}(s_j^+ + s_j^-) + (1-\sigma)\sum_{i\in I'}\Delta_i$$

is called the *slack part* of the objective function.

7.2 A Primal Heuristic for MINLP with Indicator Constraints

In the following we describe the primal heuristic for the MINLP with indicator constraints (7.1.1). It is invoked for a given vector of binary values x^*. We regard (7.1.2) as an optimization problem with parameters (x^*, p). The heuristic aims to compute a vector x^* and a feasible solution for (7.1.2) with parameters $(x^*, 0)$ so that the objective value of this solution equals zero. When neglecting the slack variables we obtain a feasible solution for (7.1.1).

The outline of this section is as follows: First we give a motivation of the mathematical background in Section 7.2.1. Then we present the procedure of our heuristic in Section 7.2.2. In Section 7.2.3 we describe how to embed the heuristic in a branch-and-prune search.

7.2.1 Theoretical Motivation

Let us denote the objective function in (7.1.2a) by $\tilde{f}(x^*, z, s, \Delta)$. We consider a KKT point of (7.1.2) with primal solution $(\hat{z}, \hat{s}, \hat{\Delta})$ and dual solution $(\hat{\lambda}, \hat{\eta}, \hat{\mu}, \hat{\xi}, \hat{\zeta})$. Here the dual variables λ correspond to inequalities (7.1.2b), η to (7.1.2c) and (7.1.2d), $\mu^\pm$ to (7.1.2e) and (7.1.2f), ξ to (7.1.2h), and $\zeta^\pm$ to (7.1.2i), respectively. Our aim is to make use of equation (5.4) of Fiacco and Ishizuka (1990), which states a relation between a modification of a constraint and its impact on the objective function value via Lagrange multipliers from a corresponding KKT point. The main result that we derive from their equation is

$$\partial_{p_i} \tilde{f}(x^*, \hat{z}, \hat{s}, \hat{\Delta}, p^*) = -\hat{\eta}_i \qquad \forall i \in I', \tag{7.2.1}$$

where the dual value $\hat{\eta}_i$ is the Lagrange multiplier of the constraints $h_i(z) - \Delta_i - p_i^* \leq 0$ or $h_i(z) - p_i^* \leq 0$ of (7.1.2).

In the following, we briefly outline the main steps of the derivation of Fiacco and Ishizuka (1990). Therefor we consider problem (7.1.2) as a parametric nonlinear optimization problem of the general form

$$\min_r F(r, p), \tag{7.2.2a}$$

$$G_i(r, p) \geq 0, \quad \forall i \in I, \tag{7.2.2b}$$

where $r \in \mathbb{R}^{N_n}$ are variables and $p \in \mathbb{R}^I$ are parameters. We note that $r = (z, s, \Delta)$ for the relaxation (7.1.2). Furthermore we assume that F and G are twice continuously differentiable, real valued functions.

A KKT point $(\hat{r}, \hat{\eta})$ of (7.2.2) with Lagrange multipliers $\eta \in \mathbb{R}^I_{\geq 0}$ depends on the chosen parameter p, hence we denote it by $(\hat{r}(p), \hat{\eta}(p))$, in order to emphasize this dependency. We select a parameter p_0 and obtain a solution $\hat{r} = \hat{r}(p_0)$ by solving (7.2.2). In order to estimate the change in the objective function when altering the parameter p in a small neighborhood of p_0, we differentiate the objective function $F(\hat{r}(p), p)$ with respect to p and obtain:

$$\frac{dF}{dp}(\hat{r}, p_0) = \nabla_r F(\hat{r}, p_0) \frac{dr}{dp}(p_0) + \frac{\partial F}{\partial p}(\hat{r}, p_0). \tag{7.2.3}$$

Writing the Lagrange function of (7.1.2) as $L(r, \eta, p) = F(r, p) - \eta G(r, p)$, it holds $\nabla_r L(\hat{r}, \hat{\eta}, p_0) = 0$ by (2.4.2) for the KKT point $(\hat{r}, \hat{\eta}, p_0)$. Hence we can rewrite the first summand on the right-hand side of (7.2.3) as

$$\nabla_r F(\hat{r}, p_0) = \eta_0 \nabla_r G(\hat{r}, p_0),$$

where Lagrange multipliers η_0 are the values $\hat{\eta}$ for the parameter choice $p = p_0$. We use this to rewrite (7.2.3) as

$$\frac{dF}{dp}(\hat{r}, p_0) = \eta_0 \nabla_r G(\hat{r}, p_0) \frac{dr}{dp}(p_0) + \frac{\partial F}{\partial p}(\hat{r}, p_0). \tag{7.2.4}$$

We denote by $I' \subseteq I$ the subset of active constraints from I for $(\hat{r}, p_0)$, i.e., $G_i(\hat{r}, p_0) = 0$ for $i \in I'$. Further define $\tilde{G} := (G_i(\hat{r}, p_0))_{i \in I'}$. By Fiacco and Ishizuka (1990) there exists a neighborhood of p_0 such that $\tilde{G}(\hat{r}(p), p) = 0$ for all p within this p_0-neighborhood. This means, the function $\tilde{G}(\hat{r}(\cdot), \cdot)$ is locally constant, which implies that $\frac{d\tilde{G}}{dp}(\hat{r}(p), p) = 0$ for all p in the neighborhood of p_0. We compute this derivation in p_0 and obtain

$$0 = \frac{d\tilde{G}}{dp}(\hat{r}, p_0) = \nabla_r \tilde{G}(\hat{r}, p_0) \frac{dr}{dp}(p_0) + \nabla_p \tilde{G}(\hat{r}, p_0). \tag{7.2.5}$$

We put (7.2.5) into (7.2.4), make use of $\eta_{0_i} = 0, i \in I \setminus I'$ for the inactive constraints and obtain

$$\frac{dF}{dp}(\hat{r}, p_0) = -\eta_0 \nabla_p G(\hat{r}, p_0) + \frac{\partial F}{\partial p}(\hat{r}, p_0). \tag{7.2.6}$$

Now we consider the derivation of the Lagrange function L with respect to p:

$$\frac{dL}{dp}(\hat{r}, \eta_0, p_0) = \underbrace{\nabla_r L(\hat{r}, \eta_0, p_0)}_{=0,\text{by KKT}} \frac{dr}{dp}(p_0) + \underbrace{\nabla_\eta L(\hat{r}, \eta_0, p_0)}_{=G(\hat{r}, \eta_0, p_0)=0} \frac{d\eta}{dp}(p_0) + \frac{\partial L}{\partial p}(\hat{r}, \eta_0, p_0).$$

From

$$\frac{\partial L}{\partial p}(\hat{r}, \eta_0, p_0) = -\eta_0 \nabla_p G(\hat{r}, p_0) + \frac{\partial F}{\partial p}(\hat{r}, p_0)$$

and (7.2.6) we conclude that

$$\frac{dF}{dp}(\hat{r}, p_0) = \frac{dL}{dp}(\hat{r}, \eta_0, p_0). \tag{7.2.7}$$

In order to make use of this equation, we first write down the Lagrange function $L(x^*, z, s, \Delta, p^*, \mu, \eta, \lambda, \xi, \zeta)$ of (7.1.2), which is defined as follows:

$$\begin{aligned} L(x^*, z, s, \Delta, p^*, \mu, \eta, \lambda, \xi, \zeta) = {} & \sigma f(x^*, z) + (1 - \sigma) \left(\sum_{j=1}^{n} (s_j^+ + s_j^-) + \sum_{i \in I''} \Delta_i \right) \\ & + \lambda\, g(x^*, z) \\ & + \sum_{i \in I'} \eta_i \,(h_i(z) - p_i^*) - \sum_{i \in I''} \eta_i \,\Delta_i \\ & + \sum_{j=1}^{n} \left(\mu_j^+ (z_j - s_j^+ - \overline{z}_j) + \mu_j^- (\underline{z}_j - z_j - s_j^-) \right) \\ & - \sum_{i \in I'} \xi_i \end{aligned}$$

$$-\sum_{j=1}^{n}(\zeta_j^+ + \zeta_j^-).$$

Making use of equation (7.2.7) we obtain from this Lagrange function:

$$\partial_{p_i} \tilde{f}(x^*, \hat{z}, \hat{s}, \hat{\Delta}, p^*) = -\hat{\eta}_i \qquad \forall i \in I'. \tag{7.2.8}$$

This equation expresses the change of the objective function $\tilde{f}$, when parameter p_i is changed.

7.2.2 The Basic Dual Value MINLP Heuristic

In the previous section we derived relation (7.2.8) between the dual value of a constraint of (7.1.2) and the impact on its objective by changing this constraint. Making use of this relation our heuristic works as follows: An iterative process is started at problem (7.1.2) based on the binary decisions x^*. In each step the heuristic selects one fixed binary variable and assigns a different binary value to it (this corresponds to a flip, from upper to lower bound, or vice versa). This is done as follows: First a KKT point of the relaxation (7.1.2) with parameters $(x^*, 0)$ is computed. Then all the constraints (7.1.2c) and (7.1.2d) are ranked according to the current values of the dual solutions, i.e., the right-hand sides in (7.2.8). This ranking yields a ranking of those binary variables x which are associated by (7.1.1c) and (7.1.1d). The variable corresponding to the constraint with the highest rank is then the most promising candidate for an assignment of a different value. That is, if $\hat{\eta}_i \geq \max\{\hat{\eta}_j : j \in I'\}$ for some $i \in I'$, then the corresponding variable x_i (which is currently fixed to $x_i^* = 1$) will be fixed to 0 in the next iteration. By most promising we mean that for the new parameter $\tilde{x}^*$ for (7.1.2), which is obtained from x^* by changing a specific value, the optimal objective value of (7.1.2) is decreased. Recall that a feasible solution for (7.1.2) with optimal objective value zero also yields a feasible solution for (7.1.1) when neglecting the slack variables. When changing x^* to $\tilde{x}^*$ we also ensure that (7.1.2) is feasible for the new parameters $(\tilde{x}^*, 0)$.

We also keep track of the variable which is changed, so that we do not repeat the same decision again in later steps, in order to prevent the heuristic from cycling. Note that cycles occur if (7.1.1) contains the relation $x_1 = 1 - x_2$ for two binary variables x_1 and x_2. In this case it might occur that both binary variables x_1 and x_2 are iteratively changed.

In the next iteration, we solve the slack model (7.1.2) again. Note that the set I' changes from one iteration to the next. The objective function values $\tilde{f}$ from the previous iteration is compared with the objective function value of the current iteration, denoted by $\tilde{f}_+$. We allow a slight increase (worsening) of at most 20 % (or any other user-defined parameter) to accept this move. If the increase is more than that, the heuristic terminates without any result. Otherwise the process is iteratively continued.

Summing up, the heuristic selects one fixed binary variable in each step, and assigns a different binary value to it. In the long run, the objective function value of the slack model (7.1.2) typically decreases. Two cases can occur: First, at some iteration, we reach a point where the slack part of the objective function value is 0. In this case we obtain a feasible solution for (7.1.1) by neglecting the slack variables and the heuristic terminates successfully. The second case that might occur is that the slack part of the objective function does not converge to 0. If after a user-defined number of rounds the slack part is still nonzero, then the heuristic terminates. Thus it failed to construct a feasible solution.

7.2.3 Embedding the Heuristic in a Branch-and-Prune Search

The heuristic outlined above can be embedded in a tree search. It can run into a dead end if the slack part of the objective does not converge to 0 after a certain number of iterations. The idea of the tree search is to restart the heuristic at an earlier stage and thus to cover a wider range of potential changes of binary variables.

To this end, we do not only consider the single best variable with respect to the dual values as before. Instead, we use a small pool (of a user-defined size, typically 5 to 10 variables), and take the best dual variable until the pool is filled. Each iteration of the heuristic is considered as a node in the tree (and the start as the root node), and for each variable in the pool a child node is created and inserted in the tree.

We then traverse the tree in a depth-first search. If the heuristic fails, the node is pruned and a back-tracking to the previous unpruned nodes takes place. An example of this tree search is discussed in Example 7.3.2.

7.3 A Specialization to the Topology Optimization Problem

We are going to apply the heuristic method to the natural gas network topology optimization problem (3.2.1). We do not insist on the assumptions because our aim is to develop a heuristic method. Recall that we assumed above that all constraints of MINLP (7.1.1) are twice continuously differentiable. This is not the case for (3.2.1) as the function $q \mapsto q|q|$, for instance, is not $\mathcal{C}^2$. Nevertheless, we are going to apply the heuristic to (3.2.1). In Section 7.3.1 we present the relaxation (7.1.2) in the special case of solving (3.2.1). This relaxation is slightly different from the domain relaxation (6.1.2). In Section 7.3.2 and Section 7.3.3 we present special adaptations of the heuristic to the topology optimization problem. We note that the variables (z, x) of (7.1.1) change their roles to (q, π, p, y, x) with $z = (q, \pi, p, y)$ when turning to (3.2.1).

7.3.1 The Relaxation

Let us write down the relaxation (7.1.2) for the topology optimization problem (3.2.1). We define the set of arcs that are selected at the current node by $A' := \{(a, i) \in A_X \mid x_{a,i} = 1, i > 0\}$, i.e., A' contains all arcs where the flow is not fixed to zero by (3.2.1d). Furthermore let the set $A'_0 := \{(a, 0) \in A_X \mid x_{a,0} = 1\}$ contain all active arcs that are in closed state. We define the index set J' appropriately such that the flow and node potential variable bounds are relaxed. Furthermore we choose I'' such that all constraints (3.2.1c), that are active for the choice of x, are relaxed, too. We introduce $\Delta_v \in \mathbb{R}_{\geq 0}$ for all $v \in V$ and $\Delta_a \in \mathbb{R}_{\geq 0}$ and $\tilde{\Delta}_a \in \mathbb{R}^{\nu_a}_{\geq 0}$ for all $a \in A'$ as slack variables. Then, for $\sigma = 0$, the relaxation (7.1.2) writes as follows:

$$\min \sum_{v \in V} \Delta_v + \sum_{a \in A'} \Delta_a + \sum_{a \in A'} \sum_{k=1}^{\nu_a} (\tilde{\Delta}_a)_k \tag{7.3.1}$$

$$\begin{aligned}
\text{s.t.} \quad \alpha_a \, q_a |q_a|^{k_a} - \beta_a y_a - (\pi_v - \gamma_a \pi_w) &= 0 && \forall\, a = (v, w) \in A', \\
q_a &= 0 && \forall\, a = (v, w) \in A'_0, \\
A_a \left(q_a, p_v, p_w\right)^T - \tilde{\Delta}_a &\leq b_a && \forall\, a = (v, w) \in A', \\
\sum_{a \in \delta^+_{A' \cup A'_0}(v)} q_a - \sum_{a \in \delta^-_{A' \cup A'_0}(v)} q_a &= d_v && \forall\, v \in V, \\
|p_v| p_v - \pi_v &= 0 && \forall\, v \in V, \\
\pi_v - \Delta_v &\leq \overline{\pi}_v && \forall\, v \in V, \\
\pi_v + \Delta_v &\geq \underline{\pi}_v && \forall\, v \in V, \\
q_a - \Delta_a &\leq \overline{q}_a && \forall\, a \in A', \\
q_a + \Delta_a &\geq \underline{q}_a && \forall\, a \in A',
\end{aligned}$$

$$
\begin{aligned}
y_a &\leq \overline{y}_a && \forall\, a \in A', \\
y_a &\geq \underline{y}_a && \forall\, a \in A', \\
\pi_v &\in \mathbb{R} && \forall\, v \in V, \\
q_a &\in \mathbb{R} && \forall\, a \in A' \cup A'_0, \\
y_a &\in \mathbb{R} && \forall\, a \in A', \\
\Delta_v &\in \mathbb{R}_{\geq 0} && \forall\, v \in V, \\
\Delta_a &\in \mathbb{R}_{\geq 0} && \forall\, a \in A', \\
\tilde{\Delta}_a &\in \mathbb{R}^{\nu_a}_{\geq 0} && \forall\, a \in A'.
\end{aligned}
$$

We note that this relaxation equals the domain relaxation of the active transmission problem (6.1.2) except the additional flow variables q_a for $a \in A'_0$ and the constraints $q_a = 0$ for these arcs. So this relaxation is always feasible by Lemma 6.1.1 which allows to proceed as already described in Section 7.2.2.

Remark 7.3.1:
Note that the domain relaxation (6.1.2) *of the active transmission problem* (6.1.1) *turned out to be more efficient than the flow conservation relaxation* (6.1.5). *Hence we do not investigate further whether a similar heuristic process based on the flow conservation relaxation* (6.1.5) *could be derived.*

7.3.2 Handling Different Modes of Active Devices

When applying the heuristic described in Section 7.2 we identify binary variables that are switched to different values. For the topology optimization problem (3.2.1) some of these binary variables correspond to different modes of the active elements in the network. Recall that a valve for instance can be open or closed. A typical situation in the network is a compressor (or a control valve) that can be active (compressing or regulating), in bypass, or closed. Let us briefly recall each state.

- If the compressor is active, then the gas can flow through the compressor. In this case, the amount of flow is restricted by lower and upper bounds. The output pressure at the exit side of the compressor is higher than the input pressure, and the compression ratio between output and input pressure has to satisfy certain bounds. The operating range of a compressor $a \in A$ is thus described by ν_a linear inequalities in the system (A_a, b_a), see (3.2.1c). They

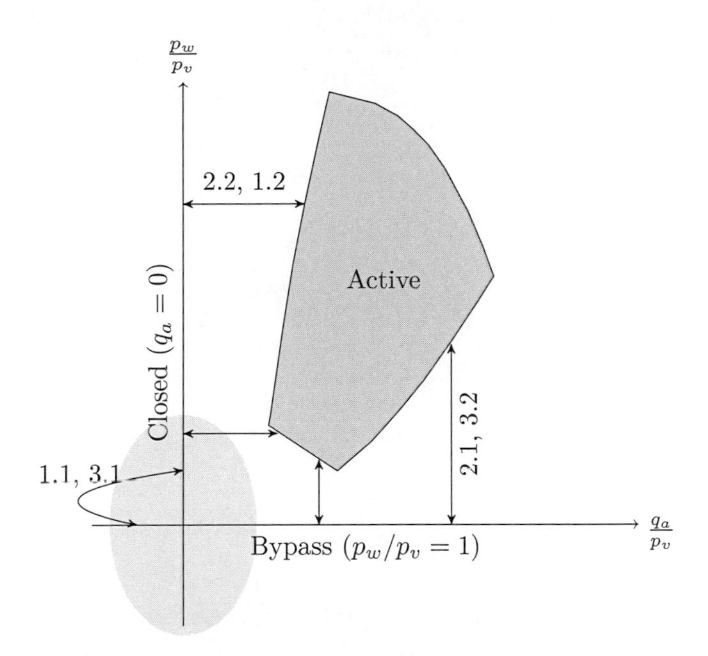

Figure 7.1: Shown are the different operation modes of a compressor. The arrows indicate the changes to binary variables for a compressor, indicated by the heuristic, labeled by cases, see Section 7.3.2. The gray shaded ellipse marks those parts of the states bypass and closed in which changes between the modes bypass and closed are possibly performed by the heuristic.

define a set of feasible points illustrated by the shaded area in the upper-right of Figure 7.1.

- If the compressor is in bypass, then again, flow through the compressor is allowed. Furthermore the pressure at the input side is identical to the pressure at the output side, and the flow can vary arbitrarily (see the horizontal line in Figure 7.1).

- Finally, if the compressor is in closed mode, then the flow is zero, and the pressures at input and output side can take arbitrary values (see the vertical line in Figure 7.1).

The situation for a control valve is similar, see Figure 7.2. The only difference is that the output pressure is lower than the input pressure in 'active' mode.

We consider a compressor or a control valve $a = (v, w) \in A$. Recall that $x_{a,1}, x_{a,2}$ are binary variables associated with the compressor (or control valve) denoting the bypass and active state respectively. The binary variable $x_{a,0}$ is used to represent the case that the active element is closed. The crucial part of the topology optimization

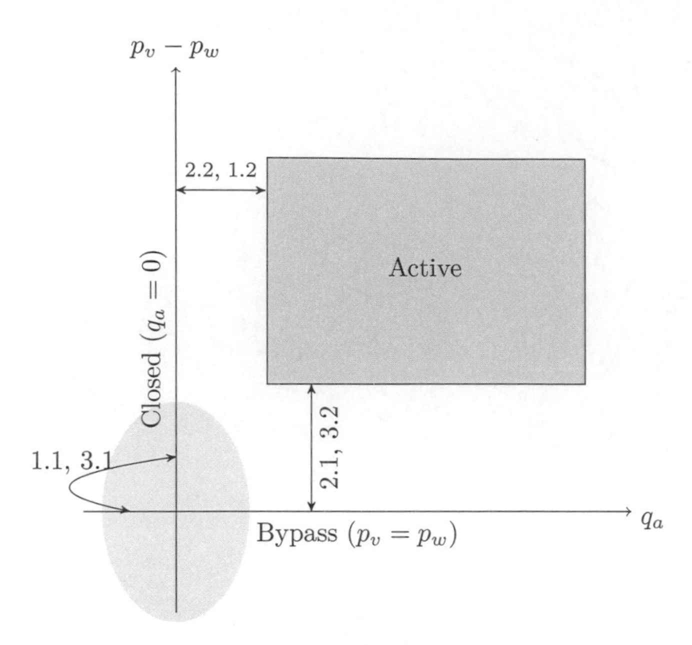

Figure 7.2: Shown are the different operation modes of a control valve. The arrows indicate the changes to binary variables for a compressor, indicated by the heuristic, labeled by cases, see Section 7.3.2. The gray shaded ellipse marks those parts of the states bypass and closed in which changes between the modes bypass and closed are possibly performed by the heuristic.

model that we will exploit in the sequel are the following three families of indicator constraints:

$$x_{a,0} = 1 \Rightarrow \begin{cases} q_{a,1} = 0 & (7.3.2a) \\ q_{a,2} = 0 & (7.3.2b) \end{cases}$$

$$x_{a,1} = 1 \Rightarrow \begin{cases} \pi_v - \pi_w = 0 & (7.3.3a) \\ q_{a,2} = 0 & (7.3.3b) \end{cases}$$

$$x_{a,2} = 1 \Rightarrow \begin{cases} A_a \left(q_{a,2}, p_v, p_w\right)^T \leq b_a & (7.3.4a) \\ q_{a,1} = 0 & (7.3.4b) \end{cases}$$

Recall that exactly one of these three cases applies because of (3.2.1e).

The heuristic creates its own branching tree $\mathcal{T}$ as described in Section 7.2.3. Its root node consists of some fixation of the discrete decision variables to binary values. We solve the slack variant (7.3.1) of model (3.2.1). We obtain a ranking among the binary variables (according to the dual solution values) as described in Section 7.2.2. This ranking indicates which binary variable should be flipped from one to zero. Changing the state of a compressor or a control valve goes along with changing the values of two binary variables. Hence another binary variable must then be flipped from zero to one in order to preserve the feasibility of (7.3.1).

Denote by ρ_f a certain user-specified threshold value for the flow (default value is 20), and by ρ_p a threshold value for the pressure difference (default value is 10). Using these values new child nodes for tree $\mathcal{T}$ are created where we distinguish the following cases.

Case 1: $x_{a,0}$ from 1 to 0. The subnet is currently “closed”. The ranking indicates that either the variable $x_{a,1}$ or $x_{a,2}$, having the present value 0, should be set to 1.

Case 1.1: “From closed to bypass.” If the pressure difference at both end nodes $|p_w - p_v|$ is below ρ_p, then a new child node is created where $x_{a,1}$ is set from 0 to 1 and $x_{a,2}$ remains at 0.

Case 1.2: “From closed to active.” If the dual value of at least one of the constraints (7.3.2a) or (7.3.2b) is positive, then in the new child we flip $x_{a,2}$ from 0 to 1 and $x_{a,1}$ remains at 0.

Case 2: $x_{a,2}$ from 1 to 0. The subnet is currently “active”. We collect all constraints of (7.3.4a) that are either fulfilled with equality or where the associated value of Δ is greater than zero. (Recall that we solve a slack model, which ensures feasibility.) Denote by $I \subseteq \{1, \ldots, \nu_a\}$ the corresponding index set. For each $i \in I$ we consider the i-th constraint in (7.3.4a) and denote by (a_1, a_2, a_3) the coefficients from the i-th row of A_a. We fix the input pressure $p_v = 1$ and check in which quadrant the remaining two-dimensional vector (a_2, a_3) lies.

Case 2.1: “From active to bypass.” If $a_2 < 0$, then we create a child node in which we set $x_{a,2}$ to 0 and $x_{a,1}$ to 1.

Case 2.2: “From active to closed.” If $a_1 < 0$, then we create a child node in which we set $x_{a,2}$ to 0 and also $x_{a,0}$ to 1.

Case 3: $x_{a,1}$ from 1 to 0. The subnet is currently in “bypass”.

Case 3.1: “From bypass to closed.” If the absolute value of flow is below ρ_f, then a new child node is created with the decision to set $x_{a,1}$ to 0, to leave $x_{a,2}$ at 0, and thus set $x_{a,0}$ to 1. Hence at this child node, the subnetwork is 'closed'.

Case 3.2: “From bypass to active.” If the dual value of the constraint (7.3.3a) is negative, then a new child node is created, in which we set $x_{a,2}$ from 0 to 1 and $x_{a,1}$ from 1 to 0.

Otherwise: Consider the next best variable according to the ranking. If no more variable exists, no other child node is created.

Note that multiple cases can occur simultaneously, hence it is possible that multiple child nodes are created.

The "otherwise" case can occur for example if the ranking favors an increase of compression for an active compressor, while the compression ratio is already at its upper limit.

Regarding control valves we create new child nodes in the very same way as we do for compressors. For simple (non-control) valves, the state "active" does not exist; they can only be in "bypass" or "closed" mode. So we apply the same child node creation routine as for compressors but we skip those parts that are concerned with "active" states.

After inserting the new child nodes to the tree $\mathcal{T}$ we perform the following node selection strategy in order to decide where to continue the search. We always proceed according to the rule: "active" first, "bypass" second, "closed" third. We always pass along from a parent node to one of its child nodes according to this rule (depth-first search). When entering the child node again the slack model (7.3.1) is solved first. This yields a new ranking of the decision variables, which leads to a new decision about which of the variables should be flipped in the next round.

The branching process terminates if at one node the slack model (7.3.1) has a zero objective function value or a certain user-defined branching depth (i.e., number of nodes in the path to the root node) is reached. Then the corresponding child node is pruned and the search procedure continues at another unfinished child node. Here we apply a simple backtracking rule, going back to the previous parent node where we came from before. An example for the branching tree $\mathcal{T}$ is given in the next section.

7.3.3 Handling Loop Extensions

Our topology optimization model (3.2.1) allows a single original arc $a = (v, w) \in A$ together with additional parallel arcs $(a, i) \in A_X, i > 1$. For these arcs the important part of the model consists of the following constraints:

$$\begin{aligned} x_{a,i} = 1 \Rightarrow \alpha_{a,i}\, q_{a,i}|q_{a,i}|^{k_a} - \beta_{a,i} y_{a,i} - (\pi_v - \gamma_a \pi_w) = 0 \quad & \forall\,(a,i) \in A_X, i \neq 0, \\ x_{a,i} = 0 \Rightarrow q_{a,i} = 0 \quad & \forall\,(a,i) \in A_X, i \neq 0, \\ \sum_{i:(a,i)\in A_X} x_{a,i} = 1. & \end{aligned}$$

Recall that the extended arc set A_X does not contain $(a, 0)$ as we consider a passive network element here. Thus, if $x_{a,i}$ is chosen to be switched from 1 to 0 by our heuristic, then the constraint $\alpha_{a,i}\, q_{a,i}|q_{a,i}|^{k_a} - \beta_{a,i} y_{a,i} - (\pi_v - \gamma_a \pi_w) = 0$, which is active in the current relaxation (7.3.1) becomes inactive in the next problem. When writing the relaxation (7.3.1) this constraint is expressed as

$$\alpha_{a,i}\, q_{a,i}|q_{a,i}|^{k_a} - \beta_{a,i} y_{a,i} - (\pi_v - \gamma_a \pi_w) \leq 0, \tag{7.3.5a}$$

$$-\alpha_{a,i}\, q_{a,i}|q_{a,i}|^{k_a} + \beta_{a,i} y_{a,i} + (\pi_v - \gamma_a \pi_w) \leq 0. \tag{7.3.5b}$$

Now we distinguish two cases:

Case 1: The dual value of constraint (7.3.5a) is positive. This means that the difference of node potential values is too large for the solution flow value. If the constant $\alpha_{a,i}$ is not at its minimum, i.e., $x_{a,j} = 0$ for a maximal index $j < i$ with $(a, j) \in A_X$, and the compression $y_{a,i}$ is maximal, i.e., $y_{a,i} = \overline{y}_{a,i}$, then we flip $x_{a,i}$ from 1 to 0 and $x_{a,j}$ from 0 to 1 in the new child.

Case 2: The dual value of constraint (7.3.5b) is positive. This reflects that the difference of potential values is too small for the solution flow value. If the constant $\alpha_{a,i}$ is not at its maximum, i.e., $x_{a,j} = 0$ for a minimal index $j > i$ with $(a, j) \in A_X$, and the compression $y_{a,i}$ is minimal, i.e., $y_{a,i} = \underline{y}_{a,i}$, we flip $x_{a,i}$ from 1 to 0 and $x_{a,j}$ from 0 to 1 in the new child node.

Otherwise: The dual values of (7.3.5a) and (7.3.5b) are zero. Consider the next best variable according to the ranking. If no more variables exist, a child node is not created.

Now let us explain the branching tree $\mathcal{T}$ as an example.

Example 7.3.2:

An example for the branching process and the corresponding branching tree $\mathcal{T}$ is shown in Figure 7.3. Here we consider a network that consists of one control valve (CV), one compressor station (CS), and a pipe, which can be extended by a parallel pipe (loop). Assume that either from the LP relaxation or by branching at the current node the CV is set to active mode and the CS is set to bypass mode and no extension for the pipe is selected. Assume as well that this selection is not feasible. Now the heuristic is invoked. It creates a ranking of the corresponding decision variables, which means, a ranking of the network elements. We assume that the CV is ranked higher than the CS while the pipe has the lowest rank. Hence the heuristic creates a branch (left son), where it changes the stage of the CV from active to bypass first

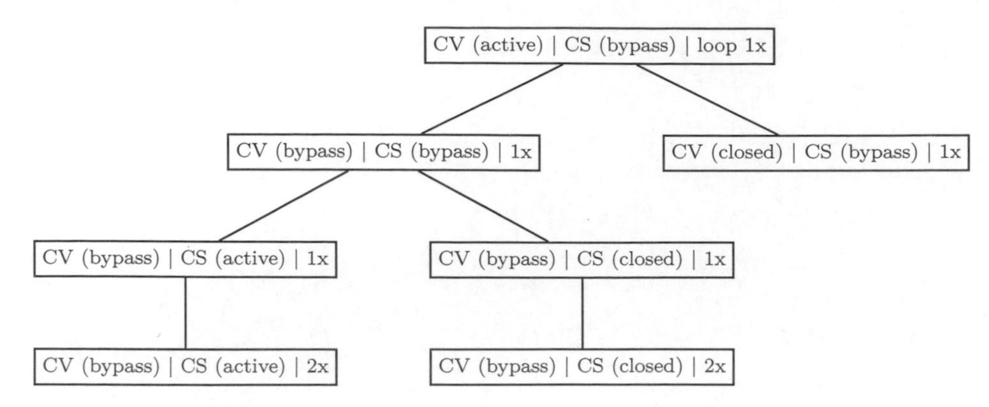

Figure 7.3: Example of the branching tree created by the heuristic for a network that contains 1 control valve (CV), 1 compressor station (CS) and a pipe that can be extended by a parallel pipe (loop). The tree is discussed in Example 7.3.2.

(according to Case 2.1 in Section 7.3.2). Another branch (right son) is created where it changes the stage of the CV from active to closed. Now we focus on the left son and solve the NLP (7.1.2) *again and obtain a new ranking. Among all elements that have not been changed before, the CS has the highest ranking now. Hence it is changed, once from bypass to active (left son) and once from bypass to closed (right son). For the left son, we solve the* NLP *relaxation* (7.1.2) *and assume that it has still a positive slack value. Now the loop is selected in a new child of the tree (according to Case 1 in this section) and the* NLP *relaxation is solved. We assume that it still has a positive slack value. Since we cannot alter more elements in this small network, this part of the tree is finished, and we track back to the parent node and once more to the next parent node. For the right son, we solve the* NLP *relaxation and again get an infeasible subproblem, so we also consider the child in the next step. Now we assume the corresponding* NLP *has a positive slack value again. Then we prune this node and afterwards are back at the root node of the search tree created by the heuristic. The sub tree where the CV was changed from active to bypass is finished, hence we now change the CV from active to closed (shown on the right). Then we solve the* NLP (7.1.2) *and assume, that we obtain a solution with slack zero. At this point, when neglecting the slack variables, a new primal feasible solution is found for* (7.3.1).

7.4 Implementation Details

Our heuristic is implemented as a plug-in for the solver SCIP. The following settings are offered to the user:

Maxcalls. Specifies the maximal number of nodes for the branching tree within the heuristic. (Default value is 30.)

Maxequals. Abort the heuristic without a solution, if the highest-ranked "Maxequals" fraction of binary variables all have equal ranks. (Default value is 1/3, i.e., 33.33 % of all binary variables.)

Mingap. Do not call the heuristic, if the gap is below the threshold value "Mingap". The gap is defined as: (best primal − best dual) / (best dual). (Default value is 0.05, i.e., a SCIP gap of 5 %.)

Sigma. Sets the value for σ in objective (7.1.2a). (Default value is 0, i.e., emphasis only on feasibility.)

Leaves only. Call the heuristic only at nodes of the branching tree where all binary variables are fixed. (Default value is "No".)

We apply the heuristic to the topology optimization problem (3.2.1). The heuristic is invoked at those nodes in the branch-and-bound tree where all binary decision variables have binary values (either by branching or by the solution of the LP relaxation). In order to make this happen as early as possible, we introduce a branching priority rule that first branches on integral variables, before spatial branching on continuous variables is applied. Furthermore, when changing x^* in our heuristic to x' in the next iteration, we ensure that x' is feasible for the constraints $Lx \leq 0$, see (3.2.1h). This is done as follows: Recall that the left-hand side of (3.2.1h) is a vector of linear functions. We assume that $x_{a_0,i}$ has to be changed from zero to one. Then we solve the following MIP problem to obtain the next vector x':

$$\begin{aligned} \min\ & ||x' - x^*||_1 \\ \text{s.t.}\quad & Lx' \leq t, \\ & x'_{a_0,i} = 1, \\ & x' \in \{0,1\}^{A_X}. \end{aligned} \tag{7.4.1}$$

If this problem is infeasible, then we prune the current node of the branching tree of our heuristic. Otherwise we obtain an optimal solution x' which has minimal deviations from x^* together with the change $x_{a_0,i} = 1$.

7.5 Computational Results

We implemented the heuristic described above in C and use the computational setup described in Section 3.5. For our computational study we considered the network `net7`. The initial network `net7a` consists of 4165 nodes and 4079 arcs, among them 3638 passive pipelines, 12 compressor stations, 121 control valves, and 308 valves. The industry partner defined a test set of 30 nominations for `net7a` that differ among each other in the pressure and flow bounds on the entry and exit nodes. The problem is a feasibility problem only (i.e., does there exist a feasible flow in the network). Hence it is considered as "solved" as soon as a feasible solution is constructed; no objective function values for opening valves (i.e., adding further topology extensions) are set. We imposed a time limit of 4 hours for all instances which is still more than the cooperation partner suggests. We compare three strategies for solving the topology optimization problem (3.2.1).

1. "SCIP/default": The first strategy is to use SCIP without any adaptations on the solver settings. All branching decisions are up to the solver. We enforce a certain branching priority rule, so that SCIP first branches on binary decision variables x. Only after all discrete variables are fixed it is allowed to perform spatial branching on continuous variables.

2. "SCIP/dualval(basic)": The second strategy is to use the heuristic as described in Section 7.2. We establish the branching tree $\mathcal{T}$ as described in Section 7.3.2 and Section 7.3.3, but neglect the special analysis which sorts out some changes as described in the case discussion in Section 7.3.2.

3. "SCIP/dualval(adapted)": The third strategy is to use the heuristic as described in Section 7.2. Additionally we use the special adaptations described in Section 7.3.2 and Section 7.3.3.

These strategies are used to demonstrate the impact of our heuristic on the solving performance of SCIP. Additionally, we use BARON and ANTIGONE to solve the instances.

A brief summary of the computational results is shown in Table 7.1. A detailed presentation of the results is available in Tables A.14-A.18. The third and fourth column of Table 7.1 show that neither BARON nor ANTIGONE solved any of the test instances (see also Table A.1 and Table A.3). The fifth column of Table 7.1 shows the number of instances that were solved with SCIP/default, strategy 1. The sixth column of Table 7.1 shows the number of instances that were solved by the heuristic (SCIP/dualval (basic), strategy 2). The last column of Table 7.1 shows

network	BARON	ANTIGONE	SCIP strategy 1	SCIP strategy 2	SCIP strategy 3
net7a	-	-	4	23	30
net7b	-	-	4	12	18
net7c	-	-	1	9	18
net7d	-	-	-	5	17
net7e	-	-	1	8	18

Table 7.1: Globally solved instances out of 30 nominations for each of the networks net7a-net7e. The underlying data are available in Tables A.14–A.18.

the number of instances that were solved by the heuristic exploiting the problem structure (SCIP/dualval (adapted), strategy 3). Thus, for example SCIP/dualval (basic) was able to find a feasible solution within 4 hours in 23 of the net7a-type instances.

The modified version of the heuristic is able to solve all problem instances for net7a. In later phases of our collaboration (net7b–net7e), the industrial project partner made minor alterations to the networks' topologies, and, more important, also made the instances more and more difficult by imposing tighter bounds on the minimum and maximum pressure levels at the nodes. Hence it became more difficult for any heuristic to construct feasible solutions. However, the adapted heuristic was still able to solve globally approximately 67 % of the instances (101 out of 150), whereas the standard MINLP heuristics of SCIP found less feasible solutions the more difficult the instances became (6 % of the instances of the test set are globally solved, 10 out of 150). The basic heuristic (strategy 2) that is not adapted to the problem's special structure is also pretty good in finding feasible solutions (when compared to SCIP/default, strategy 1). Here approximately 38 % of the instances of the test set are globally solved (57 out of 150).

Figure 7.4 shows a comparison of SCIP/default with default settings (in particular, with all standard heuristics) and SCIP/dualval(adapted) with the additional adapted heuristic on all instances from Table 7.1. It turns out that for those instances which SCIP/default is already able to solve, the additional time spent in the dual value heuristic does not pay off, but slows down the overall solution process instead. For many of those instances that SCIP/default could not solve, SCIP/dualval(adapted) was able to compute a feasible solution within the time limit.

We also applied the heuristic in its adapted version to network net5. Here we combined it with the solution approach indicated by strategy 3 presented in the previous chapter. The corresponding instances are topology expansion problems. Our computational results show that a primal feasible solution is available for every

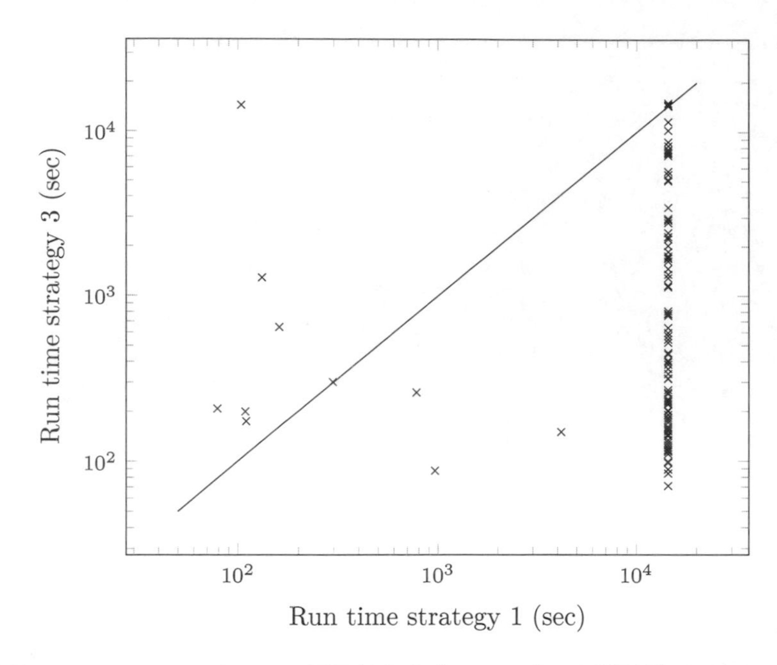

Figure 7.4: Run time comparison between SCIP/default (strategy 1) and SCIP/dualval(adapted) (strategy 3). Each cross (×) marks the run time of these two solvers. A cross above the diagonal line indicates that the strategy SCIP/default is faster, and a cross below the diagonal line indicates that SCIP/dualval(adapted) is faster. Note that there are instances, where both strategies did not find any solution. Each of these instances correspond to the same cross (×) mark in the upper right corner of Figure 7.4. The time limit was set to 4 hours.

instance. The results are available in Table A.19. Without our heuristic a feasible solution was available for approximately 45 % of the instances only, see Table A.13.

Summary

We presented a new heuristic algorithm that exploits dual information coming from KKT points in order to find feasible solutions for the mixed-integer nonlinear optimization problem (7.1.1) with indicator constraints. We applied this heuristic to the topology optimization problem (3.2.1). Using the heuristic in the standard fashion, it is better than the existing heuristics that are already available in the solver SCIP. After exploiting the problem structure, the heuristic is able to identify many more feasible solutions than any other heuristic included in SCIP that we are aware of. In total approximately 61 % more instances of the test set consisting of instances of the network `net7` are globally solved in total. We also applied the heuristic to the network `net5` and combined it with the solution approach presented in the previous chapter. The computational results showed that the number of instances with a primal feasible solution available was increased by 55 %. Recall that both networks out of those considered in this thesis remained with very few feasible solutions after the solution approaches of the previous Chapters 4-6.

Chapter 8

Conclusions

This thesis deals with gas network optimization, that is nomination validation and topology expansion problems arising from gas transport as introduced in Chapter 1. We presented a mixed-integer nonlinear program (3.2.1) that models both problems at the same time and considered it as the topology optimization problem in gas networks. Our industrial cooperation partner provided us with data of different networks in combination with corresponding nominations. Facing the fact that state-of-the-art solvers like BARON, ANTIGONE or SCIP are slow on real-world instances or unable to compute any primal feasible solution, we set our aim to improve the performance of SCIP for solving our model. Exploiting specific knowledge of our model was the key to success. Our strategy shows that an MINLP solver in combination with special-tailored adaptations offers the potential to solve globally large-scale non-convex MINLP problems of real-world size. We believe that our results presented in this thesis provide evidence for the suitability of the approach and constitute a step in the right direction of solving the topology optimization problem.

Future studies aim at solving the topology optimization problem for transient gas flows. In the focus of the study there is the identification of more general conditions for the passive transmission problem. Assuming that the convex domain relaxation of the passive transmission problem is performed, an interpretation of the dual solution of the relaxation would lead to a primal heuristic for identifying feasible solutions. Discretization of the time component in the model for transient flows allows the inclusion of the results obtained. Apart from this, an extension of the pc-regularization towards time dependency should be investigated in order to derive cuts by an augmented Benders argument. Finally, this approach used for gas transmission networks could also be extended to water or electricity networks, under some additional assumptions.

Benefits for a TSO

Let us briefly discuss different aspects which allow a TSO to improve the operation of its network. Utilizing the algorithms developed in this thesis allows the company to compute operation modes for the active network elements, arc flows, and node pressures for a given nomination. This computation requires that the flow specified by the nomination is transported through the network and all technical and physical constraints are fulfilled. Here the software can be used to assist a manual approach of expert knowledge in combination with simulation software. Given a set of extensions the model can also be used for the topology expansion problem.

Other aspects are also useful for a TSO. The company is able to get a visualization of the pressure distribution, see Figure 1.3. Given the case that the simulation software is not delivering a feasible solution the possibility of proving global infeasibility of an active transmission problem as described in Section 6.2 improves this situation. A failing simulation software does not generally imply that the current transportation situation is indeed infeasible. Especially a visualization as shown in Figure 4.2 provides assistance in comprehending the reasons of an infeasibility. The picture emphasizes those parts of the network that cause the infeasibility of a passive transmission problem as described in Section 4.2.3. Furthermore considering an infeasible passive transmission problem and solving the flow conservation relaxation yields a set of specific slack values. These values allow to circumvent infeasibilities. More precisely, an adaptation of the considered nomination can be performed such that the corresponding gas quantities are transported through the network, i.e., the passive transmission problem is feasible for the adapted nomination.

Appendix A

Tables

nom	net7a	net7b	net7c	net7d	net7e
4	limit	limit	limit	647	limit
6	limit	limit	limit	912	limit
8	limit	limit	725	706	limit
19	limit	limit	limit	limit	663
20	limit	limit	564	limit	limit
23	limit	limit	limit	3,029	limit
24	limit	limit	limit	limit	724
25	limit	limit	571	limit	996
26	limit	limit	limit	574	limit
27	limit	limit	limit	589	limit
29	limit	limit	limit	730	limit

Table A.1: Run time results in seconds using BARON to solve the topology optimization problem on 30 nominations on the network net7. The time limit was set to 4 hours and the results are discussed in Chapter 7. All nominations not depicted here ran into the time limit without a feasible solution. Those instances with a finite time limit were detected to be infeasible. No primal solution was available for all the other instances.

nom	net7a	net7b	net7c	net7d	net7e
3	limit	103	limit	limit	limit
4	limit	110	limit	limit	limit
5	78	limit	limit	limit	limit
11	limit	limit	limit	limit	132
15	limit	limit	limit	limit	limit
16	966	limit	limit	limit	limit
17	109	limit	limit	limit	limit
22	limit	297	161	limit	limit
27	779	4,171	limit	limit	limit

Table A.2: Run time results in seconds using SCIP to solve the topology optimization problem on 30 nominations on the network `net7`. The time limit was set to 4 hours and the results are discussed in Chapter 7. All nominations not depicted here ran into the time limit without a feasible solution. Those instances with a finite time limit were feasible. No primal solution was available for all the other instances.

nom	net7a	net7b	net7c	net7d	net7e
1	37	37	1,808	limit	375
2	27	27	6,665	13,825	9,221
3	62	93	711	limit	3,064
4	28	28	50	4,859	52
5	28	28	50	1,874	52
6	28	29	50	49	52
7	27	28	50	585	52
8	28	28	51	4,009	55
9	28	28	49	50	53
10	28	38	5,971	52	limit
11	37	27	131	351	304
12	27	28	49	50	51
13	28	27	51	457	52
14	28	28	50	49	51
15	28	27	50	50	52
16	28	28	49	49	52
17	28	27	50	52	52
18	7,985	38	limit	limit	51
19	37	28	49	limit	1,542
20	28	28	50	60	52
21	28	28	49	53	52
22	27	27	49	49	52
23	27	27	49	55	51
24	27	28	49	49	51
25	28	28	51	limit	52
26	28	27	51	209	52
27	28	28	50	50	53
28	27	28	49	50	51
29	33	14,205	52	51	53
30	27	28	48	52	limit

Table A.3: Run time results in seconds using ANTIGONE to solve the topology optimization problem on 30 nominations on the network net7. The time limit was set to 4 hours and the results are discussed in Chapter 7. Those instances with a finite time limit were detected to be infeasible. No primal solution was available for all the other instances.

nom	domain relaxation (strategy 3)					flow cons. relaxation (strategy 4)					SCIP prio (strategy 2)					SCIP (strategy 1)				
	gap	primal	dual	time	nodes	gap	primal	dual	time	nodes	gap	primal	dual	time	nodes	gap	primal	dual	time	nodes
1	-	0	0	1	1	-	0	0	1	1	-	0	0	1	1	-	0	0	1	1
2	-	0	0	1	1	-	0	0	1	1	-	0	0	1	1	-	0	0	1	1
3	-	0	0	1	1	-	0	0	1	1	-	0	0	1	1	-	0	0	1	1
4	-	0	0	1	1	-	0	0	1	1	-	0	0	1	1	-	0	0	1	1
5	-	0	0	1	1	-	0	0	1	1	-	0	0	1	1	-	0	0	1	1
6	-	0	0	1	1	-	0	0	1	1	-	0	0	1	1	-	0	0	1	1
7	-	0	0	1	1	-	0	0	1	1	-	0	0	1	1	-	0	0	1	1
8	-	0	0	1	1	-	0	0	1	1	-	0	0	1	1	-	0	0	1	1
9	-	0	0	1	1	-	0	0	1	1	-	0	0	1	1	-	0	0	1	1
10	-	0	0	1	1	-	0	0	1	1	-	0	0	1	1	-	0	0	1	1
11	-	95	95	6	105	-	95	95	6	105	-	95	95	7	780	-	95	95	6	1,087
12	-	95	95	1	1	-	95	95	1	1	-	95	95	1	1	-	95	95	1	1
13	-	95	95	1	1	-	95	95	1	1	-	95	95	1	1	-	95	95	1	1
14	-	95	95	1	1	-	95	95	1	1	-	95	95	1	1	-	95	95	1	1
15	-	95	95	1	1	-	95	95	1	1	-	95	95	1	1	-	95	95	1	1
16	-	249	249	7	303	-	249	249	8	303	-	249	249	1,483	763,842	-	249	249	33,270	19,155,684
17	-	320	320	9	312	-	320	320	12	312	-	320	320	68	24,322	-	320	320	12,773	7,225,869
18	-	475	475	11	445	-	475	475	32	445	-	475	475	69	25,140	636	700	95	limit	14,090,518
19	-	525	525	65	3,750	-	525	525	252	3,594			475	limit	22,919,263	160	743	285	limit	16,516,334
20	-	525	525	36	2,957	-	525	525	100	2,957	-	525	525	64	19,410	-	525	525	76	19,323
21	-	525	525	3	173	-	525	525	5	173	-	525	525	2	173	-	525	525	2	173
22	-	525	525	1	12	-	525	525	1	12	-	525	525	1	12	-	525	525	1	12
23	-	525	525	1	6	-	525	525	1	6	-	525	525	1	6	-	525	525	2	6
24	-	525	525	1	6	-	525	525	1	6	-	525	525	1	6	-	525	525	1	6
25	-	525	525	1	6	-	525	525	1	6	-	525	525	1	6	-	525	525	1	6
26	-	584	584	15	1,534	-	584	584	27	1,526	-	584	584	76	23,830	-	584	584	24,658	9,692,200
27	-	633	633	2	23	-	633	633	3	23	-	633	633	2,999	1,082,717	29	679	525	limit	13,668,016
28	-	633	633	47	3,122	-	633	633	143	3,216	-	633	633	46	11,870	72	905	525	limit	15,964,903
29	-	693	693	33	1,428	-	693	693	135	1,764	-	693	693	371	105,119	114	1,125	525	limit	15,632,260
30	-	788	788	28	893	-	788	788	154	2,382	-	788	788	6,589	2,403,190	78	938	525	limit	16,565,148
31	-	997	997	38	1,263	-	997	997	355	5,016	37	1,085	788	limit	21,938,417	143	1,279	525	limit	16,769,864
32	-	1,092	1,092	118	6,456	-	1,092	1,092	419	6,456	57	1,092	693	limit	13,526,935	180	1,476	525	limit	11,628,716
33	-	1,171	1,171	379	16,938	-	1,171	1,171	1,268	18,758	67	1,319	788	limit	7,929,185	125	1,318	584	limit	6,976,475
34	-	1,174	1,174	257	12,370	-	1,174	1,174	871	12,580	104	1,614	788	limit	9,530,654	175	1,612	584	limit	11,104,065
35	-	1,182	1,182	64	4,072	-	1,182	1,182	849	12,888	28	1,254	978	limit	16,014,817	199	1,896	633	limit	11,797,821
36	-	1,234	1,234	200	14,073	-	1,234	1,234	445	10,100	23	1,234	997	limit	8,524,150	98	1,259	633	limit	15,467,469
37	-	1,241	1,241	80	7,412	-	1,241	1,241	511	12,466	65	1,812	1,092	limit	4,921,651	95	1,241	633	limit	11,612,658
38	-	1,318	1,318	64	7,666	-	1,318	1,318	228	7,666	10	1,332	1,202	limit	26,954,607	110	1,332	633	limit	19,373,729
39	-	1,336	1,336	194	19,978	-	1,336	1,336	404	16,302	7	1,336	1,237	limit	11,395,890	241	2,164	633	limit	10,694,870
40	-	1,459	1,459	158	20,362	-	1,459	1,459	826	29,252	22	1,620	1,325	limit	14,668,683	105	1,620	788	limit	13,737,907
41	15	1,545	1,332	limit	3,223,996	-	1,545	1,545	1,269	45,195	15	1,545	1,332	limit	8,085,049	86	1,863	997	limit	15,256,088

Table A.4: Results on network `net4` and 41 nominations. A time limit of 11 h was imposed. The different strategies for solving the topology optimization problem are presented in Chapter 4.

nom	domain relaxation (strategy 3)					flow cons. relaxation (strategy 4)					SCIP prio (strategy 2)					SCIP (strategy 1)				
	gap	primal	dual	time	nodes	gap	primal	dual	time	nodes	gap	primal	dual	time	nodes	gap	primal	dual	time	nodes
1	-	555	555	152	690	-	555	555	249	690	194	1,314	446	limit	3,275,942	116	729	337	limit	4,273,098
2	-	701	701	1,090	26,834	-	701	701	3,186	27,968	-	701	701	1,531	183,041	148	836	337	limit	2,417,953
3	-	1,429	1,429	4,375	32,660	-	1,429	1,429	9,499	34,675			952	limit	5,261,200	103	1,429	701	limit	2,236,848
4	-	1,495	1,495	2,303	38,971	-	1,495	1,495	6,699	38,971			1,038	limit	4,800,505	73	1,799	1,038	limit	2,206,174
5	-	1,731	1,731	11,698	80,143	-	1,731	1,731	27,220	80,143			1,038	limit	3,953,611			1,038	limit	1,410,299
6	98	2,575	1,300	limit	298,530	107	2,575	1,242	limit	133,550			1,257	limit	4,282,280			1,038	limit	1,914,502
7	82	2,634	1,443	limit	324,788	90	2,634	1,384	limit	152,222			1,604	limit	4,225,594			1,148	limit	1,323,814
8	118	3,575	1,635	limit	664,458	166	4,130	1,548	limit	270,658			1,713	limit	5,028,474			1,384	limit	3,144,894
9	236	6,553	1,948	limit	387,517	250	6,553	1,871	limit	221,056			1,995	limit	4,244,062			1,666	limit	847,847
10	179	6,154	2,199	limit	734,553			2,147	limit	514,498			1,995	limit	3,951,594			1,995	limit	2,059,568
11			2,401	limit	2,049,433			2,337	limit	1,195,274			2,343	limit	5,293,619			2,130	limit	2,739,786

Table A.5: Results on network net5 and 11 nominations. A time limit of 11 h was imposed. The different strategies for solving the topology optimization problem are presented in Chapter 4.

scale	loops	SCIP (strategy 1)					cutoff (strategy 2)				
		gap	primal	dual	time	nodes	gap	primal	dual	time	nodes
2.0	8	-	1,500	1,500	212	176,615	-	1,500	1,500	255	212,747
2.0	9	-	1,500	1,500	113	83,832	-	1,500	1,500	157	114,824
2.0	10	-	1,500	1,500	140	100,641	-	1,500	1,500	133	94,086
2.0	11	-	1,500	1,500	130	91,718	-	1,500	1,500	139	94,076
2.0	12	-	1,500	1,500	132	91,469	-	1,500	1,500	141	88,662
2.1	8	-	1,800	1,800	2,127	1,034,782	-	1,800	1,800	1,955	1,090,390
2.1	9	-	1,800	1,800	859	625,947	-	1,800	1,800	771	524,701
2.1	10	-	1,700	1,700	719	444,753	-	1,700	1,700	807	480,961
2.1	11	-	1,700	1,700	774	504,599	-	1,700	1,700	926	552,831
2.1	12	-	1,700	1,700	728	471,606	-	1,700	1,700	912	533,415
2.2	8	-	2,200	2,200	7,069	4,521,738	-	2,200	2,200	7,267	4,211,609
2.2	9	-	2,000	2,000	3,874	2,274,304	-	2,000	2,000	4,148	2,282,214
2.2	10	-	2,000	2,000	2,594	1,441,304	-	2,000	2,000	2,339	1,339,507
2.2	11	-	2,000	2,000	2,919	1,713,374	-	2,000	2,000	2,590	1,580,747
2.2	12	-	2,000	2,000	2,699	1,494,467	-	2,000	2,000	3,562	1,866,970
2.3	8	63	3,100	1,900	limit	10,801,193	61	3,000	1,858	limit	10,007,510
2.3	9	18	2,400	2,028	limit	13,912,926	-	2,400	2,400	38,627	14,172,581
2.3	10	-	2,300	2,300	16,496	6,038,279	-	2,300	2,300	11,947	5,797,020
2.3	11	-	2,200	2,200	5,380	3,104,723	-	2,200	2,200	6,658	3,525,269
2.3	12	-	2,200	2,200	9,881	4,006,313	-	2,200	2,200	10,991	4,547,700
2.4	8	133	5,900	2,523	limit	4,451,666	155	6,200	2,425	limit	4,308,121
2.4	9	67	3,200	1,914	limit	5,744,040	60	3,100	1,933	limit	8,051,340
2.4	10	39	2,700	1,929	limit	10,522,647	40	2,700	1,916	limit	8,037,016
2.4	11	33	2,600	1,948	limit	12,099,342	33	2,600	1,952	limit	13,062,343
2.4	12	-	2,500	2,500	38,011	13,944,805	22	2,500	2,044	limit	12,970,687
2.5	9	131	5,200	2,245	limit	7,162,185	97	4,400	2,231	limit	7,122,703
2.5	10	86	4,000	2,140	limit	10,425,831	58	3,400	2,150	limit	13,198,840
2.5	11	61	3,100	1,922	limit	3,805,350	66	3,200	1,922	limit	3,860,573
2.5	12	47	3,000	2,030	limit	10,600,042	48	3,000	2,025	limit	9,488,796

Table A.6: Results on network `net1` and 30 instances. A time limit of 11 h was imposed. The different strategies for solving the topology optimization problem are presented in Chapter 5. Those instances not depicted are infeasible. The column "scale" shows the scaling of the nomination and the columns "loops" the number of available loops. The column "cuts" shows the number of generated cuts.

scale	loops	no good cuts + restarts (strategy 5) gap	primal	dual	time	nodes	with cuts (no restart) (strategy 3) gap	primal	dual	time	nodes	cuts	with cuts (restart) (strategy 4) gap	primal	dual	time	nodes	cuts
2.0	8	-	1,500	1,500	579	381,940	-	1,500	1,500	109	84,549	193	-	1,500	1,500	226	49,316	31
2.0	9	-	1,500	1,500	371	246,107	-	1,500	1,500	138	97,257	247	-	1,500	1,500	342	11,834	40
2.0	10	-	1,500	1,500	585	206,881	-	1,500	1,500	156	93,624	339	-	1,500	1,500	425	23,428	61
2.0	11	-	1,500	1,500	736	272,253	-	1,500	1,500	137	76,879	337	-	1,500	1,500	777	22,293	45
2.0	12	-	1,500	1,500	287	93,061	-	1,500	1,500	80	42,624	161	-	1,500	1,500	557	12,269	41
2.1	8	-	1,800	1,800	3,020	1,864,188	-	1,800	1,800	1,923	952,577	1,345	-	1,800	1,800	288	100,391	130
2.1	9	-	1,800	1,800	2,955	1,708,259	-	1,800	1 800	724	537,401	612	-	1,800	1,800	478	72,130	119
2.1	10	-	1,700	1,700	2,928	1,742,387	-	1,700	1,700	720	495,061	449	-	1,700	1,700	674	72,263	91
2.1	11	-	1 700	1,700	2,157	1,251,067	-	1,700	1,700	997	643,326	713	-	1,700	1,700	590	34,296	91
2.1	12	-	1,700	1,700	2,385	1,286,824	-	1,700	1,700	739	478,849	427	-	2,100	2,100	961	22,659	31
2.2	8	-	2,200	2,200	38.724	15,439,053	-	2,200	2,200	17,063	7,498,640	1,548	-	2,200	2,200	392	127,909	91
2.2	9	-	2,000	2,000	12,033	7,257,944	-	2,000	2,000	4,109	2,731,815	1,003	-	2,000	2,000	287	80,895	91
2.2	10	-	2,000	2,000	5,055	2,774,703	-	2,000	2,000	3,204	2,059,187	1,207	-	2,000	2,000	704	175,816	91
2.2	11	-	2,000	2,000	8,502	4,051,322	-	2,000	2,000	3,619	2,344,021	1,466	-	2,000	2,000	574	107,757	127
2.2	12	-	2,000	2,000	5,713	3,012,721	-	2,000	2,000	3,225	1,912,371	1,217	-	2,000	2,000	924	166,555	61
2.3	8	92	3,100	1,609	limit	11,695,102	70	3,100	1,815	limit	8,467,054	1,497	13	3,000	2,644	limit	2,622,733	5,224
2.3	9	40	2,500	1,775	limit	16,540,306	-	2,400	2,400	38,746	14,393,567	1,430	-	2,400	2,400	474	194,864	91
2.3	10	-	2,300	2,300	28,348	14,080,352	-	2,300	2,300	14,596	7,812,121	2,026	-	2,300	2,300	568	217,980	66
2.3	11	-	2,200	2,200	14,635	7,731,582	-	2,200	2,200	4,502	2,438,022	999	-	2,200	2,200	925	211,591	136
2.3	12	-	2,200	2,200	17,920	7,763,367	-	2,200	2,200	11,676	4,476,996	2,080	-	2,200	2,200	1,168	272,979	153
2.4	8	144	5,700	2,332	limit	14,554,022	120	5,800	2,628	limit	7,970,320	4,862	78	5,600	3,139	limit	9,595,085	1,032
2.4	9	73	3,200	1,843	limit	10,190,515	53	3,100	2,025	limit	11,958,426	1,891	-	3,100	3,100	620	281,801	204
2.4	10	55	2,800	1,806	limit	15,469,392	44	2,800	1,937	limit	11,329,023	2,802	-	2,700	2,700	1,054	421,319	166
2.4	11	40	2,600	1,854	limit	17,939,546	32	2,600	1,958	limit	16,327,371	1,790	-	2,600	2,600	1,122	457,510	136
2.4	12	35	2,500	1,845	limit	14,579,946	37	2,600	1,887	limit	11,569,901	2,340	-	2,500	2,500	1,697	439,561	204
2.5	9	126	4,600	2,035	limit	12,010,467	120	4,700	2,133	limit	7,206,449	1,317	39	4,500	3,221	limit	4,502,599	3,483
2.5	10	66	3,300	1,981	limit	14,218,724	61	3,400	2,107	limit	9,107,376	1,419	11	3,300	2,964	limit	3,978,200	3,753
2.5	11	68	3,100	1,842	limit	14,247,921	44	3,000	2,081	limit	14,940,544	1,729	-	3,000	3,000	2,524	1,017,901	354
2.5	12	57	3,000	1,900	limit	11,001,843	42	2,900	2,037	limit	12,776,311	1,900	-	2,900	2,900	1,967	921,701	136

Table A.7: Results (continued) on network `net1` and 30 instances. A time limit of 11 h was imposed. The different strategies for solving the topology optimization problem are presented in Chapter 5. Those instances not depicted are infeasible. The column "scale" shows the scaling of the nomination and the columns "loops" the number of available loops. The column "cuts" shows the number of generated cuts.

scale	loops	SCIP (strategy 1) gap	primal	dual	time	nodes	cutoff (strategy 2) gap	primal	dual	time	nodes
2.0	3	-	100	100	2	469	-	100	100	2	127
2.0	4	-	100	100	2	412	-	100	100	3	141
2.0	5	-	100	100	3	527	-	100	100	5	220
2.1	3	-	100	100	1	297	-	100	100	2	139
2.1	4	-	100	100	2	394	-	100	100	2	188
2.1	5	-	100	100	2	342	-	100	100	3	223
2.2	3	-	200	200	2	421	-	200	200	2	131
2.2	4	-	200	200	2	558	-	200	200	6	453
2.2	5	-	200	200	5	993	-	200	200	7	532
2.3	3	-	200	200	1	293	-	200	200	2	261
2.3	4	-	200	200	4	1,013	-	200	200	10	1,089
2.3	5	-	200	200	2	313	-	200	200	4	280
2.4	3	-	300	300	2	1,108	-	300	300	3	872
2.4	4	-	300	300	3	966	-	300	300	4	283
2.4	5	-	300	300	5	1,155	-	300	300	4	731
2.5	3	-	400	400	4	1,620	-	400	400	10	3,848
2.5	4	-	400	400	4	1,280	-	400	400	9	2,044
2.5	5	-	400	400	6	1,751	-	400	400	6	1,209
2.6	3	-	500	500	11	6,584	-	500	500	4	1,629
2.6	4	-	500	500	9	5,154	-	500	500	23	9,447
2.6	5	-	500	500	5	2,919	-	500	500	7	3,057
2.7	3	-	700	700	26	21,115	-	700	700	38	26,640
2.7	4	-	600	600	38	28,762	-	600	600	15	9,740
2.7	5	-	600	600	52	34,265	-	600	600	21	12,649
2.8	3	-	800	800	164	140,745	-	800	800	210	143,142
2.8	4	-	800	800	252	193,589	-	800	800	233	173,792
2.8	5	-	800	800	141	118,166	-	800	800	167	120,963
2.9	3	-	1,000	1,000	895	690,477	-	1,000	1,000	1,183	789,484
2.9	4	-	900	900	1,190	805,692	-	900	900	1,947	950,030
2.9	5	-	900	900	1,427	869,203	-	900	900	1,778	981,023
3.0	3	-	1,300	1,300	5,243	4,327,789	-	1,300	1,300	7,870	5,773,387
3.0	4	-	1,100	1,100	6,380	2,557,433	-	1,100	1,100	4,830	2,141,380
3.0	5	-	1,100	1,100	7,994	5,176,313	-	1,100	1,100	9,591	5,208,192
3.1	3	53	1,700	1,107	limit	24,381,440	82	2,000	1,098	limit	25,098,997
3.1	4	57	1,300	827	limit	7,358,275	58	1,300	819	limit	6,957,830
3.1	5	-	1,200	1,200	30,775	10,316,726	-	1,200	1,200	23,027	8,665,593
3.2	3	125	3,000	1,330	limit	40,470,382	126	3,000	1,323	limit	38,022,524
3.2	4	108	1,700	816	limit	6,878,911	120	1,800	816	limit	6,646,229
3.2	5	39	1,400	1,006	limit	24,473,376	63	1,500	917	limit	18,459,280
3.3	3	400	7,400	1,479	limit	24,672,027	396	7,400	1,490	limit	26,207,732
3.3	4	125	2,100	931	limit	9,094,229	107	1,900	915	limit	7,202,437
3.3	5	105	1,800	877	limit	8,842,062	99	1,700	851	limit	7,189,779
3.4	4	136	2,600	1,098	limit	13,464,692	145	2,700	1,100	limit	14,353,164
3.4	5	116	2,000	924	limit	7,219,501	146	2,300	933	limit	7,122,962
3.5	4	153	3,000	1,182	limit	14,580,836	165	3,100	1,166	limit	13,819,689
3.5	5	198	3,000	1,006	limit	11,273,438	115	2,200	1,020	limit	13,411,095
3.6	4	149	3,200	1,280	limit	17,477,233	190	3,800	1,306	limit	26,036,497
3.6	5	164	2,800	1,057	limit	15,527,932	162	2,800	1,066	limit	15,283,010
3.7	4	284	5,200	1,352	limit	13,760,208	284	5,300	1,379	limit	14,050,957
3.7	5	206	3,600	1,174	limit	17,115,145	158	3,100	1,197	limit	18,264,790
3.8	4	409	7,500	1,471	limit	10,677,370	417	7,500	1,449	limit	9,927,261
3.8	5	237	4,100	1,213	limit	14,844,993	219	3,900	1,222	limit	15,409,491
3.9	5	237	4,200	1,243	limit	12,183,208	330	5,400	1,254	limit	13,395,223

Table A.8: Results on network `net2` and 60 instances. A time limit of 11 h was imposed. The different strategies for solving the topology optimization problem are presented in Chapter 5. Those instances not depicted are infeasible. The column "scale" shows the scaling of the nomination and the columns "loops" the number of available loops. The column "cuts" shows the number of generated cuts.

scale	loops	no good cuts + restarts (strategy 5)					with cuts (no restart) (strategy 3)						with cuts (restart) (strategy 4)					
		gap	primal	dual	time	nodes	gap	primal	dual	time	nodes	cuts	gap	primal	dual	time	nodes	cuts
2.0	3	-	100	100	6	183	-	100	100	1	16	2	-	100	100	1	2	1
2.0	4	-	100	100	3	29	-	100	100	2	41	12	-	100	100	1	11	2
2.0	5	-	100	100	1	10	-	100	100	2	51	15	-	100	100	3	40	4
2.1	3	-	100	100	1	1	-	100	100	2	101	17	-	100	100	1	2	1
2.1	4	-	100	100	10	252	-	100	100	1	62	5	-	100	100	2	50	3
2.1	5	-	100	100	19	404	-	100	100	3	158	11	-	100	100	13	222	14
2.2	3	-	200	200	5	238	-	200	200	2	94	20	-	200	200	7	284	23
2.2	4	-	200	200	24	1,013	-	200	200	5	240	37	-	200	200	32	893	12
2.2	5	-	200	200	24	701	-	200	200	8	445	57	-	200	200	37	1,021	16
2.3	3	-	200	200	18	1,039	-	200	200	9	469	91	-	200	200	5	96	19
2.3	4	-	200	200	32	1,181	-	200	200	5	578	24	-	200	200	38	1,242	22
2.3	5	-	200	200	55	2,013	-	200	200	9	610	47	-	200	200	72	1,821	16
2.4	3	-	300	300	19	1,186	-	300	300	3	199	13	-	300	300	24	1,138	16
2.4	4	-	300	300	45	2,347	-	300	300	9	927	62	-	300	300	64	2,723	19
2.4	5	-	300	300	68	2,191	-	300	300	13	1,079	84	-	300	300	91	2,699	30
2.5	3	-	400	400	17	1,591	-	400	400	17	3,518	131	-	400	400	3	152	3
2.5	4	-	400	400	44	3,605	-	400	400	8	1,299	32	-	400	400	56	2,470	22
2.5	5	-	400	400	63	3,441	-	400	400	9	1,381	40	-	400	400	71	3,546	7
2.6	3	-	500	500	20	2,639	-	500	500	8	1,729	60	-	500	500	6	726	32
2.6	4	-	500	500	47	4,432	-	500	500	8	2,201	38	-	500	500	58	2,844	41
2.6	5	-	500	500	82	6,839	-	500	500	9	2,291	26	-	500	500	83	4,490	18
2.7	3	-	700	700	101	47,908	-	700	700	80	25,554	572	-	700	700	36	2,842	67
2.7	4	-	600	600	58	16,759	-	600	600	19	6,465	101	-	600	600	47	3,288	41
2.7	5	-	600	600	88	21,317	-	600	600	21	9,188	79	-	600	600	77	8,535	41
2.8	3	-	800	800	321	259,584	-	800	800	444	103,865	3,893	-	800	800	101	20,504	306
2.8	4	-	800	800	641	402,363	-	800	800	224	106,306	1,230	-	800	800	86	8,690	91
2.8	5	-	800	800	511	340,700	-	800	800	145	94,302	490	-	800	800	82	11,344	28
2.9	3	-	1,000	1,000	2,229	1,310,819	-	1,000	1,000	958	344,129	5,182	-	1,000	1,000	149	28,936	417
2.9	4	-	900	900	8,465	3,901,049	-	900	900	1,514	638,805	5,600	-	900	900	129	38,188	204
2.9	5	-	900	900	3,271	2,361,298	-	900	900	1,381	858,811	3,446	-	900	900	118	24,799	91
3.0	3	-	1,300	1,300	33,215	15,303,215	-	1,300	1,300	7,993	4,268,476	19,253	-	1,300	1,300	394	95,624	1,002
3.0	4	-	1,100	1,100	11,235	5,408,097	-	1,100	1,100	4,127	1,819,166	9,411	-	1,100	1,100	205	83,185	204
3.0	5	-	1,100	1,100	14,619	9,351,222	-	1,100	1,100	2,872	1,957,876	5,347	-	1,100	1,100	519	139,101	566
3.1	3	125	1,900	840	limit	17,659,666	69	1,700	1,005	limit	16,055,176	124,498	16	1,700	1,460	limit	1,320,841	13,121
3.1	4	90	1,400	735	limit	15,299,955	-	1,300	1,300	35,008	7,668,150	23,436	-	1,300	1,300	1,353	222,636	1,548
3.1	5	46	1,200	818	limit	23,515,936	-	1,200	1,200	36,281	14,962,333	23,035	-	1,200	1,200	460	148,351	459
3.2	3	152	2,300	912	limit	25,980,239	147	3,000	1,211	limit	22,952,077	142,038	26	2,200	1,739	limit	1,537,141	11,754
3.2	4	100	1,600	798	limit	17,818,411	65	1,500	906	limit	11,585,021	22,924	-	1,500	1,500	2,464	487,931	1,730
3.2	5	96	1,500	762	limit	23,043,580	44	1,400	971	limit	18,874,203	13,320	-	1,400	1,400	1,792	520,520	1,032
3.3	3	225	4,000	1,229	limit	38,453,552	268	5,300	1,438	limit	30,121,605	67,404	47	3,300	2,235	limit	1,553,748	17,631
3.3	4	146	2,200	893	limit	16,903,153	101	1,900	944	limit	10,384,128	15,039	22	1,800	1,468	limit	1,396,263	11,754
3.3	5	99	1,700	852	limit	25,467,262	83	1,700	924	limit	21,068,388	11,460	-	1,600	1,600	3,962	748,767	2,033
3.4	4	168	2,600	969	limit	21,328,194	112	2,300	1,083	limit	21,156,982	40,985	43	2,400	1,676	limit	1,404,166	11,754
3.4	5	148	1,900	763	limit	25,743,706	99	1,900	952	limit	18,087,948	3,929	-	1,800	1,800	13,578	1,805,176	3,495
3.5	4	205	3,100	1,014	limit	24,477,965	159	3,000	1,155	limit	21,401,578	30,400	37	2,700	1,960	limit	1,414,853	11,754
3.5	5	152	2,400	950	limit	22,388,336	133	2,400	1,027	limit	17,266,232	8,457	23	2,100	1,701	limit	2,628,007	7,836
3.6	4	233	3,300	1,137	limit	19,267,344	194	3,700	1,256	limit	18,175,173	35,035	28	3,100	2,420	limit	1,709,756	11,754
3.6	5	219	3,100	969	limit	23,991,442	152	2,700	1,067	limit	15,352,407	7,940	36	2,600	1,906	limit	2,807,075	7,836
3.7	4	382	5,400	1,118	limit	17,811,354	272	4,900	1,313	limit	9,690,771	18,186	31	3,800	2,900	limit	1,782,097	11,754
3.7	5	193	3,200	1,088	limit	23,594,610	145	2,900	1,183	limit	16,222,287	8,113	37	2,900	2,105	limit	2,180,909	7,836
3.8	4	351	5,600	1,239	limit	22,488,013	345	6,400	1,437	limit	12,589,804	14,127	40	4,900	3,480	limit	1,696,707	11,754
3.8	5	245	4,100	1,187	limit	21,967,929	188	3,600	1,249	limit	16,366,510	8,054	50	3,600	2,392	limit	2,305,573	7,836
3.9	5	299	4,900	1,227	limit	25,521,263	199	3,900	1,301	limit	14,552,546	4,267	40	3,900	2,774	limit	2,554,368	7,787

Table A.9: Results (continued) on network net2 and 60 instances. A time limit of 11 h was imposed. The different strategies for solving the topology optimization problem are presented in Chapter 5. Those instances not depicted are infeasible. The column "scale" shows the scaling of the nomination and the columns "loops" the number of available loops. The column "cuts" shows the number of generated cuts.

nom	strategy 1 time	strategy 1 nodes	strategy 2 time	strategy 2 nodes	strategy 3 time	strategy 3 nodes
1	7	131	2	112	2	112
2	3	39	1	26	1	26
3	1	44	1	34	1	34
4	4	199	1	30	1	30
5	3	162	2	35	3	35
6	327	7,022	3	235	3	235
7	1	26	10	205	93	205
8	2	72	1	50	1	50
9	1	1	1	8	1	8
10	2	18	1	17	1	17
11	2	17	1	20	1	20
12	2	55	1	28	1	28
13	3	236	3	53	3	53
14	12	788	1	26	1	26
15	1	44	1	71	1	71
16	limit	31,562	35	4,033	35	4,033
17	limit	53,277	56	8,012	55	8,012
18	16	575	5	327	4	327
19	1	19	1	12	1	12
20	2	27	1	19	1	19
21	1	39	1	20	1	20
22	251	22,587	127	3,042	36	798
23	2	35	7	129	57	129
24	2	46	4	41	7	41
25	1	1	0	1	1	1
26	1	14	1	54	1	54
27	2	41	1	33	1	33
28	4	304	2	24	2	24
29	limit	24,619	5	192	6	192
30	1	52	10	153	77	153
31	1	16	2	119	3	119
32	1	1	1	82	1	82
33	1	35	9	971	9	971
34	5	80	9	131	59	131
35	2	61	1	22	2	22
36	limit	28,498	1	40	1	40
37	1	25	2	232	2	232
38	3	226	13	2,651	12	2,651
39	3	132	6	161	5	161
40	2	134	9	140	28	140
41	58	1,183	1	8	1	8
42	1	15	2	121	2	121
43	2	22	3	202	3	202

Table A.10: Results on network `net6` and 43 instances. A time limit of 1 h was imposed. The different strategies for solving the topology optimization problem are presented in Chapter 6.

nom	SCIP (strat. 1)					SCIP with priorities (strat. 2)					domain relaxation heuristic (strat. 4)					domain relaxation and check (strat. 3)				
	gap	primal	dual	time	nodes	gap	primal	dual	time	nodes	gap	primal	dual	time	nodes	gap	primal	dual	time	nodes
1	-	0	0	1	1	-	0	0	1	1	-	0	0	1	1	-	0	0	1	1
2	-	0	0	1	1	-	0	0	1	1	-	0	0	1	1	-	0	0	1	1
3	-	34	34	131	41,225	-	34	34	49	4,538	-	34	34	63	3,343	-	34	34	53	3,120
4	-	34	34	128	41,225	-	34	34	49	4,538	-	34	34	63	3,343	-	34	34	53	3,120
5	-	125	125	9,529	5,445,216	-	125	125	425	239,810	-	125	125	632	223,460	-	125	125	959	218,147
6	-	125	125	9,523	5,445,216	-	125	125	427	239,810	-	125	125	630	223,460	-	125	125	945	218,147
7	-	150	150	217	159,121	-	150	150	144	46,594	-	150	150	263	115,702	-	150	150	166	43,836
8	-	150	150	216	159,121	-	150	150	144	46,594	-	150	150	263	115,702	-	150	150	707	225,457
9	-	0	0	1	1	-	0	0	1	1	-	0	0	1	1	-	0	0	1	1
10	-	0	0	1	1	-	0	0	1	1	-	0	0	1	1	-	0	0	1	1
11	-	19	19	14	3,403	-	19	19	34	2,473	-	21	21	14	1,429	-	19	19	215	2,247
12	-	19	19	14	3,403	-	19	19	34	2,473	-	21	21	14	1,429	-	19	19	215	2,247
13	44	143	99	limit	9,636,729	-	130	130	657	338,707	-	130	130	2,441	465,459	-	130	130	1,269	352,263
14	44	143	99	limit	9,630,988	-	130	130	648	338,707	-	130	130	2,437	465,459	-	130	130	1,282	352,263
15	-	181	181	1,800	1,340,255	-	181	181	330	169,996	-	181	181	371	145,275	-	181	181	406	182,279
16	-	181	181	1,793	1,340,255	-	181	181	330	169,996	-	181	181	368	145,275	-	181	181	403	182,279
17	-	146	146	161	114,789	-	146	146	74	41,384	-	146	146	99	43,720	-	146	146	133	43,025
18	-	146	146	161	114,789	-	146	146	74	41,384	-	146	146	99	43,720	-	146	146	132	43,025
19	-	193	193	185	84,772	-	193	193	556	103,197	-	193	193	418	55,527	-	193	193	476	59,532
20	-	193	193	186	84,772	-	193	193	557	103,197	-	193	193	419	55,527	-	193	193	477	59,532
21	9	345	316	limit	12,424,524	-	345	345	6,009	1,911,912	-	345	345	8,490	2.333,807	-	345	345	6,739	2,189,269
22	9	345	316	limit	12,424,117	-	345	345	5,983	1,911,912	-	345	345	8,439	2,333,807	-	345	345	6,736	2,189,269
23	23	467	378	limit	14,846,550	20	467	388	limit	2,487,515	-	467	467	8,945	4,969,890	-	467	467	10,177	5,521,093
24	23	467	378	limit	14,882,249	20	467	388	limit	2,489,140	-	467	467	8,920	4,969,890	-	467	467	10,213	5,521,093

Table A.11: Results on network **net3** and 24 instances. A time limit of 11 h was imposed. The different strategies for solving the topology optimization problem are presented in Chapter 6.

nom	SCIP (strat. 1) gap	primal	dual	time	nodes	SCIP with priorities (strat. 2) gap	primal	dual	time	nodes	domain relaxation heuristic (strat. 4) gap	primal	dual	time	nodes	domain relaxation and check (strat. 3) gap	primal	dual	time	nodes
1	-	0	0	1	1	-	0	0	1	1	-	0	0	1	1	-	0	0	1	1
2	-	0	0	1	1	-	0	0	1	1	-	0	0	1	1	-	0	0	1	1
3	-	0	0	1	1	-	0	0	1	1	-	0	0	1	1	-	0	0	1	1
4	-	0	0	1	1	-	0	0	1	1	-	0	0	1	1	-	0	0	1	1
5	-	0	0	1	1	-	0	0	1	1	-	0	0	1	1	-	0	0	1	1
6	-	0	0	1	1	-	0	0	1	1	-	0	0	1	1	-	0	0	1	1
7	-	0	0	1	1	-	0	0	1	1	-	0	0	1	1	-	0	0	1	1
8	-	0	0	1	1	-	0	0	1	1	-	0	0	1	1	-	0	0	1	1
9	-	0	0	1	1	-	0	0	1	1	-	0	0	1	1	-	0	0	1	1
10	-	0	0	1	1	-	0	0	1	1	-	0	0	1	1	-	0	0	1	1
11	-	95	95	166	7,798	-	95	95	39	1,834	-	95	95	104	1,587	-	95	95	116	1,042
12	-	95	95	13	872	-	95	95	17	796	-	95	95	97	1,458	-	95	95	33	665
13	-	95	95	17	1,207	-	95	95	65	3,291	-	95	95	202	2,787	-	95	95	22	1,247
14	-	95	95	12	1,290	-	95	95	13	1,189	-	95	95	66	1,179	-	95	95	17	807
15	-	95	95	1	1	-	95	95	1	1	-	95	95	1	1	-	95	95	1	1
16	-	249	249	6,315	246,873	-	249	249	97	5,009	61	249	154	limit	298,007	-	249	249	173	3,318
17	12	320	285	limit	1,691,917	12	320	285	limit	1,354,643	-	320	320	26,805	200,944	-	320	320	330	3,185
18	1	475	466	limit	1,496,516	-	475	475	23,777	916,290	8	475	439	limit	508,811	-	475	475	986	3,120
19	10	525	475	limit	1,479,491	-	525	525	634	23,754	-	525	525	3,893	70,054	-	525	525	661	6,377
20	-	525	525	8	701	-	525	525	8	719	-	525	525	8	719	-	525	525	8	725
21	-	525	525	3	246	-	525	525	3	246	-	525	525	3	246	-	525	525	3	246
22	-	525	525	1	28	-	525	525	1	28	-	525	525	1	28	-	525	525	1	28
23	-	525	525	2	6	-	525	525	2	6	-	525	525	1	6	-	525	525	2	6
24	-	525	525	1	5	-	525	525	1	5	-	525	525	1	5	-	525	525	1	5
25	-	525	525	1	5	-	525	525	2	5	-	525	525	1	5	-	525	525	1	5
26	-	584	584	6	135	11	584	525	limit	1,928,200	11	584	525	limit	369,106	-	584	584	4	13
27	-	633	633	10,076	376,520	8	633	584	limit	1,472,909	-	633	633	896	5,907	-	633	633	69	536
28	-	633	633	9	185	-	633	633	9	185	-	633	633	27	185	-	633	633	8	21
29	78	938	525	limit	1,648,593	-	693	693	199	5,681	-	693	693	942	7,944	-	693	693	717	2,842
30	78	938	525	limit	1,549,956	24	788	633	limit	1,175,948	15	788	679	limit	218,167	-	788	788	1,710	9,182
31	89	997	525	limit	1,667,251	26	997	788	limit	659,384	57	997	633	limit	256,190	-	997	997	1,618	15,045
32	131	1,218	525	limit	4,325,792	38	1,092	788	limit	476,112	38	1,092	788	limit	195,827	-	1,092	1,092	2,018	7,051
33	133	1,227	525	limit	1,410,512	49	1,174	788	limit	278,890	27	1,171	918	limit	197,865	-	1,171	1,171	5,129	25,100
34	153	1,479	584	limit	1,389,987	25	1,174	938	limit	762,033	49	1,174	788	limit	187,185	-	1,174	1,174	5,133	26,092
35	86	1,182	633	limit	1,043,948	50	1,182	788	limit	340,425	50	1,182	788	limit	212,598	-	1,182	1,182	4,580	25,073
36	94	1,234	633	limit	1,175,254	26	1,237	978	limit	337,358	26	1,237	978	limit	320,287	-	1,234	1,234	4,463	30,296
37	103	1,286	633	limit	1,305,666	24	1,241	997	limit	513,348	22	1,241	1,015	limit	201,148	-	1,241	1,241	4,856	40,554
38	133	1,479	633	limit	908,286	11	1,318	1,182	limit	594,167	9	1,318	1,202	limit	252,838	-	1,318	1,318	5,013	50,056
39	92	1,336	693	limit	707,675	8	1,336	1,234	limit	407,863	8	1,336	1,230	limit	286,698	-	1,336	1,336	3,149	35,690
40	85	1,465	788	limit	537,315	17	1,459	1,245	limit	647,954	14	1,459	1,270	limit	294,554	5	1,459	1,381	limit	430,424
41	66	1,545	926	limit	1,326,633	20	1,545	1,286	limit	837,664	19	1,545	1,294	limit	310,438	4	1,545	1,481	limit	547,743

Table A.12: Results on network **net4** and 41 nominations. A time limit of 11 h was imposed. The different strategies for solving the topology optimization problem are presented in Chapter 6.

nom	SCIP (strat. 1)					SCIP with priorities (strat. 2)					domain relaxation heuristic (strat. 4)					domain relaxation and check (strat. 3)				
	gap	primal	dual	time	nodes	gap	primal	dual	time	nodes	gap	primal	dual	time	nodes	gap	primal	dual	time	nodes
1		996	0	limit	332,054		1,999	0	limit	251,088		1,556	0	limit	142,814		1,239	0	limit	116,281
2	5,499	2,270	40	limit	739,421			40	limit	100,414			40	limit	102,524	14,755	6,024	40	limit	110,307
3	1,972	1,908	92	limit	1,918,145	1,181	1,180	92	limit	888,972	4,725	4,444	92	limit	1,568,040			92	limit	1,871,730
4	63	1,148	701	limit	342,856	63	1,148	701	limit	376,552	873	6,832	701	limit	145,366	228	2,306	701	limit	134,085
5			726	limit	2,646,703			728	limit	3,192,607			727	limit	2,843,775			726	limit	2,423,993
6			772	limit	2,955,237			759	limit	3,787,138	1,070	8,890	759	limit	2,158,999	833	7,071	757	limit	2,422,856
7			1,038	limit	3,004,645			987	limit	2,446,627			1,038	limit	3,059,269			970	limit	3,024,349
8			1,019	limit	3,700,692			994	limit	3,242,856			959	12,660*	995,717			994	limit	3,243,034
9			1,099	limit	3,520,812			1,099	limit	3,522,377			1,099	limit	3,519,282			1,099	limit	3,530,143
10			1,252	limit	2,552,061			1,250	limit	2,887,158			1,257	limit	2,598,481			1,266	limit	2,774,184
11			1,187	limit	3,153,578			1,187	limit	3,160,494			1,187	limit	3,142,580			1,187	limit	3,162,582

Table A.13: Results on network net5 and 11 nominations. A time limit of 11 h was imposed. The different strategies for solving the topology optimization problem are presented in Chapter 5. The instance marked with * had numerical troubles due to a CPLEX LP error.

	strategy 1				strategy 2			strategy 3		
	no priorities		with priorities		dualval (basic)			dualval (adapted)		
nom	time	nodes	time	nodes	time	nodes	calls	time	nodes	calls
1	limit	241,297	limit	144,404	159	79	1	172	79	1
2	limit	109,869	limit	611,865	limit	744	118	134	308	1
3	limit	122,915	limit	112,174	limit	731	118	177	295	1
4	limit	138,123	limit	27,594	2,784	249	18	315	88	1
5	78	58	101	701	limit	195	50	207	64	1
6	limit	119,661	limit	592,352	272	64	1	254	64	1
7	limit	87,125	limit	162,034	193	296	1	114	296	1
8	limit	134,228	limit	503,061	90	134	1	84	134	1
9	limit	481,537	limit	128,105	1,036	328	10	1,758	328	10
10	limit	231,002	limit	335,555	limit	1,205	79	145	62	1
11	limit	113,277	limit	110,301	647	206	6	117	180	1
12	limit	140,155	limit	711,685	214	51	1	385	51	1
13	limit	114,814	limit	129,237	limit	714	108	5,759	356	45
14	limit	103,021	limit	388,074	limit	648	99	222	189	1
15	limit	141,208	limit	134,031	207	63	2	110	61	1
16	966	7,999	limit	87,825	84	268	0	87	268	0
17	109	183	limit	618,599	515	95	4	198	76	1
18	limit	105,106	limit	82,548	6,173	234	36	99	65	1
19	limit	134,483	limit	193,098	4,186	634	27	117	90	1
20	limit	133,759	limit	116,794	limit	704	141	2,036	88	6
21	limit	248,095	limit	399,486	198	177	2	120	172	1
22	limit	651,493	limit	771,975	603	313	4	269	85	1
23	limit	277,121	101	437	202	37	1	183	37	1
24	limit	637,225	limit	1,108,129	487	117	3	161	62	1
25	limit	133,181	limit	111,325	3,572	202	40	89	68	1
26	limit	105,742	limit	121,007	1,385	82	12	1,638	85	14
27	779	7,314	limit	130,146	222	72	2	259	72	2
28	limit	119,299	109	459	70	39	0	71	39	0
29	limit	110,801	limit	112,277	130	70	1	99	70	1
30	limit	112,109	limit	178,885	1,406	282	15	154	201	1

Table A.14: Results on network `net7a` and 30 nominations. A time limit of 4 h was imposed. The different strategies for solving the topology optimization problem are presented in Chapter 7. Those instances with a finite time limit were feasible. The column "calls" shows the number of calls of the primal heuristic based on dual information.

	strategy 1				strategy 2			strategy 3		
	no priorities		with priorities		dualval (basic)			dualval (adapted)		
nom	time	nodes	time	nodes	time	nodes	calls	time	nodes	calls
1	limit	140,940	limit	68,362	limit	2,812	130	limit	2,665	102
2	limit	90,836	limit	381,513	143	261	1	142	261	1
3	103	442	limit	96,709	limit	36,705	118	limit	36,249	112
4	110	903	limit	59,288	limit	464	83	173	188	1
5	limit	140,474	86	274	482	69	5	151	54	1
6	limit	412,625	limit	639,087	limit	12,284	115	573	11,786	1
7	limit	117,310	limit	98,116	limit	3,451	159	limit	3,233	131
8	limit	312,566	limit	129,571	limit	932	157	limit	752	121
9	limit	152,721	limit	113,130	563	606	4	158	593	1
10	limit	105,796	limit	126,752	limit	63,450	80	7,546	66,510	21
11	limit	156,664	limit	68,237	218	121	1	545	273	3
12	limit	234,332	limit	450,738	limit	1,649	145	limit	1,179	94
13	limit	771,320	limit	1,220,105	2,964	112	12	limit	481	64
14	limit	491,487	limit	128,450	1,865	326	24	116	56	1
15	limit	144,534	limit	144,477	limit	1,093	156	8,260	639	61
16	limit	613,588	limit	116,138	limit	1,876	208	limit	1,265	124
17	limit	412,334	limit	392,347	5,019	559	34	811	262	5
18	limit	148,791	limit	94,420	limit	9,055	138	125	138	1
19	limit	337,272	limit	131,345	limit	1,304	137	limit	1,145	99
20	limit	490,877	limit	125,256	4,186	429	44	97	115	1
21	limit	249,526	limit	236,225	1,854	194	10	232	56	1
22	297	10,988	limit	158,298	limit	610	102	300	169	1
23	limit	103,000	limit	611,098	279	1,236	1	254	1,236	1
24	limit	137,949	limit	123,919	limit	1,381	139	limit	881	74
25	limit	106,484	limit	124,315	limit	2,572	108	limit	2,473	98
26	limit	127,864	limit	126,024	limit	1,784	152	limit	1,384	110
27	4,171	175,096	limit	613,955	limit	919	123	149	68	1
28	limit	147,134	limit	125,502	limit	1,561	157	limit	883	94
29	limit	118,015	limit	179,521	2,653	160	24	2,305	129	18
30	limit	210,212	limit	124,893	1,561	562	12	406	471	3

Table A.15: Results on network `net7b` and 30 nominations. A time limit of 4 h was imposed. The different strategies for solving the topology optimization problem are presented in Chapter 7. Those instances with a finite time limit were feasible. The column "calls" shows the number of calls of the primal heuristic based on dual information.

	strategy 1				strategy 2			strategy 3		
	no priorities		with priorities		dualval (basic)			dualval (adapted)		
nom	time	nodes	time	nodes	time	nodes	calls	time	nodes	calls
1	limit	209,792	limit	175,127	5,081	1,554	28	7,574	1,558	30
2	limit	235,184	limit	293,344	544	1,233	3	202	1,225	1
3	limit	95,608	limit	194,961	limit	1,674	59	1,359	673	4
4	limit	725,232	limit	455,667	limit	4,942	86	limit	5,047	101
5	limit	453,246	limit	409,942	limit	741	69	7,398	661	45
6	limit	730,765	limit	681,565	1,725	7,254	9	442	5,201	1
7	limit	410,226	limit	544,259	269	699	1	317	699	1
8	limit	115,252	limit	592,581	limit	594	88	7,498	399	49
9	limit	116,342	14,088	104,447	limit	104,776	33	5,513	361	33
10	limit	225,347	limit	490,606	limit	1,060	90	limit	911	64
11	limit	96,301	limit	113,693	limit	2,254	86	190	174	1
12	limit	549,066	limit	789,363	419	384	2	228	375	1
13	limit	126,058	limit	119,756	limit	610	101	limit	548	85
14	limit	569,523	limit	375,609	limit	2,178	118	limit	2,190	121
15	limit	252,135	limit	237,202	limit	5,624	104	limit	5,116	78
16	limit	553,331	limit	119,828	3,864	349	33	2,806	204	20
17	limit	236,160	limit	145,606	limit	786	78	limit	753	68
18	limit	333,974	limit	329,769	limit	1,241	68	limit	1,214	58
19	limit	373,279	limit	213,921	limit	773	69	5,136	299	17
20	limit	165,352	limit	146,906	limit	4,872	77	limit	2,972	64
21	limit	133,832	limit	174,777	6,991	6,962	47	142	263	1
22	161	1,193	limit	672,644	1,328	217	10	645	190	4
23	limit	479,245	limit	367,011	limit	2,031	65	761	512	3
24	limit	700,383	limit	535,530	limit	13,133	30	2,234	5,525	4
25	limit	137,073	limit	116,393	limit	3,925	73	limit	339	47
26	limit	425,587	limit	92,517	limit	870	90	200	58	1
27	limit	104,986	limit	315,932	limit	764	95	limit	697	77
28	limit	128,744	limit	909,649	321	4,709	1	337	4,709	1
29	limit	101,648	limit	110,769	limit	663	59	limit	680	64
30	limit	256,820	limit	83,425	limit	979	82	limit	933	70

Table A.16: Results on network `net7c` and 30 nominations. A time limit of 4 h was imposed. The different strategies for solving the topology optimization problem are presented in Chapter 7. Those instances with a finite time limit were feasible. The column "calls" shows the number of calls of the primal heuristic based on dual information.

	strategy 1				strategy 2			strategy 3		
	no priorities		with priorities		dualval (basic)			dualval (adapted)		
nom	time	nodes	time	nodes	time	nodes	calls	time	nodes	calls
1	limit	92,986	limit	53,134	limit	1,911	75	limit	1,832	64
2	limit	107,559	limit	221,183	limit	1,525	81	limit	1,462	61
3	limit	294,798	limit	213,185	limit	16,201	97	2,305	10,172	3
4	limit	264,519	limit	407,828	limit	575	87	error		5
5	limit	481,910	limit	119,520	limit	6,422	95	1,716	3,110	10
6	limit	686,619	limit	695,384	limit	533	59	1,291	303	5
7	limit	495,048	limit	471,546	1,727	1,011	8	8,721	1,246	43
8	limit	125,606	limit	327,295	limit	92,402	0	limit	94,489	0
9	limit	52,482	limit	452,027	limit	1,789	99	limit	1,454	63
10	limit	194,612	limit	296,502	limit	681	91	2,981	436	19
11	limit	117,377	limit	98,475	limit	133,894	10	limit	18,324	52
12	limit	482,468	limit	465,670	6,544	975	25	1,470	852	6
13	limit	114,820	limit	140,953	limit	593	67	limit	439	37
14	limit	560,563	limit	378,138	limit	1,350	60	7,847	1,247	29
15	limit	258,260	limit	268,361	limit	358	89	237	81	1
16	limit	536,908	170	1,511	7,616	524	61	5,044	276	36
17	limit	341,725	limit	210,353	limit	1,502	117	397	372	1
18	limit	175,509	limit	102,087	limit	359,541	14	limit	374,860	14
19	limit	197,670	limit	288,764	9,091	1,114	41	11,581	1,112	43
20	limit	136,959	limit	129,193	limit	226	36	limit	669	60
21	limit	114,578	limit	466,233	8,580	12,477	56	3,476	11,595	16
22	limit	392,281	limit	402,646	limit	388	94	361	57	2
23	limit	545,123	limit	480,047	limit	7,061	50	limit	7,765	75
24	limit	764,936	limit	464,159	limit	577	78	2,949	350	20
25	limit	161,312	limit	629,419	limit	1,721	100	7,751	318	40
26	limit	119,817	limit	166,047	limit	5,484	76	limit	3,542	36
27	limit	91,989	limit	242,744	limit	2,525	91	limit	2,500	71
28	limit	423,135	limit	682,647	limit	5,590	81	785	98	4
29	limit	131,923	limit	230,005	limit	12,891	85	642	239	4
30	limit	257,926	limit	162,310	limit	3,119	54	limit	2,972	42

Table A.17: Results on network `net7d` and 30 nominations. A time limit of 4 h was imposed. The different strategies for solving the topology optimization problem are presented in Chapter 7. The error originates from numerical troubles in the LP solver. Those instances with a finite time limit were feasible. The column "calls" shows the number of calls of the primal heuristic based on dual information.

	strategy 1				strategy 2			strategy 3		
	no priorities		with priorities		dualval (basic)			dualval (adapted)		
nom	time	nodes	time	nodes	time	nodes	calls	time	nodes	calls
1	limit	47,244	limit	195,673	limit	1,623	49	limit	1,640	53
2	limit	265,266	limit	167,224	limit	27,063	46	1,869	17,775	3
3	limit	193,970	limit	154,650	limit	1,422	57	limit	1,513	77
4	limit	425,537	limit	388,587	6,013	691	38	2,438	296	7
5	limit	154,950	limit	362,629	limit	616	53	259	381	1
6	limit	311,869	limit	148,075	limit	48,493	62	7,196	48,133	28
7	limit	306,996	limit	335,788	limit	2,139	55	limit	2,160	59
8	limit	311,992	limit	162,684	limit	459	76	114	275	1
9	limit	55,740	limit	33,938	limit	824	67	limit	828	65
10	limit	397,544	limit	305,236	limit	776	49	limit	849	70
11	132	466	limit	291,708	limit	489	60	1,292	168	6
12	limit	52,270	limit	526,302	558	2,133	2	779	2,188	4
13	limit	59,516	limit	244,825	limit	513	73	10,266	449	53
14	limit	295,171	limit	197,110	limit	22,572	77	limit	4,895	64
15	limit	356,782	limit	254,441	954	258	6	220	239	1
16	limit	85,951	limit	119,517	895	428	6	1,161	428	6
17	limit	362,794	limit	382,453	limit	48,787	67	limit	45,209	31
18	limit	269,555	limit	79,645	limit	3,752	54	452	3,228	1
19	limit	335,511	limit	263,183	limit	3,981	47	2,905	3,625	11
20	limit	138,307	limit	46,787	limit	929	86	129	133	1
21	limit	100,438	limit	59,936	1,697	2,462	10	522	1,449	3
22	limit	256,607	limit	344,673	8,336	1,193	41	limit	1,386	59
23	limit	415,409	limit	512,692	limit	879	64	limit	760	47
24	limit	363,684	limit	536,135	451	6,806	1	1,698	10,269	6
25	limit	77,653	limit	127,088	limit	254	67	limit	169	42
26	limit	58,964	limit	26,786	limit	915	58	limit	998	82
27	limit	58,944	limit	235,868	limit	18,841	81	limit	17,303	63
28	limit	422,310	limit	462,954	998	19,059	1	1,138	19,081	2
29	limit	117,490	limit	51,026	limit	554	93	98	43	1
30	limit	62,883	limit	309,384	limit	18,383	56	598	469	3

Table A.18: Results on network `net7e` and 30 nominations. A time limit of 4 h was imposed. The different strategies for solving the topology optimization problem are presented in Chapter 7. Those instances with a finite time limit were feasible. The column "calls" shows the number of calls of the primal heuristic based on dual information.

nom	heuristic dualval (adapted) & domain relaxation and check gap	primal	dual	time	nodes
1		1,007	0	limit	84,984
2	3,759	1,565	40	limit	90,292
3	2,954	2,813	92	limit	155,623
4	1,867	13,801	701	limit	164,731
5	174	1,997	728	limit	248,268
6	886	7,513	761	limit	109,671
7	630	6,673	913	limit	125,628
8	1,089	9,915	833	limit	469,076
9	883	10,994	1,117	limit	158,182
10	619	12,450	1,731	limit	190,015
11	674	13,199	1,703	limit	137,384

Table A.19: Results on network `net5` and 11 nominations. A time limit of 11 h was imposed. The strategy for solving the topology optimization problem is presented in Chapter 7.

Bibliography

K. Abhishek, S. Leyffer, and J. T. Linderoth. FilMINT: An Outer-Approximation-Based Solver for Nonlinear Mixed Integer Programs. In: *INFORMS Journal On Computing* 22, No. 4, 2010, pp. 555–567 (cit. on p. 17).

T. Achterberg. *SCIP - a framework to integrate Constraint and Mixed Integer Programming.* Tech. rep. ZR-04-19. Zuse Institute Berlin, 2004 (cit. on p. 14).

T. Achterberg. SCIP: Solving Constraint Integer Programs. In: *Mathematical Programming Computation* 1, No. 1, 2009, pp. 1–41 (cit. on pp. 10, 13, 17).

T. Ahadi-Oskui, S. Vigerske, I. Nowak, and G. Tsatsaronis. Optimizing the design of complex energy conversion systems by Branch and Cut. In: *Computers & Chemical Engineering* 34, No. 8, 2010, pp. 1226–1236 (cit. on p. 14).

J. André, J. F. Bonnans, and L Cornibert. "Planning reinforcement of gas transportation networks with optimization methods." In: *Proceedings 19th Mini EURO Conference on the Energy Sector.* 2006 (cit. on p. 7).

F. Babonneau, Y. Nesterov, and J.-P. Vial. Design and Operations of Gas Transmission Networks. In: *Operations Research* 60, No. 1, 2012, pp. 34–47 (cit. on p. 8).

P. Belotti, C. Kirches, S. Leyffer, J. Linderoth, J. Luedtke, and A. Mahajan. *Mixed-Integer Nonlinear Optimization.* Preprint ANL/MCS-P3060-1121. Argonne National Laboratory, 2012 (cit. on p. 13).

T. Berthold. "Heuristic algorithms in global MINLP solvers." PhD thesis. Technische Universität Berlin, 2014 (cit. on p. 17).

P. Bonami and J. Gonçalves. Heuristics for convex mixed integer nonlinear programs. In: *Computational Optimization and Applications* 51 (2), 2012, pp. 729–747 (cit. on p. 17).

P. Bonami, L. T. Biegler, A. R. Conn, G. Cornuéjols, I. E. Grossmann, C. D. Laird, J. Lee, A. Lodi, F. Margot, N. Sawaya, and A. Wächter. An algorithmic framework for convex mixed integer nonlinear programs. In: *Discrete Optimization* 5, No. 2, 2008, pp. 186–204 (cit. on p. 17).

J. Bonnans, G. Spiers, and J.-L. Vie. *Global optimization of pipe networks by the interval analysis approach: the Belgium network case.* Rapport de Recherche RR 7796. INRIA, 2011 (cit. on pp. 7 sq.).

C. Borraz-Sánchez and R. Z. Ríos-Mercado. "A Tabu Search Approach for Minimizing Fuel Consumption on Cyclic Natural Gas Pipeline Systems." In: *Proceedings of 12th CLAIO Conference.* 2004 (cit. on p. 7).

C. Borráz-Sánchez and R. Ríos-Mercado. A Non-Sequential Dynamic Programming Approach for Natural Gas Network Optimization. In: *WSEAS Transactions on Systems* 3, No. 4, 2004, pp. 1384–1389 (cit. on p. 7).

I. D. Boyd, P. D. Surry, and N. J. Radcliffe. *Constrained Gas Network Pipe Sizing with Genetic Algorithms.* Tech. rep. EPCC-TR94-11. Edinburgh Parallel Computing Centre, 1994 (cit. on p. 7).

S. Boyd and L. Vandenberghe. *Convex Optimization.* Cambridge University Press, 2004 (cit. on p. 17).

C. Bragalli, C. D'Ambrosio, J. Lee, A. Lodi, and P. Toth. On the optimal design of water distribution networks: a practical MINLP approach. In: *Optimization and Engineering* 13 (2), 2012, pp. 219–246 (cit. on p. 35).

BGR 2013: Bundesanstalt für Geowissenschaften und Rohstoffe. *Energiestudie 2013 - Reserven, Ressourcen und Verfügbarkeit von Energierohstoffen.* 2013. URL: `http://www.bgr.bund.de/DE/Themen/Energie/Produkte/energiestudie2013_Zusammenfassung.html`. Downloaded on August 26, 2014 (cit. on p. 1).

GasNZV 2005: Bundesministerium der Justiz. *Verordnung über den Zugang von Gasversorgungnetzen (Gasnetzzugangsverordnung – GasNZV).* 2005. URL: `http://www.bmwi.de/BMWi/Redaktion/PDF/Gesetz/energiewirtschaftsgesetz-verordnung-gasnetz-gasnzv-april-2005,property=pdf,bereich=bmwi2012,sprache=de,rwb=true.pdf`. Version released Apr. 14, 2005 (cit. on p. 1).

J. Burgschweiger, B. Gnädig, and M. C. Steinbach. Optimization Models for Operative Planning in Drinking Water Networks. In: *Optimization and Engineering* 10, No. 1, 2009, pp. 43–73 (cit. on p. 35).

M. R. Bussieck and S. Vigerske. MINLP Solver Software. In: *Wiley Encyclopedia of Operations Research and Management Science*. Ed. by J. J. C. et.al. Wiley & Sons, Inc., 2010 (cit. on p. 13).

R. G. Carter. *Compressor Station Optimization: Computational Accuracy and Speed.* Tech. rep. PSIG 9605. Pipeline Simulation Interest Group, 1996 (cit. on p. 28).

R. G. Carter. "Pipeline Optimization: Dynamic Programming after 30 Years." In: *Proceedings of the 30th PSIG Annual Meeting.* 1998 (cit. on p. 7).

R. G. Carter, D. Schroeder, and T. Harbick. *Some Causes and Effects of Discontinuities in Modeling and Optimizing Gas Transmission Networks.* Tech. rep. PSIG 9308. Pipeline Simulation Interest Group, 1993 (cit. on p. 28).

L. Castillo and A. Gonzáleza. Distribution network optimization: Finding the most economic solution by using genetic algorithms. In: *European Journal of Operational Research* 108, No. 3, 1998, pp. 527–537 (cit. on p. 7).

G. Cerbe. *Grundlagen der Gastechnik: Gasbeschaffung – Gasverteilung – Gasverwendung.* Hanser Verlag, Leipzig, Germany, 2008 (cit. on pp. 1, 30).

A. Chebouba, F. Yalaoui, A. Smati, L. Amodeo, K. Younsi, and A. Tairi. Optimization of Natural Gas Pipeline Transportation Using Ant Colony Optimization. In: *Computers & Operations Research* 36, No. 6, 2009, pp. 1916–1923 (cit. on p. 7).

M. Collins, L. Cooper, R. Helgason, J. Kennington, and L. LeBlanc. Solving the pipe network analysis problem using optimization techniques. In: *Management Science* 24, No. 7, 1978, pp. 747–760 (cit. on pp. 54 sq., 73).

A. Conn, N. Gould, and P. Toint. *Trust-Region Methods.* MPS-SIAM Series on Optimization. Society for Industrial and Applied Mathematics, 2000 (cit. on p. 19).

C. D'Ambrosio and A. Lodi. Mixed integer nonlinear programming tools: a practical overview. In: *4OR* 9, No. 4, 2011, pp. 329–349 (cit. on p. 13).

C. D'Ambrosio and A. Lodi. Mixed integer nonlinear programming tools: an updated practical overview. In: *Annals of Operations Research* 204, No. 1, 2013, pp. 301–320 (cit. on p. 13).

E. Danna, E. Rothberg, and C. L. Pape. Exploring relaxation induced neighborhoods to improve MIP solutions. In: *Mathematical Programming* 102, No. 1, 2004, pp. 71–90 (cit. on p. 17).

G. B. Dantzig. "Maximization of a linear function of variables subject to linear inequalities." In: *Activity Analysis of Production and Allocation.* Ed. by T. Koopmanns. John Wiley & Sons, New York, 1951, pp. 339–347 (cit. on p. 15).

D. De Wolf and B. Bakhouya. *Optimal dimensioning of pipe networks: the new situation when the distribution and the transportation functions are disconnected.* Tech. rep. 07/02. Ieseg, Université catholique de Lille, HEC Ecole de Gestion de l'ULG, 2008 (cit. on p. 45).

D. De Wolf and Y. Smeers. Optimal dimensioning of pipe networks with application to gas transmission networks. In: *Operations Research* 44, No. 4, 1996, pp. 596–608 (cit. on p. 8).

D. De Wolf and Y. Smeers. The Gas Transmission Problem Solved by an Extension of the Simplex Algorithm. In: *Management Science* 46, No. 11, 2000, pp. 1454–1465 (cit. on pp. 7, 28, 45).

K. Ehrhardt and M. C. Steinbach. "Nonlinear Optimization in Gas Networks." In: *Modeling, Simulation and Optimization of Complex Processes.* Ed. by H. G. Bock, E. Kostina, H. X. Phu, and R. Ranacher. Springer Berlin Heidelberg, 2005, pp. 139–148 (cit. on p. 7).

K. Ehrhardt and M. C. Steinbach. KKT Systems in Operative Planning for Gas Distribution Networks. In: *Proceedings in Applied Mathematics and Mechanics* 4, No. 1, 2004, pp. 606–607 (cit. on p. 7).

M. Feistauer. *Mathematical Methods in Fluid Dynamics.* Vol. 67. Pitman Monographs and Surveys in Pure and Applied Mathematics Series. Longman Scientific & Technical, Harlow, 1993 (cit. on p. 24).

A. V. Fiacco and Y. Ishizuka. Sensitivity and Stability Analysis for Nonlinear Programming. In: *Annals of Operations Research* 27, No. 1, 1990, pp. 215–236 (cit. on pp. 184 sqq.).

FICO. *Xpress-Optimizer Reference manual.* 2009. URL: `http://www.fico.com/xpress` (cit. on p. 17).

C. A. Floudas. *Nonlinear and Mixed Integer Optimization: Fundamentals and Applications.* Oxford University Press, New York, 1995 (cit. on p. 14).

A. Fügenschuh, B. Hiller, J. Humpola, T. Koch, T. Lehman, R. Schwarz, J. Schweiger, and J. Szabó. Gas Network Topology Optimization for Upcoming Market Requirements. In: *IEEE Proceedings of the 8th International Conference on the European Energy Market (EEM)*, 2011, pp. 346–351 (cit. on pp. iv, viii).

A. Fügenschuh, H. Homfeld, H. Schülldorf, and S. Vigerske. "Mixed-Integer Nonlinear Problems in Transportation Applications." In: *Proceedings of the 2nd International Conference on Engineering Optimization (CD-ROM).* Ed. by H. Rodrigues et al. 2010 (cit. on p. 15).

A. Fügenschuh, B. Geißler, R. Gollmer, C. Hayn, R. Henrion, B. Hiller, J. Humpola, T. Koch, T. Lehmann, A. Martin, R. Mirkov, A. Morsi, J. Rövekamp, L. Schewe, M. Schmidt, R. Schultz, R. Schwarz, J. Schweiger, C. Stangl, M. C. Steinbach, and B. M. Willert. Mathematical Optimization for Challenging Network Planning Problems in Unbundled Liberalized Gas Markets. In: *Energy Systems*, 2013 (cit. on pp. iv, viii).

A. Fügenschuh, B. Geißler, R. Gollmer, A. Morsi, M. E. Pfetsch, J. Rövekamp, M. Schmidt, K. Spreckelsen, and M. C. Steinbach. Physical and technical fundamentals of gas networks. In: *Evaluating Gas Network Capacities.* Ed. by T. Koch, B. Hiller, M. E. Pfetsch, and L. Schewe. SIAM-MOS series on Optimization. SIAM, 2015. Chap. 2, pp. 17–44 (cit. on p. 24).

B. Furey. A Sequential Quadratic Programming-Based Algorithm for Optimization of Gas Networks. In: *Automatica* 29, No. 6, 1993, pp. 1439–1450 (cit. on p. 7).

GAMS Model Library. *General Algebraic Modeling System (GAMS) Model Library.* URL: `http://www.gams.com/modlib/modlib.htm` (cit. on p. 45).

B. Geißler. “Towards Globally Optimal Solutions for MINLPs by Discretization Techniques with Applications in Gas Network Optimization.” PhD thesis. Friedrich-Alexander-Universität Erlangen-Nürnberg, 2011 (cit. on pp. 9, 24).

I. Gentilini, F. Margot, and K. Shimada. The Traveling Salesman Problem with Neighborhoods: MINLP Solution. In: *Optimization Methods and Software* 28, 2013, pp. 364–378 (cit. on p. 51).

A. M. Geoffrion. Generalized Benders Decomposition. In: *Journal of Optimization Theory and Applications* 10, No. 4, 1972, pp. 237–260 (cit. on pp. 12, 103).

B. Gilmour, C. Luongo, and D. Schroeder. *Optimization in Natural Gas Transmission Networks: A Tool to Improve Operational Efficiency.* Tech. rep. Stoner Associates Inc., 1989 (cit. on p. 7).

A. M. Gleixner, H. Held, W. Huang, and S. Vigerske. Towards globally optimal operation of water supply networks. In: *Numerical Algebra, Control and Optimization* 2, No. 4, 2012, pp. 695–711 (cit. on p. 35).

I. E. Grossmann and Z. Kravanja. Mixed-Integer nonlinear programming: A survey of algorithms and applications. In: *Large-Scale Optimization with Applications, Part II: Optimal Design and Control.* Ed. by A. R. Conn, L. T. Biegler, T. F. Coleman, and F. N. Santosa. Vol. 93. The IMA Volumes in Mathematics and its Applications. Springer, New York, 1997, pp. 73–100 (cit. on p. 14).

Y. Hamam and A. Brameller. Hybrid Method for the Solution of Piping Networks. In: *Proceedings of the Institution of Electrical Engineers* 118, No. 11, 1971, pp. 1607–1612 (cit. on p. 7).

C. T. Hansen, K. Madsen, and H. B. Nielsen. Optimization of pipe networks. In: *Mathematical Programming* 52, No. 1-3, 1991, pp. 45–58 (cit. on p. 8).

M. Hübner. “Druckebenenübergreifende Grundsatzplanung von Gasverteilungsnetzen.” PhD thesis. RWTH Aachen, 2009 (cit. on p. 8).

J. Humpola and A. Fügenschuh. Convex reformulations for solving a nonlinear network design problem. In: *Computational Optimization and Applications* 62, No. 3, 2015, pp. 717–759 (cit. on pp. iv, viii).

J. Humpola and F. Serrano. Sufficient pruning conditions for MINLP in gas network design. In: *EURO Journal on Computational Optimization* 5, No. 1, 2017, pp. 239–261 (cit. on pp. iv, viii).

J. Humpola, A. Fügenschuh, and T. Lehmann. A primal heuristic for optimizing the topology of gas networks based on dual information. In: *EURO Journal on Computational Optimization* 3, No. 1, 2015, pp. 53–78 (cit. on pp. iv, viii).

J. Humpola, A. Fügenschuh, B. Hiller, T. Koch, T. Lehmann, R. Lenz, R. Schwarz, and J. Schweiger. The specialized MINLP approach. In: *Evaluating Gas Network Capacities.* Ed. by T. Koch, B. Hiller, M. E. Pfetsch, and L. Schewe. SIAM-MOS series on Optimization. SIAM, 2015. Chap. 7, pp. 123–144 (cit. on pp. iv, viii, 101).

J. Humpola, A. Fügenschuh, and T. Koch. Valid inequalities for the topology optimization problem in gas network design. In: *OR Spectrum* 38, No. 3, 2016, pp. 597–631 (cit. on pp. iv, viii).

IBM. *CPLEX.* URL: `http://www.cplex.com` (cit. on pp. 10, 17, 44).

T. Jeníček. "Steady-State Optimization of Gas Transport." In: *Proceedings of 2nd International Workshop SIMONE on Innovative Approaches to Modeling and Optimal Control of Large Scale Pipeline Networks.* Prague, 1993, pp. 26–38 (cit. on p. 7).

W. Karush. Minima of Functions of Several Variables with Inequalities as Side Conditions. In: *Traces and Emergence of Nonlinear Programming.* Ed. by G. Giorgi and T. H. Kjeldsen. Springer Basel, 2014, pp. 217–245 (cit. on p. 20).

S. Kim, R. Z. Ríos-Mercado, and E. A. Boyd. Heuristics for Minimum Cost Steady-State Gas Transmission Networks. In: *Computing Tools for Modeling, Optimization and Simulation.* Ed. by M. Laguna and J. L. González-Velarde. Vol. 12. Interfaces in Computer Science and Operations Research. Kluwer, Boston, USA, 2000. Chap. 11, pp. 203–213 (cit. on p. 7).

T. Koch, B. Hiller, M. E. Pfetsch, and L. Schewe, eds. *Evaluating Gas Network Capacities.* SIAM-MOS series on Optimization. SIAM, 2015 (cit. on pp. 9, 24).

B. Korte and J. Vygen. *Combinatorial Optimization: Theory and Algorithms.* Springer Verlag, Berlin, 2007 (cit. on pp. 2, 23, 35, 58).

H. W. Kuhn and A. W. Tucker. Nonlinear programming. In: *Proceedings of the 2nd Berkley Symposium on Mathematical Statistics and Probability.* Ed. by J. Neyman. Uinversity Press, Berkley, California, 1951, pp. 481–493 (cit. on p. 20).

H. Lall and P. Percell. A dynamic programming based Gas Pipeline Optimizer. In: *Analysis and Optimization of Systems.* Ed. by A. Bensoussan and J. Lions. Vol. 144. Lecture Notes in Control and Information Sciences. Springer Berlin Heidelberg, 1990, pp. 123–132 (cit. on p. 7).

Lamatto++. *A Framework for Modeling and Solving Mixed-Integer Nonlinear Programming Problems on Networks.* URL: `http://www.mso.math.fau.de/edom/projects/lamatto.html` (cit. on p. 44).

C. Li, W. Jia, Y. Yang, and X. Wu. Adaptive Genetic Algorithm for Steady-State Operation Optimization in Natural Gas Networks. In: *Journal of Software* 6, No. 3, 2011, pp. 452–459 (cit. on p. 7).

LIWACOM. *Benutzerhandbuch.* Version 5.5. LIWACOM Informations GmbH and SIMONE Research Group s.r.o. 2005 (cit. on p. 8).

M. V. Lurie. *Modeling of Oil Product and Gas Pipeline Transportation.* Wiley-VCH, Weinheim, 2008 (cit. on pp. 24, 27).

D. Mahlke, A. Martin, and S. Moritz. A Simulated Annealing Algorithm for Transient Optimization in Gas Networks. In: *Mathematical Methods of Operations Research* 66, No. 1, 2007, pp. 99–116 (cit. on p. 7).

J. Mallinson, A. Fincham, S. Bull, J. Rollet, and M. Wong. Methods for optimizing gas transmission networks. In: *Annals of Operations Research* 43, 1993, pp. 443–454 (cit. on p. 7).

O. Mariani, F. Ancillai, and E. Donati. *Design of a Gas Pipeline: Optimal Configuration.* Tech. rep. PSIG 9706. Pipeline Simulation Interest Group, 1997 (cit. on p. 8).

A. Martin, M. Möller, and S. Moritz. Mixed Integer Models for the Stationary Case of Gas Network Optimization. In: *Mathematical Programming* 105, No. 2, 2006, pp. 563–582 (cit. on p. 7).

A. Martin, B. Geißler, C. Hayn, J. Humpola, T. Koch, T. Lehman, A. Morsi, M. Pfetsch, L. Schewe, M. Schmidt, R. Schultz, R. Schwarz, J. Schweiger, M. Steinbach, and B. Willert. Optimierung Technischer Kapazitäten in Gasnetzen. In: *VDI-Berichte: Optimierung in der Energiewirtschaft* 2157, 2011, pp. 105–115 (cit. on pp. iv, viii).

J. J. Maugis. Etude de réseaux de transport et de distribution de fluide. In: *RAIRO Operations Research* 11, No. 2, 1977, pp. 243–248 (cit. on pp. 54 sq., 58, 73).

J. Mischner. Notices about hydraulic calculations of gas pipelines. In: *GWF–Gas/Erdgas* 4, 2012, pp. 158–273 (cit. on p. 25).

R. Misener and C. Floudas. A Framework for Globally Optimizing Mixed-Integer Signomial Programs. In: *Journal of Optimization Theory and Applications* 161, No. 3, 2014, pp. 905–932 (cit. on p. 10).

M. Möller. "Mixed Integer Models for the Optimisation of Gas Networks in the Stationary Case." PhD thesis. Technische Universität Darmstadt, Fachbereich Mathematik, 2004 (cit. on p. 7).

A. Morsi. "Solving MINLPs on Loosely-Coupled Networks with Applications in Water and Gas Network Optimization." PhD thesis. Friedrich-Alexander-Universität Erlangen-Nürnberg, 2013 (cit. on p. 9).

G. L. Nemhauser and L. A. Wolsey. Integer Programming. In: *Optimization.* Ed. by G. L. Nemhauser, A. H. G. Rinnooy Kan, and M. J. Todd. Elsevier, 1989. Chap. 6, pp. 447–527 (cit. on pp. 14 sqq.).

J. Nikuradse. *Strömungsgesetze in rauhen Rohren.* Forschungsheft auf dem Gebiete des Ingenieurwesens. VDI-Verlag, Düsseldorf, 1933 (cit. on p. 25).

J. Nikuradse. *Laws of Flow in Rough Pipes.* Vol. Technical Memorandum 1292. National Advisory Committee for Aeronautics Washington, 1950 (cit. on p. 25).

V. S. Nørstebø, F. Rømo, and L. Hellemo. Using operations research to optimise operation of the Norwegian natural gas system. In: *Journal of Natural Gas Science and Engineering* 2, 2010, pp. 153–162 (cit. on p. 7).

J. Oldham. "Combinatorial approximation algorithms for generalized flow problems." In: *Proceedings of the tenth annual ACM-SIAM symposium on Discrete algorithms SODA '99.* 1999, pp. 704–714 (cit. on p. 66).

A. Osiadacz and M. Górecki. *Optimization of Pipe Sizes for Distribution Gas Network Design.* Tech. rep. PSIG 9511. Pipeline Simulation Interest Group, 1995 (cit. on p. 8).

A. J. Osiadacz. *Simulation and analysis of gas networks.* E. & F.N. Spon Ltd, London, 1987 (cit. on p. 4).

P. B. Percell and M. J. Ryan. *Steady-State Optimization of Gas Pipeline Network Operation.* Tech. rep. PSIG 8703. Pipeline Simulation Interest Group, 1987 (cit. on p. 7).

M. E. Pfetsch, A. Fügenschuh, B. Geißler, N. Geißler, R. Gollmer, B. Hiller, J. Humpola, T. Koch, T. Lehmann, A. Martin, A. Morsi, J. Rövekamp, L. Schewe, M. Schmidt, R. Schultz, R. Schwarz, J. Schweiger, C. Stangl, M. C. Steinbach, S. Vigerske, and B. M. Willert. Validation of Nominations in Gas Network Optimization: Models, Methods, and Solutions. In: *Optimization Methods and Software,* 2014 (cit. on pp. iv, viii, 24).

J. D. Pintér, ed. *Global Optimization: Scientific and Engineering Case Studies.* Vol. 85. Nonconvex Optimization and Its Applications. Springer, 2006 (cit. on p. 14).

A. U. Raghunathan. Global Optimization of Nonlinear Network Design. In: *SIAM Journal on Optimization* 23, No. 1, 2013, pp. 268–295 (cit. on pp. 51, 55).

R. Z. Ríos-Mercado and C. Borraz-Sánchez. Optimization Problems in Natural Gas Transmission Systems: A State-of-the-Art Survey. Submitted. 2012 (cit. on p. 4).

R. Z. Ríos-Mercado, S. Wu, L. R. Scott, and E. A. Boyd. A Reduction Technique for Natural Gas Transmission Network Optimization Problems. In: *Annals of Operations Research* 117, No. 1, 2002, pp. 217–234 (cit. on p. 7).

R. Z. Ríos-Mercado, S. Kim, and E. A. Boyd. Efficient Operation of Natural Gas Transmission Systems: A Network-Based Heuristic for Cyclic Structures. In: *Computers & Operations Research* 33, No. 8, 2006, pp. 2323–2351 (cit. on p. 7).

D. Scheibe and A. Weimann. Dynamische Gasnetzsimulation mit GANESI. In: *GWF–Gas/Erdgas* 9, 1999, pp. 610–616 (cit. on p. 8).

M. Schmidt, M. C. Steinbach, and B. M. Willert. High detail stationary optimization models for gas networks. In: *Optimization and Engineering*, 2014, pp. 1–34 (cit. on p. 9).

D. Shaw. *Pipeline System Optimization: A Tutorial.* Tech. rep. PSIG 9405. Pipeline Simulation Interest Group, 1994 (cit. on p. 4).

C. Stangl. "Modelle, Strukturen und Algorithmen für stationäre Flüsse in Gasnetzen." PhD thesis. Universität Duisburg-Essen, 2014 (cit. on p. 9).

M. C. Steinbach. On PDE Solution in Transient Optimization of Gas Networks. In: *Journal of Computational and Applied Mathematics* 203, No. 2, 2007, pp. 345–361 (cit. on p. 7).

C. K. Sun, V. Uraikul, C. W. Chan, and P. Tontiwachwuthikul. An integrated expert system/operations research approach for the optimization of natural gas pipeline operations. In: *Engineering Applications of Artificial Intelligence* 13, No. 4, 2000, pp. 465–475 (cit. on p. 7).

M. Tawarmalani and N. Sahinidis. A polyhedral branch-and-cut approach to global optimization. In: *Mathematical Programming* 103, No. 2, 2005, pp. 225–249 (cit. on p. 10).

M. Tawarmalani and N. V. Sahinidis. *Convexification and Global Optimization in Continuous and Mixed-Integer Nonlinear Programming: Theory, Algorithms, Software, and Applications.* Vol. 65. Nonconvex Optimization and Its Applications. Kluwer Academic Publishers, 2002 (cit. on p. 14).

A. Tomasgard, F. Rømo, M. Fodstad, and K. Midthun. Optimization Models for the Natural Gas Value Chain. In: *Geometric Modelling, Numerical Simulation, and Optimization.* Ed. by G. Hasle, K.-A. Lie, and E. Quak. Springer Verlag, New York, 2007, pp. 521–558 (cit. on p. 7).

FNBGas 2013: Vereinigung der Fernleitungsnetzbetreiber Gas e. V. *Netzentwicklungsplan Gas 2013. Konsultationsdokument der deutschen Fernleitungsnetzbetreiber.*

URL: http://www.netzentwicklungsplan-gas.de/. Downloaded on August 26, 2014 (cit. on p. 1).

S. Vigerske. "Decomposition in Multistage Stochastic Programming and a Constraint Integer Programming Approach to Mixed-Integer Nonlinear Programming." PhD thesis. Humboldt-Universität zu Berlin, 2012 (cit. on pp. 10, 13 sq., 17 sqq.).

Z. Vostrý. "Transient Optimization of Gas Transport and Distribution." In: *Proceedings of 2nd International Workshop SIMONE on Innovative Approaches to Modeling and Optimal Control of Large Scale Pipeline Networks.* Prague, 1993, pp. 53–62 (cit. on p. 7).

A. Wächter and L. T. Biegler. On the Implementation of a Primal-Dual Interior Point Filter Line Search Algorithm for Large-Scale Nonlinear Programming. In: *Mathematical Programming* 106, No. 1, 2006, pp. 25–57 (cit. on pp. 10, 20, 44).

J. F. Wilkinson, D. V. Holliday, E. H. Batey, and K. W. Hannah. *Transient Flow in Natural Gas Transmission Systems.* American Gas Association, New York, 1964 (cit. on p. 25).

P. J. Wong and R. E. Larson. Optimization of Natural-Gas Pipeline Systems via Dynamic Programming. In: *IEEE Transactions Automatic Control* 15, No. 5, 1968, pp. 475–481 (cit. on p. 7).

P. J. Wong and R. E. Larson. Optimization of Tree-Structured Natural-Gas Transmission Networks. In: *Journal of Mathematical Analysis and Applications* 24, 1968, pp. 613–626 (cit. on p. 7).

S. Wright, M. Somani, and C. Ditzel. *Compressor Station Optimization.* Tech. rep. PSIG 9805. Pipeline Simulation Interest Group, 1998 (cit. on p. 7).

S. Wu, R. Z. Ríos-Mercado, E. A. Boyd, and L. R. Scott. Model Relaxations for the Fuel Cost Minimization of Steady-State Gas Pipeline Networks. In: *Mathematical and Computer Modelling* 31, No. 2, 2000, pp. 197–220 (cit. on pp. 7, 28).

J. Zhang and D. Zhu. A Bilevel Programming Method for Pipe Network Optimization. In: *SIAM Journal on Optimization* 6, No. 3, 1996, pp. 838–857 (cit. on p. 8).

Q. Zheng, S. Rebennack, N. Iliadis, and P. Pardalos. Optimization Models in the Natural Gas Industry. In: *Handbook of Power Systems I.* Springer, 2010, pp. 121–148 (cit. on p. 4).

H. Zimmer. *Calculating Optimum Pipeline Operations.* Tech. rep. SAND2009-5066C. El Paso Natural Gas Company, 1975 (cit. on p. 7).